2⊕14CAR

UNITED STATES CLIMATE ACTION REPORT 2014

First Biennial Report
of the United States of America

Sixth National Communication
of the United States of America

Under the United Nations Framework
Convention on Climate Change

Climate change is one of the most urgent and profoundly complex challenges we face. That's why, everywhere I travel as Secretary of State – in every meeting, here at home and across the more than 280,000 miles I've traveled since I raised my hand and took the oath to serve in this office – I have made this issue a top priority.

Today, all the scientific evidence is telling us that we cannot afford to delay the reckoning with climate change. With each passing day, the case grows more compelling and the costs of inaction grow beyond anything that anyone with conscience or common sense should be willing to contemplate.

The IPCC's Fifth Assessment Report is another wakeup call. It marshals unassailable evidence of the perils of inaction: Summertime Arctic sea ice volume has shrunk by 70 percent since 1979, 12 of the hottest 13 years on record have occurred since 2000, and the oceans are 30 percent more acidic than they were a century ago. Bottom line: Climate change is real, it's happening now, and human beings are the cause.

In the face of these risks and these warnings, it is time for all of us to do what the science tells us we must, to do what our faiths require of us, and to do what our fragile planet demands of us: It's time to take strong action to combat a truly life-and-death challenge.

Today, people all over the world are demanding action on climate change, and those of us in positions of authority globally have a responsibility to lead the way toward progress. The United States is committed to doing its part.

That's why I am pleased to present this 2014 Climate Action Report to the United Nations Framework Convention on Climate Change (UNFCCC). This report contains our national communication – a quadrennial report detailing actions the United States is taking at home and internationally to mitigate, adapt to, and assist others in addressing climate change as part of our commitments under the UNFCCC.

The report builds on the most authoritative assessments of climate change. It outlines U.S. efforts to promote the research, development, and deployment of technologies to reduce greenhouse gas emissions. It highlights our substantial and growing efforts to support developing countries in the global response to climate change. It also details the financial assistance, education programs, and policies and measures we've implemented both to reduce greenhouse gases and to adapt to the effects of climate change.

I am especially pleased that the 2014 Climate Action Report contains the United States' first-ever Biennial Report, which outlines our plan to reduce greenhouse gas emissions trends even further through 2020. The agreement within the UNFCCC to submit biennial reports represents one of the most significant outcomes from recent negotiations. It is a critical means of ensuring that the parties are implementing the pledges they made under the Cancun decisions. The U.S. Biennial Report shows we are working toward meeting our Cancun commitments by taking action to reduce emissions across our energy economy, as well as in the land sector.

The path to progress has been long. But I'm proud to say that we are closer than we've ever been to a breakthrough.

Under President Obama's leadership, we have doubled wind and solar electricity generation; adopted the toughest fuel economy standards for passenger vehicles in U.S. history; advanced environmental standards to expedite the transition to cleaner and more efficient fuels in power plants; and increased the energy efficiency of our homes, industries, and businesses.

We know from history that fundamental change never comes easily or without a fight. But we're already seeing results. Just look at the facts: Since 2005, our emissions have fallen 6.5 percent, even as our economy continues to grow. What's more, we significantly scaled up our financial assistance to help developing countries mitigate and adapt to the effects of climate change.

We know we must do more, and believe me: we are. President Obama's Climate Action Plan will keep the United States on track to reach our goal of reducing U.S. greenhouse gas emissions in the range of 17 percent below 2005 levels by 2020. Commitments like this are an important signal to the world that America is ready to act.

This is a test of our leadership in the century ahead. We are not just the "indispensable nation" – today we must be the indispensable stewards of our shared planet. Strong, transparent action from all countries contributing to climate change is necessary to solve the global climate challenge. I am pleased to present this report, which demonstrates this Administration's commitment to leading the fight to confront climate change head-on, for our children and generations to come.

John F. Kerry
Secretary of State

2014CAR
UNITED STATES CLIMATE ACTION REPORT 2014

First Biennial Report
of the United States of America

Sixth National Communication
of the United States of America

Under the United Nations Framework
Convention on Climate Change

Published by the U.S. Department of State.
You may electronically download this document from the U.S. Department of State Web site:
http://www.state.gov/e/oes/rls/rpts/car6/index.htm

Foreword

This *U.S. Climate Action Report 2014* (2014 CAR) contains two documents that respond to reporting requirements under the United Nations Framework Convention on Climate Change (UNFCCC): (1) the *Sixth National Communication of the United States of America*, which is provided in accordance with Articles 4 and 12 of the UNFCCC and accompanying decisions,[1] and (2) the *First Biennial Report of the United States of America*. The Biennial Report outlines how U.S. action on climate change puts the United States on a path to reach its commitments in Copenhagen, Cancún, and Durban, covering the period up to 2020, and contains additional reporting information as specified in decisions 1/CP.16, 2/CP.17 (Annex I), and 19/CP.18.

These two reports are separate, but complementary, communications to the UNFCCC. Some of the information in the 2014 CAR and Biennial Report is duplicative, in order for the United States to meet its reporting requirements and ensure that each document is complete.

This 2014 CAR also reflects extensive public comments, as well as edits from more than 21 federal agencies during four rounds of interagency review.

---

[1] The following decisions provide rules and guidance to assist Parties to the UNFCCC in preparing their National Communications, among other things: 3/CP.1, 2/CP.1, 9/CP.2, 6/CP.3, 11/CP.4, 6/CP.5, 5/CP.5, 4/CP.5, 3/CP.5, 34/CP.7, 33/CP.7, 4/CP.8, 1/CP.9, 10/CP.13, 9/CP.16, 20/CP.18.

First Biennial Report
of the United States of America

Table of Contents

Acronyms and Abbreviations

AB	Assembly Bill
AIP	Africa Infrastructure Program
Btu	British thermal unit
ºC	degree Centigrade
CAP	2013 *Climate Action Plan*
CAR	*U.S. Climate Action Report*
CH$_4$	methane
CO$_2$	carbon dioxide
CO$_2$e	carbon dioxide equivalents
COP	Conference of the Parties
CSP	Climate Services Partnership
CTF	Clean Technology Fund
DC	direct current
DOI	U.S. Department of the Interior
DOS	U.S. Department of State
EC-LEDS	Enhancing Capacity for Low-Emission Development Strategies
EOP	Executive Office of the President
EPA	U.S. Environmental Protection Agency
EUPP	Energy Utility Partnership Program
Ex-Im	Export-Import Bank of the United States
ºF	degree Fahrenheit
FCMC	Forest Carbon, Markets and Communities
FEWS NET	Famine Early Warning Systems Network
FSF	fast start finance
FY	fiscal year
G-20	Group of Twenty
GHG	greenhouse gas
GWP	global warming potential

HCFC	hydrochlorofluorocarbon
HFC	hydrofluorocarbon
HIAMAP	High Mountains Adaptation Partnership
IPCC	Intergovernmental Panel on Climate Change
kg	kilogram
lb	pound
LEAP	Lighting and Energy Access Partnership
LULUCF	land use, land-use change, and forestry
MCC	Millennium Challenge Corporation
MM	million
MMBtu	million British thermal units
MW	megawatts
MWh	megawatt-hour
MY	model year
N_2O	nitrous oxide
NASA	National Aeronautics and Space Administration
NCDC	National Clean Diesel Campaign
NF_3	nitrogen trifluoride
OAP	Office of Atmospheric Programs
OPIC	Overseas Private Investment Corporation
PFAN	Private Financing Advisory Network
PFC	perfluorocarbon
PPD	Presidential Policy Directive on Global Development
REDD+	reducing emissions from deforestation and forest degradation
RGGI	Regional Greenhouse Gas Initiative
RPS	renewable energy portfolio standard
SAR	Second Assessment Report
SEAD	Super-Efficient Equipment and Appliances Deployment
SERVIR	Regional Visualization and Monitoring System
SF_6	sulfur hexafluoride
SNAP	Significant New Alternatives Policy Program
Tg	teragram
UNFCCC	United Nations Framework Convention on Climate Change
U.S.-ACEF	U.S.-Africa Clean Energy Finance
USAID	U.S. Agency for International Development
USFS	U.S. Forest Service
VOCTEC	Vocational Training and Education for Climate Energy

U.S. Biennial Report

The *U.S. Biennial Report*, as part of the 2014 *U.S. Climate Action Report*, outlines how U.S. action on climate change puts the United States on a path to reach the ambitious but achievable goal of reducing U.S. greenhouse gas (GHG) emissions in the range of 17 percent below 2005 levels by 2020.

During 2009-2011, average U.S. GHG emissions fell to the lowest level for any three-year period since 1994-1996, due to contributions from both economic factors and government policies. The United States has made significant efforts over the past five years, and our progress can be attributed in part to these efforts, including stringent, long-term standards for vehicle GHG emissions and efficiency, increased building and appliance efficiency, and doubling electricity generation from wind and solar.

The President's Climate Action Plan (EOP 2013a), released in June 2013, builds upon the progress of the past five years and outlines significant additional actions that are necessary to maintain the downward trend in U.S. GHG emissions, such as putting in place new rules to cut carbon pollution from the power sector, increasing energy efficiency, and reducing methane (CH_4) and hydrofluorocarbon (HFC) emissions. The plan also initiates efforts to bolster the capacity of our forests and other lands to continue sequestering carbon in the face of a changing climate and other pressures. We expect that implementation of these actions will achieve substantial additional emission reductions.

This report is a first step toward tracking our progress toward meeting the U.S. 2020 emission reduction goal. It represents an assessment of the range of GHG emission reductions that implementation of a collection of actions across sectors of the economy, consistent with those included in *The President's Climate Action Plan*, can achieve. Over the coming years, as standards and policies are put in place, we will sharpen our estimates of achievable emission reductions.

In addition, this report discusses U.S. actions to assist developing countries in their efforts to mitigate and adapt to climate change. The United States is engaging the full range of institutions—bilateral, multilateral, development finance, and export credit—to mobilize private finance and invest strategically in building lasting resilience to unavoidable climate impacts; to reduce emissions from deforestation and land degradation; and to support low-carbon development strategies and the transition to a sustainable, clean energy economy.

1. FACING THE CLIMATE CHALLENGE

The most significant long-term environmental challenge facing the United States and the world is climate change that results from anthropogenic emissions of GHGs. The scientific consensus, as reflected in the most recent Assessment Reports of the Intergovernmental Panel on Climate Change (IPCC) is that anthropogenic emissions of GHGs are causing changes in the climate that include rising average national and global temperatures, warming oceans, rising average sea levels, more extreme heat waves and storms, extinctions of species, and loss of biodiversity (IPCC 2007, 2013).[1]

[1] The Working Group I contribution (*The Physical Science*) to the IPCC's Fifth Assessment was approved and accepted by governments in Stockholm, Sweden, and ultimately released in September 2013. It is available online at www.climatechange2013.org. The Working Group II (*Impacts, Adaptation and Vulnerability*) and Working Group III (*Mitigation*) reports are scheduled for government approval and release in March and April 2014, respectively.

Climate change is no longer a distant threat. Average U.S. temperature has increased by about 0.8°C (1.5°F) since 1895; more than 80 percent of this increase has occurred since 1980. The warmest year ever recorded in the contiguous United States was 2012, when about one-third of all Americans experienced 10 days or more of 38°C (100°F) temperatures. Globally, the 12 hottest years on record have all come in the last 15 years (NOAA/NCDC 2012b).

These changes come with far-reaching consequences and real economic costs. In 2012 alone, there were 11 different weather and climate disaster events across the United States, with estimated losses exceeding $1 billion each (NOAA/NCDC 2012a). Taken together, these 11 events resulted in more than $110 billion in estimated damages, which made 2012 the second-costliest year on record, affecting many regions of the country and virtually all economic sectors. Although no individual event can be attributed to climate change alone, rising global temperatures are increasing the severity and costs associated with extreme weather events.

We have an obligation to current and future generations to take action to meet this challenge. By building on important progress achieved during the President's first term, the United States plans to meet its commitment to cut GHGs in the range of 17 percent below 2005 levels by 2020 and make additional progress toward forging a robust international response to this global challenge. We will also improve our ability to manage the climate impacts that are already being felt at home and around the world. Preparing for increasingly extreme weather and other consequences of climate change will save lives now and help to secure long-term American and global prosperity and security.

2. A COMMITMENT TO ACT

Key Pillars of *The President's Climate Action Plan*

On June 25, 2013, President Obama laid out a comprehensive plan to reduce GHG pollution, prepare the country for the impacts of climate change, and lead global efforts to fight climate change (EOP 2013a). *The President's Climate Action Plan*, which consists of a variety of executive actions grounded in existing legal authorities, has three key pillars.

Reduce U.S. GHG Emissions

During 2009–2011, average U.S. GHG emissions fell to the lowest level for any three-year period since 1994-1996, due to contributions from both economic factors and government policies. To build on this progress, the Obama administration is putting in place robust new rules to cut GHG emissions. The plan includes such steps as developing the first-ever national carbon pollution standards for both new and existing power plants, under the Clean Air Act; establishing post-2018 advanced fuel efficiency and GHG emission standards for heavy-duty vehicles; setting a new goal to double electricity generation from wind and solar power; boosting energy efficiency in appliances, homes, buildings, and industries; reducing emissions of highly potent hydrofluorocarbons (HFCs); developing a comprehensive methane emissions reduction strategy; and advancing efforts to protect our forests and other critical landscapes.

Prepare the United States for the Impacts of Climate Change

Even as we take new steps to reduce carbon pollution, we must also prepare for the impacts of a changing climate that are already being felt across the country. Building on its ongoing efforts to strengthen America's climate resilience, the Obama administration will continue to work with state and local governments to prepare for the unavoidable impacts of climate change by establishing policies that promote national resilience; supporting science and research that allow climate risk to be integrated into decision making; and protecting critical infrastructure and natural resources, to better protect people's homes, businesses, and ways of life from severe weather. In November, 2013, President Obama signed Executive Order 13653, *Preparing the United States for the Impacts of Climate Change* (EOP 2013b), and created the Task Force of Governors, Mayors, Tribal Leaders, and local officials to share approaches and advise the federal government on building preparedness and resilience across the United States.

Lead International Efforts to Combat Global Climate Change and Prepare for Its Impacts

Just as no country is immune from the impacts of climate change, no country can meet this challenge alone. That is why it is imperative for the United States to couple action at home with leadership internationally. America is working to help forge a truly global solution to this global challenge by galvanizing international action to significantly reduce emissions, prepare for climate impacts, and drive progress through international negotiations.

Building on Success

The President's Climate Action Plan builds on the successes achieved in the first five years of the Obama administration and initiates additional actions that will put the United States on a course to meet its goal of reducing emissions in the range of 17 percent below 2005 levels by 2020. The rest of this report is organized as follows: section 3 outlines the U.S. 2020 emission reduction goal and how progress toward it will be measured; section 4 explains U.S. GHG emission trends from 1990 through 2011 and key emission drivers; section 5 summarizes significant actions taken in the first term of the Obama administration to reduce GHG emissions; section 6 outlines the suite of new major actions in *The President's Climate Action Plan* to tackle this anticipated growth in emissions; section 7 presents projections of the emission reductions that could be achievable through a range of additional actions, consistent with implementation of the Climate Action Plan and measured against the U.S. 2020 goal; and section 8 summarizes international climate finance the United States has provided to developing countries.

3. 2020 GOAL: TRACKING PROGRESS

In 2009, the United States made a commitment to reduce U.S. GHG emissions in the range of 17 percent below 2005 levels by 2020. The President remains firmly committed to that ambitious goal and to building on the progress of his first term to help put the nation and the world on a sustainable long-term emissions trajectory. Although there is more work to do, the United States has already made significant progress, including doubling generation of electricity from wind and solar power and establishing historic new fuel economy standards. Building on these achievements, *The President's Climate Action Plan* lays out additional executive actions the administration will take, in partnership with states, communities, and the private sector, to continue on a path toward meeting the U.S. 2020 goal (EOP 2013a). (Section 7 lays out in detail the full scope of executive actions contained in the President's plan.)

The United States is committed to providing regular, transparent updates on progress toward meeting its 2020 goal. Progress will be tracked and reported annually, using the official national GHG inventory, prepared using IPCC and United Nations Framework Convention on Climate Change (UNFCCC) inventory guidelines (IPCC 2006, UNFCCC 2006). These reports provide annual information on the full scope of our 2020 goal, based on emissions and removals (taking into account emissions absorbed by U.S. forests and other lands) resulting from all sectors of the economy, and including all primary GHGs (carbon dioxide [CO_2], CH_4, nitrous oxide [N_2O], HFCs, perfluorocarbons [PFCs], sulfur hexafluoride [SF_6], and nitrogen trifluoride [NF_3]) (Table 1). This inventory-based accounting approach means that the U.S. goal is truly comprehensive, including the full scope of emissions included under the UNFCCC inventory that contribute to global climate change.

The institutional arrangements for measuring progress toward the goal are explained in more detail in the *Inventory of U.S. Greenhouse Gas Emissions and Sinks 1990–2011*, in Section 1.2 on Institutional Arrangements (U.S. EPA/OAP 2013). The U.S. Environmental Protection Agency (EPA), in cooperation with other U.S. government agencies, prepares the annual U.S. GHG inventory. A range of agencies and individuals are involved in supplying data to, reviewing, or preparing portions of the inventory, including federal and state government authorities, research and academic institutions, industry associations, and private consultants. Information on methods and arrangements for tracking progress on individual policies and measures implemented or planned by agencies across the U.S. government are provided in Chapter 4 of the *Sixth National Communication*.

4. U.S. GREENHOUSE GAS EMISSIONS AND TRENDS

According to the most recent national GHG inventory, in 2011 U.S. GHG net emissions—including land use, land-use change, and forestry (LULUCF)—were 5,797 teragrams (Tg) of CO_2 equivalents (CO_2e). This represents a 6.5 percent reduction below 2005 levels. Even with continued economic growth, annual net emissions have declined annually by 1.1 percent on average since 2005, a reversal of past trends of average annual increases of 1.0 percent per year from 1990 to 2005. In 2011, net emissions were down 2.0 percent from 2010 levels (Figure 1).

Carbon Dioxide Emissions

U.S. emissions of carbon dioxide (CO_2), the primary GHG emitted by human activities in the United States, have significantly declined. In 2011, CO_2 emissions represented more

Table 1   **Key Parameters of the U.S. Economy-wide Emission Reduction Targets**

Parameters	Targets
Base Year	2005
Target Year	2020
Emission Reduction Target	In the range of 17 percent below 2005 levels.
Gases Covered	CO_2, CH_4, N_2O, HFCs, PFCs, SF_6, and NF_3.
Global Warming Potential	100-year values from the IPCC Fourth Assessment Report (IPCC 2007).
Sectors Covered	All IPCC sources and sectors, as measured by the full annual inventory (i.e., energy, transport, industrial processes, agriculture, LULUCF, and waste).
Land Use, Land-Use Change, and Forests (LULUCF)	Emissions and removals from the LULUCF sector will be accounted using a net-net approach and a 2005 base year, including a production approach to account for harvested wood products. The United States is considering approaches for identifying the impact of natural disturbances on emissions and removals.
Other	To be in conformity with U.S. law.

Notes:

• Consistent with the formal UNFCCC inventory reporting guidelines for developed countries (IPCC 2006), the Inventory of U.S. Greenhouse Gas Emissions and Sinks, which will be submitted to the UNFCCC in April 2015, will utilize 100-year global warming potential values from the IPCC Fourth Assessment Report (IPCC 2007).

• CH_4 = methane; CO_2 = carbon dioxide; HFCs = hydrofluorocarbons; IPCC = Intergovernmental Panel on Climate Change; N_2O = nitrous oxide; NF_3 = nitrogen trifluoride; PFCs = perfluorocarbons; SF_6 = sulfur hexafluoride.

Figure 1   **1990–2011 U.S. Greenhouse Gas Emissions and Removals by Source**

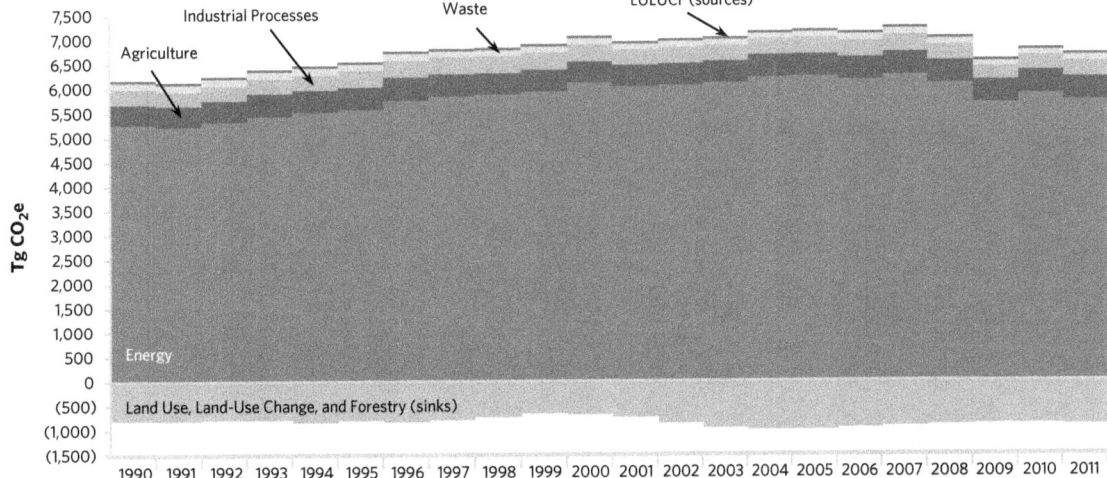

Source: U.S. EPA/OAP 2013.

Note: The 2013 U.S. GHG inventory is calculated using global warming potential values from the IPCC Second Assessment Report (IPCC 1996).

[2] U.S. DOE/EIA 2013i, Table 1.3.

than 80 percent of total U.S. GHG emissions (Figure 2). From 1990 through the mid-2000s, energy-related CO_2 emissions increased from approximately 5,100 Tg to a peak of just over 6,100 Tg in 2007. CO_2 emissions fell sharply, to approximately 5,500 Tg in 2011, down 8.0 percent from 2005 levels.

Emissions from fossil fuel combustion, the largest source of CO_2 emissions (94 percent, excluding removals from LULUCF) and of overall gross GHG emissions (79 percent, excluding removals from LULUCF) decreased at an average annual rate of 1.4 percent from 2005 through 2011. Historically, changes in emissions from fossil fuel combustion have been the dominant factor affecting U.S. emission trends. According to the U.S. Energy Information Administration, in 2012, approximately 82 percent of the energy consumed in the United States (on a British thermal unit [Btu] basis) was produced through the combustion of fossil fuels.[2] The remaining 18 percent came from other energy sources, such as hydropower, biomass, and nuclear, wind, and solar energy (Figure 3).

The five major fuel-consuming sectors contributing to CO_2 emissions from fossil fuel combustion are electricity generation and the transportation, industrial, residential, and commercial "end-use" sectors. The electricity generation sector produces CO_2 emissions as it consumes fossil fuel to provide electricity to one of the other four sectors. For the following discussion, emissions from electricity generation have been distributed to each end-use sector on the basis of each sector's share of aggregate electricity consumption.

Electricity Generation

The United States relies on electricity to meet a significant portion of its energy demands. Electricity generators consumed 36 percent of U.S. energy from fossil fuels and emitted 41 percent of the CO_2 from fossil fuel combustion in 2011. Principally due to a shift from coal to natural gas, as well as rapidly growing deployment of renewable sources of energy, CO_2 emissions from electricity generation **decreased by 10 percent below 2005 levels in 2011**.

Transportation End-Use Sector

Transportation activities (excluding international bunker fuels) accounted for 33 percent of CO_2 emissions from fossil fuel combustion in 2011. Virtually all of the energy consumed in this end-use sector came from petroleum products. Nearly 63 percent of the emissions resulted from gasoline consumption for personal vehicle use. The remaining emissions came from

Figure 2
2011 Greenhouse Gas Emissions by Gas

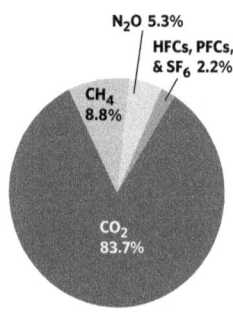

Note: Percentages based on Tg CO_2e. The 2013 U.S. GHG inventory is calculated using global warming potential values from the IPCC Second Assessment Report (IPCC 1996).

Source: U.S. EPA/OAP 2013.

Figure 3   **U.S. Primary Energy Profile Highlights: 2005–2012**

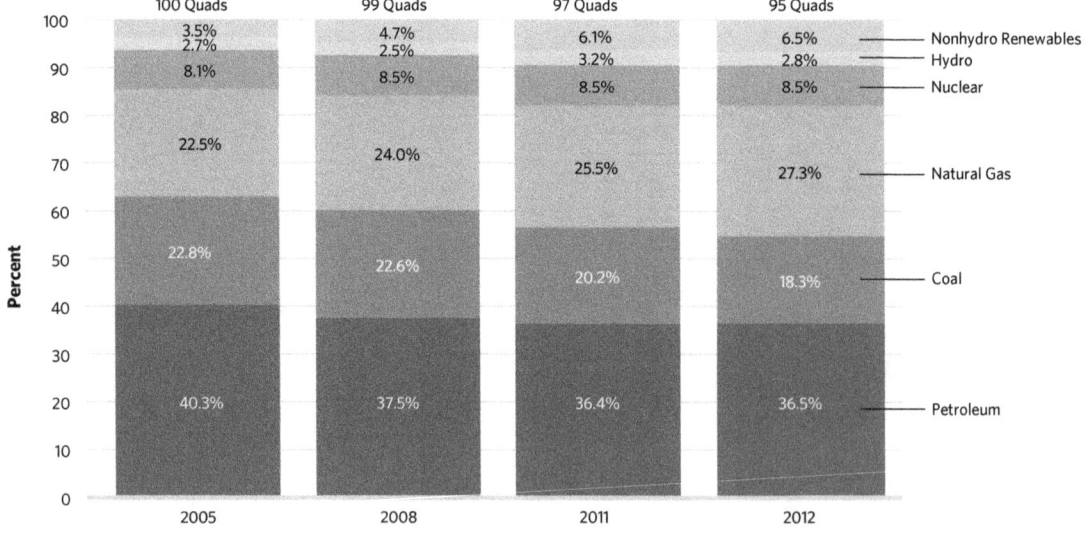

Source: U.S. DOE/EIA 2013h.

other transportation activities, including the combustion of diesel fuel in heavy-duty vehicles and jet fuel in aircraft. From 2005 through 2011, transportation emissions **dropped by 8 percent** due, in part, to increased fuel efficiency across the U.S. vehicle fleet, as well as higher fuel prices, and an associated decrease in the demand for passenger transportation.

Industrial End-Use Sector

Industrial CO_2 emissions, resulting both directly from the combustion of fossil fuels and indirectly from the generation of electricity that is consumed by industry, accounted for 26 percent of CO_2 from fossil fuel combustion in 2011. Emissions from industry have steadily **declined since 2005 (11.2 percent)**, due to structural changes in the U.S. economy (e.g., shifts from a manufacturing-based to a service-based economy), fuel switching, and efficiency improvements.

Residential and Commercial End-Use Sectors

The residential and commercial end-use sectors accounted for 21 and 18 percent, respectively, of CO_2 emissions from fossil fuel combustion in 2011, including each sector's "indirect" emissions from electricity consumption. Both sectors relied heavily on electricity to meet energy demands; 71 and 77 percent, respectively, of residential and commercial emissions were attributable to electricity consumption for lighting, heating, cooling, and operating appliances. Emissions from the residential and commercial end-use sectors, including direct and indirect emissions from electricity consumption, have **decreased by 7.3 percent and 6.5 percent since 2005**, respectively.

Methane Emissions

CH_4 emissions **decreased by 1.1 percent since 2005**, primarily resulting from the following sources: natural gas systems, enteric fermentation associated with domestic livestock, and decomposition of wastes in landfills. Emissions from natural gas systems, the largest anthropogenic source of CH_4 emissions, have **decreased by 9 percent since 2005**, due largely to a decrease in emissions from field production.

Nitrous Oxide Emissions

Agricultural soil management, mobile source fuel combustion, and stationary fuel combustion were the major sources of N_2O emissions, which **increased slightly from 2005 levels**. Making up 70 percent of total N_2O emissions, highly variable agricultural sector factors—including weather, crop production decisions, and fertilizer application patterns—are the main factors that influence overall N_2O levels.

Hydrofluorocarbon, Perfluorocarbon, and Sulfur Hexafluoride Emissions

Despite being emitted in smaller quantities relative to the other principal GHGs, emissions of HFCs, PFCs, and SF_6 are a significant and growing share of U.S. emissions because many of these gases have extremely high global warming potentials and, in the cases of PFCs and SF_6, long atmospheric lifetimes. Emissions of substitutes for ozone-depleting substances and emissions of HFC-23 during the production of hydrochlorofluorocarbon (HCFC)-22 were the primary contributors to aggregate HFC emissions, which as a class of fluorinated gases **increased by 12.2 percent since 2005**. PFC emissions **rose by 13 percent**, resulting from semiconductor manufacturing and as a by-product of primary aluminum production. Electrical transmission and distribution systems accounted for most SF_6 emissions, which **were down 37 percent** from 2005 levels in 2011.

Land Use, Land-Use Change, and Forestry

LULUCF activities in 2011 resulted in a net carbon sequestration of 905 Tg CO_2e, which, in aggregate, **offset 13.5 percent of total U.S. GHG emissions**. Forest management practices, tree planting in urban areas, the management of agricultural soils, and growth in other carbon pools resulted in a net uptake (sequestration) of carbon in the United States. Forests (including vegetation, soils, and harvested wood) accounted for 92 percent of total 2011 net CO_2 flux; urban trees accounted for 8 percent; and mineral and organic soil carbon stock changes combined with landfilled yard trimmings and food scraps together accounted for less than 1 percent. The net forest sequestration is a result of net forest growth and increasing forest area, as well as a net accumulation of carbon stocks in harvested wood pools. Forest carbon estimates, with the exception of CO_2 fluxes from wood products and urban trees, are calculated annually based on activity data collected through forest and land-use surveys conducted at multiple-year intervals ranging from 1 to 10 years.

5. FIVE YEARS OF SIGNIFICANT NEW ACTION

The past five years have seen a remarkable turnaround in U.S. GHG emissions, due in part to the unprecedented action taken by the Obama administration to tackle climate change. During the past five years, the United States has taken a series of important steps that not only reduce the harmful emissions that contribute to climate change, but also improve public health, while protecting America's water and air.

In the past five years, the United States has pursued a combination of near- and long-term, regulatory and voluntary activities to reduce GHG emissions. Policies and measures are being implemented across the economy, including in the transportation, energy supply, energy end-use, industrial, agricultural, land use and forestry, and waste sectors, and in federal facilities. These cross-cutting policies and measures encourage cost-effective reductions across multiple sectors. Chapter 4 of the *Sixth National Communication* outlines in more detail the full set of policies and measures adopted and implemented since 2010, organized by sector and by gas. Table 4-2 of the chapter includes measured GHG emission reductions achieved in 2011, and estimated emission reductions expected from each policy and measure in 2020.

National Achievements

Increased the Efficiency of Cars and Trucks

The United States is aggressively working to reduce GHG pollution from America's vehicles. The Obama administration has adopted the toughest fuel economy and GHG emission standards for passenger vehicles in U.S. history, requiring an average performance equivalent of 54.5 miles per gallon by 2025, if achieved through fuel economy improvements alone. These standards are projected to reduce oil consumption by more than 2 million barrels per day in 2025 and will cut 6 billion metric tons of GHGs over the lifetime of model year (MY) 2012–2025 vehicles. The administration has also finalized the first-ever national fuel efficiency and GHG emission standards for commercial trucks, vans, and buses for MYs 2014–2018. Under President Obama's leadership, the nation has also made critical investments in advanced vehicle and fuel technologies, public transit, and high-speed rail.

Delivered on a Commitment to Double Generation of Electricity from Wind and Solar Sources

Since 2008, the United States has doubled renewable generation from wind and solar sources, helping to develop nearly 50,000 new clean energy projects that are supporting jobs throughout the country. In 2012, the President set a goal to permit 10,000 megawatts (MW) of renewable energy sources on public lands—a goal the U.S. Department of the Interior (DOI) has achieved. America is now home to some of the largest wind and solar farms in the world.

Cut Pollution and Saved Money for Consumers through Energy Efficiency

During President Obama's first term, significant progress was made in cutting domestic carbon pollution and reducing consumer energy bills by setting appliance efficiency standards for nearly 40 products; weatherizing more than 1 million homes; recognizing superior energy savings across more than 65 product categories, new single and multifamily homes, 16 commercial building space types, and 12 manufacturing plant types that can earn the ENERGY STAR label; and forging additional private and public partnerships to drive investments in energy efficiency across sectors.

Issued Federal Air Standards for the Oil and Natural Gas Industry

In 2012, EPA issued cost-effective regulations to reduce harmful air pollution from the oil and natural gas industry, while allowing continued, responsible growth in U.S. oil and natural gas production. The final rules include the first national air standards for natural gas wells that are hydraulically fractured. The final rules are expected to yield a nearly 95 percent reduction in volatile organic compound emissions from regulated emission sources and, as a co-benefit, significant methane emission reductions, estimated at 32.6 Tg CO_2e in 2015 and 39.9 Tg CO_2e in 2020.

Cut Federal Government Carbon Pollution

In 2010, President Obama announced that the federal government would reduce its direct GHG emissions by 28 percent from 2008 levels by 2020. Agencies are also meeting the directive to enter into performance-based contracts to achieve substantial energy savings at no net cost to American taxpayers.

Regional, State, and Local Achievements

Within the United States, several regional, state, and local policies and initiatives complement federal efforts to reduce GHG emissions. These include actions that directly regulate GHG emissions, as well as policies that indirectly reduce emissions. The Obama administration supports state and local government actions that reduce GHG emissions by sponsoring policy dialogues, issuing technical documents, facilitating consistent measurement approaches and model policies, and providing direct technical assistance. Such federal support helps state and local governments learn from each other to leverage best practice approaches, helping reduce overall time and costs for both policy adoption and implementation. A full discussion of state and local efforts can be found in Chapter 4 of the *Sixth National Communication*. Following is a sample of major regional, state, and local efforts currently underway.

State Emission Targets

As of August 2013, 29 states had adopted some form of state GHG reduction targets or limits, which vary in stringency, timing, and enforceability. Statewide GHG targets are nonregulatory commitments to reduce GHG emissions to a specified level in a certain timeframe (e.g., 1990 levels by 2020). Such targets can be included in legislation, but are more typically established by the state's governor in an executive order or a state advisory board in a climate change action plan. Statewide GHG *limits* reduce emissions within a certain timeframe, but are regulatory in nature and more comprehensive than emission targets. These policies can include regulations to require GHG emission reporting and verification, and may establish authority for monitoring and enforcing compliance.

Regional Greenhouse Gas Initiative (RGGI)

Launched on January 1, 2009, RGGI is the first U.S. mandatory market-based cap-and-trade program to reduce GHG emissions. RGGI currently applies to 168 electricity-generation facilities in nine Northeast and Mid-Atlantic states, which account for approximately 95 percent of CO_2 emissions from electricity generation in the region. In February 2013, the participating states agreed to make significant revisions to the program, capping CO_2 emissions at 91 million short tons per year in 2014—a 45 percent reduction from the previous cap of 165 million short tons. The cap will then be reduced by 2.5 percent each year from 2015 through 2020.

Under the initiative, nearly 90 percent of allowances are distributed through auction. As of March 2013, cumulative auction proceeds exceeded $1.2 billion. Participating states have invested approximately 80 percent of auction proceeds in consumer benefit programs, including investments in end-use energy efficiency and renewable energy deployment programs at the state and local levels.[3]

California's Global Warming Solutions Act (AB 32)

Signed into law in 2006, AB 32 established a statewide GHG emissions limit of 1990 levels to be achieved by 2020. As part of a portfolio of measures implemented to achieve this statewide GHG emissions limit, the California Air Resources Board adopted cap-and-trade regulations in 2011. The regulations established a declining cap on sources responsible for approximately 85 percent of statewide GHG emissions, including refineries, power plants, industrial facilities, and transportation fuels. In addition, the portfolio of programs implemented to achieve the statewide GHG emissions limit under AB 32 includes a mandatory GHG emissions reporting program for large emitters, a renewable portfolio standard, and various energy efficiency measures and incentives.[4]

Power Sector Standards

As of February 2013, three states (New York, Oregon, and Washington) have GHG emission standards for electric-generating utilities, requiring power plants to have emissions equivalent to or lower than the established standard. For example, in New York, new or expanded

[3] See www.rggi.org.

[4] See http://www.arb.ca.gov/cc/ab32/ab32.htm.

baseload plants (25 MW and larger) must meet an emission rate of either 925 pounds (lb) of CO_2 per megawatt-hour (MWh) (output based) or 120 lb CO_2/per million British thermal units (MMBtu) (input based). Non-baseload plants (25 MW and larger) must meet an emission rate of either 1,450 lb CO_2/MWh (output based) or 160 lb CO_2/MMBtu (input based).

Three states (California, Oregon, and Washington) also have standards that apply to electric utilities that provide electricity to retail customers. These standards place conditions on the emission attributes of electricity procured by electric utilities. And as of January 2013, 29 states had a renewable portfolio standard, which requires utilities to supply a certain amount of electricity to customers from renewable energy sources or install a certain amount of electricity-generating capacity from renewable energy sources in a set time frame. Standards can vary, with annual or cumulative targets.

Energy Efficiency Programs and Standards

As of August 2013, 18 states have mandatory energy efficiency resource standards in place, which require utilities to reduce energy use by a certain percentage or amount each year. Many of these utilities use public benefit funds to invest in energy efficiency projects. Also, as of August 2013, 19 states, Washington, D.C., and Puerto Rico have some form of public benefit fund policy in place, in which utility consumers pay a small charge to a common fund that is then used to invest in energy efficiency and renewable energy projects and programs. In addition, many state and local governments lead by example by establishing programs to reduce energy bills and emissions in their own operations and buildings.

6. LOOKING AHEAD—*THE PRESIDENT'S CLIMATE ACTION PLAN*

During the President's first term, the United States made significant progress in several key sectors, through federal as well as state and local actions, in reducing U.S. GHG emissions. Significant new measures will be required to stay on track to reach the U.S. goal of achieving reductions in the range of 17 percent below 2005 levels by 2020. By building on the success of the first term, the United States can achieve substantial further emission reductions consistent with this ambitious goal.

In his 2013 State of the Union Address, President Obama called on Congress to pursue a bipartisan, market-based approach to combating climate change.[5] In the absence of congressional action to date, the President has laid out a comprehensive *Climate Action Plan* of executive actions, grounded in existing legal authorities, that will be implemented across U.S. government agencies to reduce GHGs, prepare our cities and nation for the worsening effects of climate change, and accelerate the transition to more sustainable sources of energy (EOP 2013a).

The first pillar of the President's plan focuses on tackling U.S. emissions of GHGs by taking the following actions.

Cutting Carbon Pollution from Power Plants

The President has directed EPA to work closely with states and other stakeholders to establish carbon pollution standards for both new and existing power plants. EPA is moving forward on the President's plan. For newly built power plants, EPA issued a new proposal on September 20, 2013. Issuance of the new proposal, together with the ensuing rulemaking process, will advance adoption of carbon pollution standards for new power plants reflect recent developments and trends in the power sector. The new proposal, comment period, and public hearings will allow an open and transparent review and robust input on the broad range of technical and legal issues contained among the more than 2.5 million comments generated by the first proposal submitted by EPA in April 2012. For existing power plants, the plan directs EPA to issue a draft rule by June 2014 and a final rule by June 2015.

Promoting American Leadership in Renewable Energy

During the President's first term, the United States more than doubled generation of electricity from wind and solar sources. To continue U.S. leadership in clean energy, President Obama has set a goal to double renewable electricity generation from wind and solar once again by 2020. To meet this ambitious target, the President directed DOI to permit enough renewable energy projects on public lands by 2020 to power more than 6 million homes; designated the

[5] See http://www.whitehouse.gov/ state-of-the-union-2013.

first-ever hydropower project for priority permitting; and set a new goal to install 100 MW of renewable power in federally assisted housing by 2020, while expanding and modernizing the electric grid to make electricity more reliable and promote clean energy sources.

Unlocking Long-Term Investment in Clean Energy Innovation

The plan furthers the President's commitment to keeping the United States at the forefront of clean energy research, development, and deployment. The President's fiscal year (FY) 2014 budget requested increasing funding for clean energy technology across all government agencies by 30 percent, to approximately $7.9 billion. This includes investment in a range of energy technologies, from advanced biofuels and emerging nuclear technologies to clean coal.

Expanding the President's Better Buildings Challenge

Focused on helping American commercial and industrial buildings become at least 20 percent more energy efficient by 2020, the Better Buildings Challenge is already showing results: first-year results show that the Challenge Partners are on track to meet the 2020 goal. To continue this success, the Obama administration has expanded the program to multifamily housing, partnering with private and affordable building owners and public housing agencies to cut energy waste, and launched the Better Buildings Accelerators to support state and local government-led efforts to reduce energy waste.

Establishing a New Goal for Energy Efficiency Standards

The plan sets a new goal to establish efficiency standards for appliances and federal buildings, which will reduce carbon pollution by at least 3 billion metric tons cumulatively by 2030—more than half of the annual carbon pollution from the U.S. energy sector.

Advancing Vehicle Fuel Efficiency and Greenhouse Gas Emission Standards

In 2011, the Obama administration finalized the first-ever fuel efficiency and GHG emission standards for heavy-duty trucks, buses, and vans, specifically for MYs 2014–2018. The administration will seek input from industry and stakeholders as it develops fuel efficiency and GHG emission standards for heavy-duty vehicles beyond 2018.

Curbing Emissions of Hydrofluorocarbons

The United States will lead efforts to curb global HFC emissions through both international diplomacy as well as domestic actions, building on its success in addressing HFC leakage from vehicle air conditioning systems through flexible approaches within the U.S. vehicle GHG standards. Moving forward, EPA will use its authority through the Significant New Alternatives Policy (SNAP) Program to encourage private-sector investment in low-emission technology by identifying and approving climate-friendly chemicals, while prohibiting certain uses of the most harmful chemical alternatives. In addition, the President has directed the federal government to purchase cleaner alternatives to HFCs whenever feasible, and to transition over time to equipment that uses safer and more sustainable alternatives.

Reducing Methane Emissions

Methane emissions will be addressed by developing a comprehensive, interagency methane strategy, focusing on assessing current emissions data, addressing data gaps, identifying technologies and best practices for reducing emissions, and identifying existing authorities and incentive-based opportunities to reduce methane emissions. As part of this strategy, the administration will also work collaboratively with state governments, as well as the private sector, to reduce emissions across multiple sectors.

Preserving the Role of Forests in Mitigating Climate Change

Mitigation across the forest sector will be addressed by identifying new approaches to protect and restore our forests, as well as other critical landscapes, including grasslands and wetlands, in the face of a changing climate.

Phasing Out Subsidies That Encourage Wasteful Consumption of Fossil Fuels

At the 2009 G-20 meeting in Pittsburgh, Pennsylvania, the United States successfully advocated for a commitment to phase out fossil fuel subsidies, and the administration has since won similar commitments in other fora, such as the Asia-Pacific Economic Cooperation forum. President Obama has called for the elimination of U.S. fossil fuel tax subsidies in his FY 2014 budget, and the administration will continue to collaborate with partners around the world toward this goal.

Instituting a Federal Quadrennial Energy Review

The administration will conduct a Quadrennial Energy Review to ensure that U.S. federal energy policy meets its economic, environmental, and security goals in this changing landscape. This first-ever review will focus on infrastructure challenges, and will identify the threats, risks, and opportunities for U.S. energy and climate security, enabling the federal government to translate policy goals into a set of analytically based, clearly articulated, sequenced, and integrated actions and proposed investments.

Leading at the Federal Level

President Obama believes that the federal government must be a leader in clean energy and energy efficiency. Federal agencies have a goal of reducing GHG emissions by 28 percent by 2020, and have already reduced them by more than 15 percent between 2008 and 2012. As outlined in the plan, in December 2013, the President issued a Presidential Memorandum on Federal Leadership on Energy Management directing federal agencies to lead by example in acting on climate and increasing the nation's renewable energy use. The memorandum strengthens established administration efforts to increase government-wide energy efficiency and sustainability, and sets a goal of 20 percent renewable energy use by the federal government by 2020—nearly triple the previous goal of 7.5 percent. In addition, the federal government will continue to pursue greater energy efficiency and GHG emission reductions in its operations.

7. SIGNIFICANT REDUCTIONS ACHIEVABLE IN 2020

The administration is already hard at work implementing *The President's Climate Action Plan*. Moreover, all of the actions outlined above are grounded in existing authorities and build on policies and programs already in place, and many of the specific measures that scale up and expand existing efforts are already underway.

However, several of the actions will require U.S. government agencies to develop recommendations, propose new rules, augment existing activities, and undertake processes that entail significant stakeholder outreach and public comment before final rules and programs are in place. Although the purpose of each action is clear, the exact shape and details of each will be developed over time. Until recommendations, rulemakings, and other administrative activities for these specific actions are complete, it will not be possible to estimate the exact scale of emission reductions that will be achieved by each specific action.

Nevertheless, at this early stage, the potential range of GHG reductions achievable by 2020 toward the ultimate goal of achieving economy-wide emission reductions in the range of 17 percent below 2005 levels can be assessed. Light can be shed on the potential scale of additional reductions through 2020 by assessing the broad categories of actions contained in the plan, using integrated models to the extent possible to ensure no double counting of reductions within each category.

Key Categories of Action for Achieving Additional Emission Reductions

Starting with projections of U.S. emissions based on policies enacted before 2012 (the "2012 Policy Baseline"), the additional reductions that are achievable by 2020 were estimated for three key categories of actions: energy CO_2, HFCs, and methane (Table 2).[6]

Energy CO_2

Estimates for energy CO_2 are based on a range of potential actions, including increasing levels of clean electricity generation, extension of energy efficiency standards and actions affecting residential and commercial buildings, and enhanced measures addressing industrial efficiency and transportation. Although these estimates do not explicitly measure projected emission reductions from specific rules, standards, and other efforts laid out in the *Climate Action Plan* but not yet implemented, they do provide a range of potential reductions that can be achieved across the relevant sectors (see Biennial Report Methodologies Appendix for further

[6] Unless otherwise stated, all GHGs in this document are reported in teragrams of CO_2 equivalents (Tg CO_2e), using the 100-year global warming potentials (GWPs) listed in the IPCC's Second Assessment Report (SAR) (IPCC 1996) to convert non-CO_2 gases to CO_2e. UNFCCC guidelines for inventories and national communications require that emissions be reported using SAR GWP values.

information). As reflected in Table 2, this analysis shows that, taken together, additional actions across the energy sector have the potential to reduce CO_2 emissions by an additional 485-800 Tg relative to the 2012 Policy Baseline or, equivalently, to reduce emissions from 2005 levels by an additional 8-12 percent.

Hydrofluorocarbons

Estimates for potential achievable U.S. reductions for HFCs, reflected in Table 2, are based on analysis conducted by EPA for a proposal for a global commitment to phase down production and consumption of HFCs under the Montreal Protocol on Substances That Deplete the Ozone Layer (U.S. EPA/OAP 2013). The United States can, and will, take several steps domestically as it moves toward an international agreement, including using EPA authority through the SNAP Program and leveraging federal government purchasing power to promote cleaner alternatives. These actions can set the United States on firm ground for reaching reductions proposed under the Montreal Protocol.

Methane

There are many options for continued and further actions to address U.S. methane emissions. The President has called for U.S. agencies to develop a comprehensive interagency methane strategy, and work on this strategy is already underway. Until such a strategy is complete, however, assessing the potential achievable reductions of methane emissions in 2020 involves considerable uncertainty, as reflected in the estimate of potential methane abatement in Table 2.

Taken together, these additional reductions have the potential to bring emissions within the range of 17 percent below 2005 levels. In the coming months and years, as the administration works to implement the *Climate Action Plan*, the scope and scale of each policy and measure will become clearer, allowing a more detailed and in-depth assessment of the potential emission reductions than this initial analysis provides. As rules and standards become finalized and programs and partnerships are rolled out, we will be able to assess their expected impacts over time with more accuracy, and thus will narrow the range of potential emission reductions displayed in Figure 4.

Table 2 2020 Ranges of Potential Emission Reductions Relative to Emissions in the 2012 Policy Baseline Scenario (Tg CO_2e)

Pollutant	Potential Reductions
Energy Sector CO_2	485-800
HFCs	100-135
CH_4	25-90
Total	**610-1,025**

Note: HFC values listed for potential abatement in 2020 were calculated using GWP values from the IPCC Fourth Assessment Report (IPCC 2007). CH_4 values listed for potential abatement in 2020 were calculated using GWP values from the IPCC Second Assessment Report (IPCC 1996).

Figure 4 **U.S. Emission Projections—2012 Policy Baseline Compared with Potential Reductions from Additional Measures Consistent with the *Climate Action Plan***

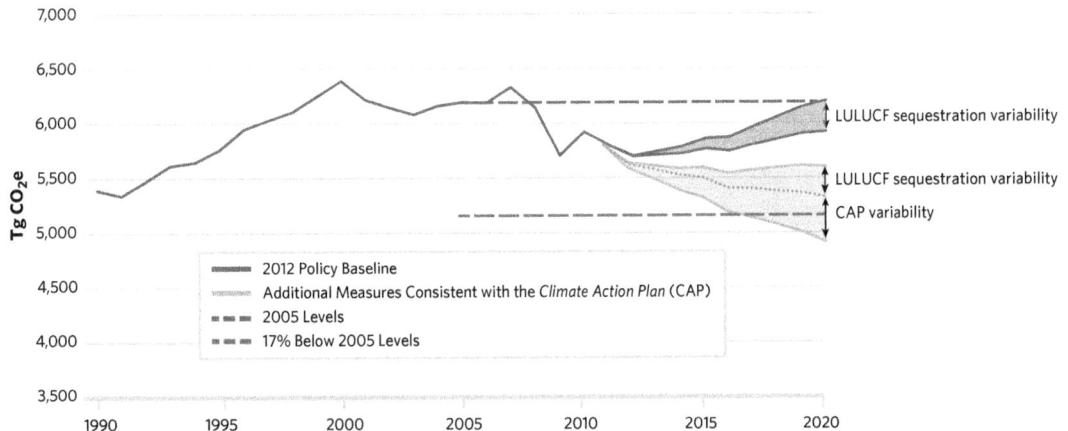

Notes: Figure 4 shows the range of projected emissions for both (1) the 2012 Policy Baseline scenario (in blue), which assumes that no additional measures are implemented after 2012; and (2) a scenario (in green) that incorporates post-2012 implementation of Additional Measures Consistent with the *Climate Action Plan*. The range (in blue) for the 2012 Policy Baseline scenario reflects variability in projected net sequestration rates from land use, land-use change, and forestry (LULUCF), much of which will be determined by factors that cannot be directly influenced by policies and measures. The range (in green) for the Additional Measures Consistent with the *Climate Action Plan* scenario reflects both LULUCF sequestration variability, as well as uncertainty regarding projected emission reductions from measures that will be implemented consistent with the *Climate Action Plan*. The dotted line delineates the share of projected variability that is attributable to LULUCF and the *Climate Action Plan*, respectively. Specifically, the portion labeled "CAP variability" illustrates the range of emission outcomes that can be directly influenced by implementation of the *Climate Action Plan*, assuming best-case LULUCF sequestration outcomes. The LULUCF sequestration variability ranges are identical in both scenarios.

The scenarios displayed in Figure 4 illustrate the ranges of projected emissions from the 2012 Policy Baseline (no additional action from 2012 onward) and from implementation of additional measures consistent with the *Climate Action Plan*. The 2012 Policy Baseline range and a portion of the Additional Measures range result from uncertainty and variability in the projected rate of net carbon sequestration from LULUCF in 2020. Specifically, the top of the range (in green) reflects the low end of the potential GHG reduction due to policy and weaker LULUCF sequestration. The bottom of the range reflects the high end of the potential reduction due to policy and stronger LULUCF sequestration. Due to the inherent uncertainty of projected emissions and removals from LULUCF, and the more limited ability to influence these outcomes relative to other sectors of the economy, both scenarios include a wide range of potential LULUCF outcomes.[7]

There are indications that in the long term, U.S. forest carbon stocks are likely to accumulate at a slower rate than in past decades, and eventually may decline as a result of forestland conversion, the maturation of land that has previously been converted to forests, and adverse impacts related to climate change and other disturbances (Haynes et al. 2007, Alig et al. 2010, Haim et al. 2011, USDA/FS 2012). The exact magnitude and timing of these changes are uncertain, but forests are unlikely to continue historical trends of sequestering additional carbon stocks in the future under current policy conditions. These changes may already be starting in U.S. forests: however, major changes in U.S. forest inventory monitoring results are not expected in the next 5–10 years. The ranges presented in the scenarios above use high and low estimates for U.S. LULUCF carbon pathways to 2020: high sequestration (which reflects lower CO_2 emissions to the atmosphere) is an extrapolation based on recent forestland and forest carbon density accumulation rate trends, and low sequestration estimates reflect possible slower accumulation of forestland and carbon density.

2012 Policy Baseline Emission Projections

Comparing the range of reductions possible under the scenarios described above (including actions consistent with the *Climate Action Plan*) with the 2012 Policy Baseline scenario provides a starting point to assess additional reductions needed to continue to make progress toward the 2020 goal (Table 3). The 2012 Policy Baseline (or "with measures" scenario)[8] takes into account only those policies adopted before September 2012; it shows that U.S. emissions start to trend upward absent additional measures. The *Climate Action Plan* initiates

[7] For more information on the methodologies used to develop the LULUCF projections, *see* Chapter 5 of the *Sixth National Communication* of the *U.S. Climate Action Report 2014*.

[8] The 2012 Policy Baseline scenario refers to the "with measures" scenario required by the UNFCCC National Communications reporting guidelines (UNFCCC 2006).

Table 3   **Historical and Projected U.S. Greenhouse Gas Emissions Baseline, by Sector: 1990–2030** (Tg CO_2e)

Sectors[b]		Historical GHG Emissions[a]				Projected GHG Emissions			
		2000	2005	2010	2011	2015	2020	2025	2030
Energy		4,258	4,321	4,104	3,981	3,936	4,038	4,141	4,207
Transportation		1,861	1,931	1,786	1,765	1,710	1,702	1,660	1,627
Industrial Processes		357	335	308	331	378	438	504	536
Agriculture		432	446	462	461	461	485	498	512
Forestry and Land Use		31	25	20	37	30	27	40	35
Waste		136	137	131	128	127	126	125	123
Total Gross Emissions		**7,076**	**7,195**	**6,812**	**6,702**	**6,643**	**6,815**	**6,967**	**7,041**
Forestry and Land Use (Sinks)[c]	high sequestration	-682	-998	-889	-905	-884	-898	-917	-937
	low sequestration					-787	-614	-573	-565
Total Net Emissions	**high sequestration**	**6,395**	**6,197**	**5,923**	**5,797**	**5,759**	**5,918**	**6,050**	**6,104**
	low sequestration					**5,856**	**6,201**	**6,394**	**6,476**

[a] Historical emissions and sinks data are from U.S. EPA/OAP 2013. Bunker fuels and biomass combustion are not included in 2013 U.S. GHG inventory calculations.

[b] Sectors correspond to 2013 U.S. GHG inventory reporting sectors, except that carbon dioxide, methane, and nitrous oxide emissions associated with mobile combustion have been moved from energy to transportation, and solvent and other product use is included within industrial processes.

[c] Sequestration is only included in the net emissions total.

Box 1   **International Impacts of Measures to Respond to Climate Change**

The most significant action the United States can take to positively impact global climate and all those affected by its changes is to mitigate emissions. As appropriate and consistent with domestic law, the United States in many instances also assesses and considers the potential impacts that certain U.S. mitigation actions themselves may have on other countries. The most effective way to maximize the positive and minimize any negative impacts on other countries as a result of U.S. mitigation action is to enhance less developed countries' capacities to transition to clean-energy, low-emission economies themselves. Three basic categories of significant U.S. government support address this cause: policy development support, public-private partnerships, and worker training. The following are examples of programs in each of these three categories.

Policy Development Support

The U.S. Enhancing Capacity for Low Emissions Development Strategies (EC-LEDS) program provides technical assistance to more than 20 partner countries to develop LEDS that grow and strengthen the economy while reducing GHG emissions over the long term. Through this program, U.S. government expertise is mobilized to provide tools, trainings, and resources to practitioners in partner countries that build capacity for these country-driven policy strategies.

Public-Private Partnerships

The Energy Utility Partnership Program (EUPP) was created by the U.S. Energy Association, a nonprofit public-private association devoted to increasing the understanding of energy issues. EUPP establishes voluntary partnerships between energy utilities, energy system operators, energy markets, and other energy service providers in countries assisted by the U.S. Agency for International Development (USAID) and their U.S. counterparts. These partnerships facilitate the sharing of experiences and best practices in the day-to-day planning, operation, and management of utilities and other energy service providers.

Worker Training

The Vocational Training and Education for Clean Energy (VOCTEC) program is a five-year global program funded by USAID and led by Arizona State University. VOCTEC aims to improve the sustainability of renewable energy infrastructure and investments in developing countries by increasing the awareness, knowledge, and capacity of local stakeholders to facilitate renewable energy investments, primarily in decentralized clean energy technologies. VOCTEC's vocational training programs for operators and technicians focus on installation, operations, and maintenance of renewable energy systems in developing countries.

additional actions that will achieve substantial emission reductions and put the United States on a course to meet the 2020 goal. For detailed information on the 2012 Policy Baseline projections, including underlying methodologies and projections "with measures" to 2030, see the *Sixth National Communication*, Chapter 5.

Projections of gross GHG emissions (not including emissions and removals from LULUCF) under the 2012 Policy Baseline case presented in this report are significantly lower than emission projections presented in previous U.S. Climate Action Reports (CARs) (Figure 5). These differences can be traced to a combination of changes in policies, energy prices, and economic growth. In the 2010 CAR, emissions were projected to increase by 4.3 percent from 2005 to 2020, versus a 14–20 percent decline from 2005 levels projected in this report under a range of actions across economic sectors consistent with those included in the 2010 CAR (U.S. DOS 2010). In the 2006 CAR, the expected growth was even higher, totaling 17 percent over the same time period. Actual emissions for 2011 are also significantly below those projected in past reports (U.S. DOS 2007).

8. INTERNATIONAL CLIMATE FINANCE

The United States is committed to assisting developing countries in their efforts to mitigate and adapt to climate change. The United States is using the full range of institutions—bilateral, multilateral, development finance, and export credit—to mobilize private finance and invest strategically in building lasting resilience to impacts; to reduce emissions from deforestation and land degradation; and to support low-carbon development strategies and the transition to a sustainable, clean energy economy. We work to ensure that our capacity-building and investment support is efficient, effective, innovative, based on country-owned plans, and focused on achieving measurable results, with a long-term view of economic and environmental sustainability.

[9] Fact Sheet: U.S. Global Development Policy. *See* http://www.whitehouse.gov/the-press-office/2010/09/22/fact-sheet-us-global-development-policy.

[10] Foreign Assistance Initiatives. *See* http://foreignassistance.gov/InitiativeLanding.aspx.

Figure 5   **Comparison of Gross GHG Emission Projections from Previous U.S. Climate Action Reports**

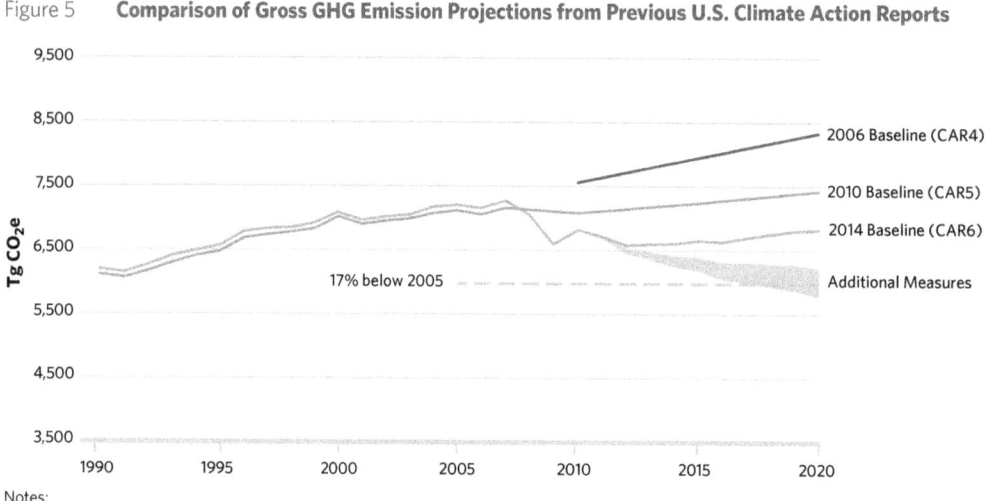

Notes:

• Emissions displayed are "Total Gross Emissions" from Table 3, and do not include CO_2 sinks from land use, land-use change, and forestry. Projections from each CAR reflect a baseline or "with measures" scenario, including the effect of policies and measures implemented at the time that the projections were prepared, but not future additional measures.

• Each year, emission and sink estimates are recalculated and revised for all years in the *Inventory of U.S. Greenhouse Gas Emissions and Sinks,* as attempts are made to improve both the analyses themselves, through the use of better methods or data, and the overall usefulness of the report. In this effort, the United States follows the 2006 IPCC Guidelines (IPCC 2006), which states, "Both methodological changes and refinements over time are an essential part of improving inventory quality."

Climate change has become a major focus of U.S. diplomatic and development assistance efforts and has been integrated into the core operations of all major U.S. foreign assistance agencies. The 2010 Presidential Policy Directive on Global Development[9] identified the Global Climate Change Initiative as one of three priority U.S. development initiatives.[10] In addition, the 2012 U.S. Agency for International Development (USAID) *Climate Change and Development Strategy* sets out principles, objectives, and priorities for USAID climate change assistance from 2012 through 2016 (USAID 2012). This strategy prioritizes not only clean energy, sustainable landscapes, and adaptation, but also integration—factoring climate change knowledge and practice into all USAID programs to ensure that all sector portfolios are climate resilient and, where possible, reduce GHG emissions.

At the 15th Conference of the Parties (COP-15) in Copenhagen, the United States committed to working with other developed countries to collectively provide resources approaching $30 billion in the "fast start" finance (FSF) period 2010–2012 to support developing countries in their mitigation and adaptation efforts. The United States also agreed, in conjunction with other developed country Parties, to the goal of collectively mobilizing $100 billion in climate finance per year by 2020, from a wide variety of public and private sources, to address the needs of developing countries in the context of meaningful mitigation actions and transparency on implementation.

As noted in Decision 1 of COP-18 in Doha, developed country Parties successfully achieved the FSF goal. U.S. climate finance was $7.5 billion[11] from FYs 2010 through 2012 and reached more than 120 countries through bilateral and multilateral channels, meeting the President's commitment to provide our fair share of the collective pledge.

This section of the *Biennial Report* provides details on U.S. climate finance by channels and instruments, thematic pillar, and region for FYs 2011 and 2012. It also describes U.S. efforts to mobilize private climate finance, and illustrates examples of U.S. contributions to capacity building and transfer of technology. For additional information on U.S. climate finance, including further examples of U.S. activities, see Chapter 7 of the *Sixth National Communication.*

U.S. Climate Finance and International Leadership to Address Climate Change in FY 2011 and FY 2012

In FYs 2011 and 2012, U.S. climate finance was $5.5 billion, which is comprised of approximately $3.1 billion in congressionally appropriated assistance, $496 million of export credit,

---

[11] The totals reported here reflect slight revisions to previously reported levels, based on updated information received since the release of the November 2012 Fast Start Finance report (U.S. DOS 2012).

and $1.8 billion of development finance. The United States organizes its support according to three pillars: adaptation, clean energy, and sustainable landscapes. Signature initiatives for each of the three pillars follow; they are not intended to be comprehensive.

Adaptation

For adaptation, dedicated U.S. climate assistance prioritizes countries, regions, and populations that are highly vulnerable to the impacts of climate change. By increasing resilience in key sectors, such as food security, water, coastal management, and public health, U.S. programs help vulnerable countries prepare for and respond to increasing climate- and weather-related risks. Assistance identifies and disseminates adaptive strategies, makes accessible the best available projected climate change impact and weather data to counterparts, and builds the capacity of partner governments and civil society partners to respond to climate change risks. This is why the Obama administration has made significant investments in bolstering the capacity of countries to respond to climate change risks. In FYs 2011 and 2012, the United States committed[12] $960 million in promoting climate resilience in developing countries.

Even in its early stages, U.S. adaptation work has made significant impacts:

* The SERVIR[13] global program has vastly increased access to and ability to use climate science and data through its three regional knowledge-sharing hubs in MesoAmerica, Africa, and the Himalaya Hindu-Kush region. SERVIR is part of a broader commitment to support climate data and services for meteorological offices and other agencies around the world.

* On-the-ground action is needed to learn what adaptation approaches will work best in different environments. USAID supports the launch of projects, programs, consultations, and planning processes around the world, with an emphasis on country and community ownership. Among these efforts, communities in Peru and Nepal are exploring multiple approaches to adapt to glacier melt in high-mountain areas, while Eastern Caribbean communities are testing water catchment areas, greenhouses, rainwater harvesting systems, and other adaptive practices to deal with increased flooding and drought. Pilot projects in Ethiopia, Senegal, and the Dominican Republic are helping local pastoralists, farmers, and insurance companies experiment with low-cost weather index insurance products, based on a model that reduced hunger following severe drought in neighboring Kenya. The Coral Triangle Initiative has provided tools, training, and projects to help the nine countries of this important region assess risks and increase the resiliency and adaptive capacity of marine resources and the communities that depend on them.

Clean Energy—For clean energy, dedicated U.S. climate assistance focuses on countries and sectors offering significant emission reduction potential over the long term, as well as countries that offer the potential to demonstrate leadership in sustained, large-scale deployment of clean energy. In terms of sector coverage, clean energy includes renewable energy and energy efficiency and excludes natural gas and other fossil fuel power plant retrofits. The United States also supports regional energy programs that improve the enabling environments for regional energy grids to distribute clean energy, as well as global programs that focus chiefly on information sharing and building coalitions for action on clean energy technologies and practices.

Expanding Clean Energy Use and Energy Efficiency—In the past three years, we have reached agreements with more than 20 countries around the world through the Enhancing Capacity for Low Emission Development Strategies program. EC-LEDS supports developing countries' efforts to pursue low-emission, climate-resilient economic development and growth. The program now has official partnerships with more than 20 countries.

Combating Short-Lived Climate Pollutants—Pollutants, such as methane, black carbon, and many HFCs, are relatively short-lived in the atmosphere, but have more potent greenhouse effects than CO_2. In February 2012, the United States launched the Climate and Clean Air Coalition to Reduce Short-Lived Climate Pollutants. The coalition has grown to include more than 30 state partners and nearly an equal number of nonstate partners, such as the World Bank, the United Nations Environment Programme, and civil society. Major efforts include reducing methane and black carbon from waste and landfills, oil and gas, diesel vehicles and engines, brick kilns, and cookstoves, and promoting activities aimed at enabling climate-

[12] While the U.S. fast start finance reports use the term "provided" to describe our support, the term "committed" is used in this report to be consistent with the new Biennial Report Common Tabular Format guidelines (UNFCCC 2012).

[13] SERVIR is a Spanish language acronym for Regional Visualization and Monitoring System.

friendly alternatives to high-global-warming-potential HFC use and reducing HFC emissions. The United States is also leading through the Global Methane Initiative, which works with 42 partner countries and an extensive network of more than 1,100 private-sector participants to reduce methane emissions.

High-Carbon Energy

Although climate finance generally refers to investing in low-carbon infrastructure, it is equally important from a climate impact point of view to address financing for high-carbon forms of energy. In June 2013, President Obama called for an end to U.S. government support for public financing of new coal power plants overseas, except for (1) the most efficient coal technology available in the world's poorest countries in cases where no other economically feasible alternatives exist, or (2) facilities deploying carbon capture and sequestration technologies (EOP 2013a). As part of this new commitment, the United States is working to secure the agreement of other countries, export credit agencies, development finance institutions, and multilateral development banks to adopt similar policies as soon as possible.

Sustainable Landscapes

For activities related to land-use mitigation (or "sustainable landscapes"), including reducing emissions from deforestation and forest degradation (REDD+), dedicated U.S. climate change assistance works to combat unsustainable forest clearing—for example, for agriculture and illegal logging—and helps ensure good governance at local and national levels to support the sustainable management of forests. U.S. support prioritizes mitigation potential; countries with the political will to implement large-scale efforts to reduce emissions from deforestation, forest degradation, and other land-use activities; and potential for investments in monitoring, reporting, and verification of forest cover and GHG emission reductions.

Reducing Emissions from Deforestation and Forest Degradation—GHG emissions from

deforestation, agriculture, and other land uses constitute approximately one-third of global emissions. In some developing countries, as much as 80 percent of GHG emissions come from the land sector. To meet the challenge of reducing these emissions, the Obama administration is working with partner countries to put in place the systems and institutions necessary to significantly reduce global land-use-related emissions, creating new models for rural development that generate climate benefits, while conserving biodiversity, protecting watersheds, and improving livelihoods.

In 2012 alone, USAID's bilateral and regional forestry programs contributed to reducing more than 140 million metric tons of CO_2 emissions.[14] Support from the U.S. Department of the Treasury and U.S. Department of State (DOS) for multilateral initiatives, such as the Forest Investment Program and the Forest Carbon Partnership Facility, is building capacity and facilitating implementation of REDD+ strategies in dozens of developing countries. Together with the Consumer Goods Forum, a coalition of more than 400 global corporations, USAID and DOS launched the Tropical Forest Alliance 2020 to reduce tropical deforestation linked to major commodities and their supply chains. In Indonesia, the Millennium Challenge Corporation (MCC) is funding a five-year "Green Prosperity" program that supports environmentally sustainable, low-carbon economic development in select districts.

Channels and Instruments

U.S. climate finance is provided through multiple channels, which can be broadly grouped into bilateral climate finance, multilateral climate finance, development finance, and export credit. Congressionally appropriated assistance is delivered through both bilateral and multilateral channels.

- **Bilateral Climate Finance**—Grant-based U.S. bilateral climate assistance is programmed directly through bilateral, regional, and global programs. These programs are principally supported by USAID, but are also supported by DOS, MCC, and other U.S. government agencies. In FY 2011–2012, the United States committed more than $2.4 billion in bilateral climate finance to its developing country partners.

- **Multilateral Climate Finance**—Multilateral climate change funds feature institutional structures governed jointly by developed and developing countries. They play an important

---

[14] See http://www.afolucarbon.org/.

role in promoting a coordinated, global response to climate change. During FY 2011–2012, the United States committed more than $700 million through multilateral climate change funds.

- **Development Finance and Export Credit**—The Overseas Private Investment Corporation (OPIC) and the Export-Import Bank of the United States (Ex-Im) play a critical role by using public money to mobilize much larger sums of private investment directed at mitigation through loans, loan guarantees, and insurance in developing countries. During FY 2011–2012, OPIC committed $1.8 billion[15] and Ex-Im committed $500 million. These numbers do not include private investment leveraged.

New and Additional Climate Finance

International assistance for climate change continues to be a major priority for the United States. The U.S. administration seeks new funding from Congress on an annual basis. Since ratifying the Convention, which is where the term "new and additional" was first used, U.S. international climate finance increased from virtually zero in 1992 to an average of $2.5 billion per year during the FSF period (2010 to 2012). During the period, average annual appropriated climate assistance increased fourfold compared with 2009 funding levels. U.S. climate assistance has increased in the context of an overall increasing foreign assistance budget.

Ensuring Transparency and Promoting Effectiveness

The United States is committed to transparently tracking and reporting its climate finance in a manner that encourages accountability and effectiveness. In 2010, President Obama issued a Presidential Policy Directive on Global Development (PPD) that emphasized the importance of tracking foreign assistance.[16]

During the FY 2010–2012 FSF period, the U.S. government refined its climate finance tracking methodologies to better reflect the totality of climate finance across the full range of government agencies. Each implementing government agency or entity follows strict guidelines and eligibility criteria when collecting and reporting information on support of activities related to adaptation, clean energy, and sustainable landscapes. For instance, activity descriptions provided by USAID missions are reviewed by climate change specialists to ensure compliance with USAID climate change goals. To improve financial reporting, DOS and USAID modified their budget and activity planning database to track climate change funding and developed standardized performance indicators to capture key outputs and outcomes of each agency's programs.[17]

In counting and aggregating climate finance, the United States includes programs that have a primary mitigation and/or adaptation purpose, as well as activities with significant climate co-benefits (e.g., relevant biodiversity and food security activities). In the case of programs for which only part of the activity is targeted toward a climate objective, only the relevant financial support is counted, rather than the entire program budget.

In addition, each implementing agency engages in strategic planning to ensure that climate finance is distributed effectively and is designed to meet U.S. partner countries' needs. The Enhancing Capacity for Low Emissions Development Strategies (EC-LEDS), a key mitigation program, illustrates one such approach to ensuring partner countries' priorities are addressed (Box 2). The program supports partner countries in developing their own LEDS. Within the LEDS framework, U.S. climate change funding directly supports the country-led process by providing technical support for developing GHG inventories, conducting technical and economic analyses, and implementing activities under the LEDS. Significantly, the LEDS can be a blueprint guiding the countries' own development investments.

U.S. government funding for adaptation is also tailored to partner country needs and often works directly through country-led processes. For example, Jamaica worked closely with USAID in 2011 and 2012 to establish a national adaptation planning process owned and led by the Ministry for Water, Land, Environment and Climate Change. In West Africa, USAID is working with ministry-level officials and regional institutions to provide technical support for developing country-owned National Adaptation Plans.[18]

[15] This number includes only those OPIC projects that are related to climate change, and are therefore counted under Fast Start Finance (FSF). However, OPIC's renewable resources portfolio (renewable energy, sustainable water, and agriculture) totals exceed the FSF-eligible totals being reported here.

[16] See http://www.whitehouse.gov/the-press-office/2010/09/22/fact-sheet-us-global-development-policy/.

[17] For the three U.S. Fast Start Finance reports, see www.state.gov/faststartfinance.

[18] For additional information on assumptions and methodologies related to U.S. international climate finance, see the accompanying annex at http://www.state.gov/e/oes/rls/rpts/car6/index.htm.

Box 2   **EC-LEDS—Strategic Programming of Assistance**

Step 1—Scoping

Once a partner country declares its intent to join the EC-LEDS program, an interagency scoping team, comprised of experts in a variety of fields, travels to the country to interview government officials and other stakeholders to analyze needs and opportunities for assistance.

Step 2—Identification of Opportunities

The scoping team completes an opportunities and options report, which identifies country needs that overlap with U.S. capacities for assistance.

Step 3—Discussions with Partner Country

The U.S. officials operating in the partner country, as part of the USAID Mission or U.S. Embassy, discuss the opportunities identified in the report, and prioritize actions based on available resources and country needs.

Step 4—Formal Agreement

A formal agreement is announced that publicly lays out the work plan.

Step 5—Implementation

The agreement is implemented in partnership with the partner country.

The United States acknowledges the critical role of our partner countries in promoting the effectiveness of climate finance. The PPD declares that where our partners set in place systems that reflect high standards of transparency, good governance, and accountability, the United States will respond directly to country priorities, making new investments in line with established national strategies and country development plans based on broad consultation, and will empower responsible governments to drive development and sustain outcomes by working through national institutions, rather than around them.[19]

U.S. Efforts to Mobilize Private Finance

The United States recognizes the role that private investment must play in mitigation and adaptation in developing countries. While maintaining a strong core of public climate finance is essential, the United States also recognizes that private finance must play a key role. Private finance has been and will continue to be the dominant force driving economic growth in most economies. How it is channeled will determine whether that growth is low in carbon emissions and resilient to changes in climate.

The U.S. government is actively pursuing strategies to encourage private investment in low-carbon, climate-resilient activities, both at home and in developing countries. We are working to combine our significant, but finite, public contributions with targeted, smart policies to mobilize maximum private investment in climate-friendly activities. For example, the United States is laying the foundation for larger-scale investments (1) by encouraging OPIC's development finance and Ex-Im Bank's export credit authorities to invest in clean energy technologies and create new products tailored toward climate change solutions; and (2) by leveraging significant private-sector investments across all three pillars through bilateral and multilateral programs.

More efficient leveraging of private investment can enable the use of available public resources in areas and sectors where the private sector is less likely to invest on its own, particularly in adaptation strategies for the most vulnerable and least developed countries. Continuing to execute this vision will be especially important as developed countries, including the United States, work toward a collective goal of mobilizing $100 billion per year in climate change finance for developing countries by 2020, in the context of meaningful mitigation actions and transparency on implementation.

USAID also contributes to mobilizing private finance, using a range of approaches. For example, the Private Finance Advisory Network (PFAN) provides direct advisory services to help promising clean energy entrepreneurs in developing countries connect with private investors and secure financing. In roughly six years of support from USAID, PFAN has helped more than three dozen clean energy start-ups or small businesses secure nearly $300 million

[19] See http://www.whitehouse.gov/the-press-office/2010/09/22/fact-sheet-us-global-development-policy.

in private financing. Another approach is to leverage local, private capital through partial credit guarantees under USAID's Development Credit Authority.

USAID also supports capacity building for the government and nongovernment staff and institutions that regulate specific sectors and private investment in order to help enhance a country's private financial enabling environment. For instance, USAID's Black Sea Regulatory Initiative links power regulators from Armenia, Azerbaijan, Georgia, Moldova, and Ukraine with midwestern U.S. state regulators to support development of harmonized regulatory practices, including guidelines for renewable energy and energy efficiency, in order to spur private investment in the region.

Another approach to mobilizing private finance is the Africa Clean Energy Finance (ACEF) Initiative, launched in 2012. ACEF seeks to address sub-Saharan Africa's acute energy needs by mobilizing private investment in clean energy projects, ranging from household-level solar energy to utility-scale power plants. ACEF represents a new way of doing business that harnesses the best of the U.S. government's technical and financial expertise. By combining $20 million in grant-based financing from DOS, project planning expertise from the U.S. Trade and Development Agency, and financing and risk mitigation tools from OPIC, ACEF will catalyze hundreds of millions of dollars in financing from OPIC, which will then leverage hundreds of millions of dollars in private investment. ACEF demonstrates how a very limited amount of grant-based public resources—when surgically applied—can catalyze a much larger pool of finance that can bring climate projects to fruition at scale.

The United States contributed $714.6 million during FY 2010–2012 to support the critical work of the Clean Technology Fund (CTF). CTF catalyzes clean energy investments in emerging economies with rapidly growing emissions by helping countries achieve access to renewable energy, green growth, and energy efficiency in transport, industry, and agriculture. CTF is working with 18 countries on various projects, such as wind power in Egypt, sustainable urban transportation in the Philippines, and energy efficiency in Turkey. The funds are channeled toward projects that focus on scaling up proven technologies, thereby promoting new markets for maximum impact. To date, CTF has approved 41 projects for a total of $2.3 billion. These funds have leveraged $18.8 billion in co-financing, including $5.8 billion from the multilateral development banks and $13 billion from other sources, and have contributed to the saving of 525 million metric tons of CO_2 emissions, the equivalent of taking 99 million cars off the road for a year.

Technology Development and Transfer

Since 2010, the United States has engaged in numerous activities with developing countries and economies in transition, with the primary goal of promoting the development and deployment of climate-friendly technologies and practices. Table 4 highlights examples of U.S. involvement in technology development and transfer. Please note that this table is purely illustrative and is not a comprehensive list of U.S. technology development and transfer activities.

Capacity Building

Reflecting its belief that a long-term view of climate change and development is crucial to sustainability and results, the United States is approaching the issue of capacity building for climate change in an integrated manner. Linking capacity building directly to projects and programs helps ensure that capacity built is relevant, effective, and tied to results. Building local capacity through greater reliance on local cooperating agencies is an explicit goal of USAID. In 2012, USAID missions awarded 14.3 percent of their funding, or $1.4 billion, to local institutions. This number is expected to double by 2015.

Capacity-building needs are addressed throughout all U.S. support activities, not as separate line items or projects, and are provided as a means for taking action on a mutually shared goal. Table 5 highlights examples of U.S. capacity-building support. Please note that this table is purely illustrative and does not represent an exhaustive list of U.S. capacity-building activities.

Table 4   **Sample U.S. Technology Development and Transfer Activities**

Since 2010, the United States has engaged in numerous activities with developing countries and economies in transition, with the primary goal of promoting the development and deployment of climate-friendly technologies and practices.

Recipient Country and/or Region	Targeted Area	Measures and Activities Related to Technology Transfer	Sector	Source of Funding	Activities Undertaken by:	Status	Additional Information
Global Methane Initiative							
Global	Mitigation	Focuses on best practices/technologies for evaluating and measuring methane emissions from target sectors, and mitigation technologies/ best practices, such as coal mine gas and landfill methane capture systems, biodigestors, and technologies for reducing oil and gas sector methane emissions.	Energy	Public	Public	Implemented	Reduced methane emissions by approximately 23 million metric tons (23 Tg CO_2e) in 2012 alone; cumulative emission reductions exceed 150 Tg CO_2e.
Super-efficient Equipment and Appliance Deployment Initiative (SEAD)							
Global	Mitigation	Peer-to-peer exchange among technical and policy experts from participating governments; complementary activities that develop clear, broadly accepted test procedures for products; and collaboration with industry to ensure its participation in promoting a transition to energy-efficient products.	Energy	Public	Public	Implemented	Employing current best practices in SEAD economies can by 2030 reduce annual electricity demand by more than 2,000 billion kilowatt-hours. These measures would decrease CO_2 emissions over the next two decades by 11 billion tons (1,000 Tg CO_2e).
Global Lighting and Energy Access Partnership (Global LEAP)							
Global	Mitigation	Quality assurance activities for solar-powered lanterns for off-grid lighting, a global competition in two categories (lights and televisions) to identify the best DC-powered products in the market for use in an off-grid context, and efforts to advance commercially viable mini-grid solutions for rural energy access.	Energy	Public	Public	Implemented	An estimated 138,600 metric tons of CO_2e (0.1386 Tg CO_2e) have been avoided. The climate benefits are even more significant when the black carbon implications of kerosene lighting are considered.

Table 4 (Continued)   **Sample U.S. Technology Development and Transfer Activities**

Recipient Country and/or Region	Targeted Area	Measures and Activities Related to Technology Transfer	Sector	Source of Funding	Activities Under-taken by:	Status	Additional Information
SERVIR							
Global (Central America, East Africa, and Hindu Kush-Himalaya)	Adaptation and Mitigation	USAID and NASA collaboration to build capacity of regional institutions in developing countries to improve environmental manage-ment and climate change resilience through the application of geospatial information in decision making.	Water, agriculture, land cover, climate, disasters, biodiversity, ecosystems	Public	Public	Implemented	Decision support will aid land and forest manage-ment, monitoring, emission estima-tions, and policy improvement, leading to emission reductions, as well as disaster risk reduction and adaptation to climate variability and change.
Famine Early Warning System Network (FEWS NET)							
Afghanistan, Angola, Burkina Faso, Central African Republic, Chad, Djibouti, Ethiopia, Guatemala, Haiti, Honduras, Kenya, Madagascar, Malawi, Mali, Mauritania, Mozambique, Nicaragua, Niger, Nigeria, Rwanda, Senegal, Somalia, South Sudan, Sudan, Tajikistan, Uganda, Yemen, Zambia, Zimbabwe	Adaptation	Assesses short- to long-term vulnerability to food insecurity with environmental informa-tion from satellites and agricultural and socio-economic information from field representatives; conducts vulnerability assessments and contingency and response planning, aimed at strengthening host country food security networks.	Adaptation	Public	Public	Implemented	
SilvaCarbon							
Governments of Colombia, Peru, Ecuador, Vietnam, and Gabon. Regional training activities in South and Central America, Congo Basin, and Southeast Asia.	Mitigation	A multi-agency U.S. government effort to improve developing country capacity for forest and other terrestrial carbon measurement and monitoring, through coordinated support on tool and methodology development and training to use appropriate methods for building and implementing forest carbon monitoring systems.	Forestry	Public	Public	Implemented	Providing countries with improved capacity to measure and report on current carbon stocks and emissions and use information together with other natural resource management data to reduce emissions from future deforestation.

Note: This table is purely illustrative and is not a comprehensive list of U.S. technology development and transfer activities.

NASA = National Aeronautics and Space Administration; USAID = U.S. Agency for International Development.

Table 5   **Examples of U.S. Capacity-Building Activities**

Capacity-building needs are addressed throughout all of U.S. support activities, not as separate line items or projects, and are provided as a means for taking action on a mutually shared goal.

Recipient Country/ Region	Targeted Area	Program or Project Title	Description of Program or Project
Global	Adaptation	Climate Services Partnership (CSP)	USAID is working with the UK Met Office, the World Bank, the WMO's Global Framework for Climate Services, and developing countries to build the capacity of national weather services to deliver accurate climate information that will facilitate the efforts of government ministries, private-sector entities, and other stakeholders to take effective adaptation actions. CSP is also compiling and disseminating current climate services experiences, conducting case studies and assessments of climate services, exploring economic valuation of climate services, developing a climate information guidebook, and piloting a nation-level climate services analysis.
Peru, the Himalaya Hindu-Kush region of South Asia, and the Pamir Mountain region of Central Asia	Adaptation	High Mountains Adaptation Partnership (HIMAP)	With support from USAID and DOS, HIMAP facilitates South–South learning to understand and manage climate-related challenges in high-mountain communities. The program has pioneered rapid assessment techniques for studying the risks of glacier lakes, and has supported community-led consultation and planning to address these risks in a timely and effective fashion.
Albania, Bangladesh, Cambodia, Colombia, Costa Rica, Gabon, Guatemala, Indonesia, Jamaica, Kazakhstan, Kenya, Macedonia, Malawi, Mexico, Moldova, Peru, Philippines, Serbia, South Africa, Thailand, Vietnam, Zambia	Mitigation	Enhancing Capacity for Low Emission Development Strategies (EC-LEDS)	This program supports partner countries in developing low-emission development strategies (LEDS) and country-led national plans to promote sustainable development while reducing GHG emissions. EC-LEDS provides countries with technical assistance to develop GHG inventories, conduct a range of economic analyses, and plan and implement LEDS across multiple economic sectors. Anticipated actions stemming from LEDS include putting policies, regulations, and infrastructure in place to dramatically increase clean energy use, and energy efficiency and piloting payments for sustainable forest management, including REDD+ arrangements.
Africa	Mitigation	Africa Infrastructure Program (AIP)	AIP works with partner countries in Africa to build capacity for regulatory reforms, tariff formulation, and key analyses required to support clean energy for power grids. AIP also provides transaction advisory services and technical, financial, commercial, regulatory, legal, and environmental support to specific clean energy projects.
Global	Mitigation	Forest Carbon, Markets, and Communities (FCMC)	FCMC provides technical support and capacity building to partner country governments around the world. Capacity building supports analysis, evaluation, tools, and guidance for program design support, training materials, and other services to improve the management and conservation of natural forests.

Note: This table is purely illustrative and does not represent an exhaustive list of U.S. capacity-building activities.

DOS = U.S. Department of State; NASA = National Aeronautics and Space Administration; REDD+ = reducing emissions from deforestation and forest degradation; USAID = U.S. Agency for International Development.

BIBLIOGRAPHY

Alig, R.J., A.J. Plantinga, D. Haim, and M. Todd. 2010. *Area Changes in U.S. Forests and Other Major Land Uses, 1982 to 2002, with Projections to 2062.* Gen. Tech. Rep. PNW-GTR-815. Portland, OR: U.S. Department of Agriculture, Forest Service, Pacific Northwest Research Station. 98 p. <http://www.fs.fed.us/pnw/pubs/pnw_gtr815.pdf>

EOP (Executive Office of the President). 2013a. *The President's Climate Action Plan.* Washington, DC. June. <http://www.whitehouse.gov/sites/default/files/image/president27sclimateactionplan.pdf>

———. 2013b. *Preparing the United States for the Impacts of Climate Change.* Executive Order 13653. *Federal Register* 78(215, Nov. 6). <http://www.gpo.gov/fdsys/pkg/FR-2013-11-06/pdf/2013-26785.pdf>

Haim, D., R.J. Alig, A.J. Plantinga, and B. Sohngen. 2011. Climate change and future land use in the United States: an economic approach. *Climatic Change Economics* 2(1): 27-51. <http://treesearch.fs.fed.us/pubs/40838>

Haynes, R.W., D.M. Adams, R.J. Alig, P.J. Ince, R. John, and X. Zhou. 2007. *The 2005 RPA Timber Assessment Update.* Gen. Tech. Rep. PNW-GTR-699. Portland, OR: U.S. Department of Agriculture, Forest Service, Pacific Northwest Research Station. 212 pp. <http://www.fs.fed.us/pnw/publications/gtr699/>

IPCC (Intergovernmental Panel on Climate Change). 2006. *2006 IPCC Guidelines for National Greenhouse Gas Inventories.* Ed. S. Eggleston, L. Buendia, K. Miwa, T. Ngara, and K. Tanabe. Prepared by the IPCC Task Force on National Greenhouse Gas Inventories. Hayama, Japan: Institute for Global Environmental Strategies. <http://www.ipcc-nggip.iges.or.jp/public/2006gl/index.html>

———. 2007. *Climate Change 2007: The Physical Science Basis. Contribution of Working Group I to the Fourth Assessment Report of the Intergovernmental Panel on Climate Change.* Ed. S. Solomon, D. Qin, M. Manning, Z. Chen, M. Marquis, K.B. Averyt, M. Tignor, and H.L. Miller. Cambridge, UK: Cambridge University Press. <http://www.ipcc.ch/publications_and_data/publications_ipcc_fourth_assessment_report_synthesis_report.htm>

———. 2013. *Climate Change 2013: The Physical Science Basis.* Contribution of Working Group I to the Fifth Assessment Report of the Intergovernmental Panel on Climate Change. Ed. T.F. Stocker, D. Qin, G.-K. Plattner, M. Tignor, S.K. Allen, J. Boschung, A. Nauels, Y. Xia, V. Bex, and P.M. Midgley. Cambridge, United Kingdom, and New York, NY: Cambridge University Press, in press.

NOAA/NCDC (National Oceanic and Atmospheric Administration, National Climatic Data Center). 2012a . *Billion-Dollar U.S. Weather/Climate Disasters: 1980-2012.* Asheville, NC. December. <http://www.ncdc.noaa.gov/billions/events>

———. 2012b. *State of the Climate: National Overview—Annual 2012.* Asheville, NC. December. <http://www.ncdc.noaa.gov/sotc/national/2012/13>

UNFCCC (United Nations Framework Convention on Climate Change). 2006. *Updated UNFCCC Reporting Guidelines on Annual Inventories Following Incorporation of the Provisions of Decision 14/CP.11.* FCCC/ SBSTA/2006/9. August 18. <http://unfccc.int/resource/docs/2006/sbsta/eng/09.pdf>

———. 2012. *Common Tabular Format for UUNFXXX Biennial Reporting Guidelines for Developed Country Parties.* FCCC/ CP/2012/8/L.12. December 7. <http://unfccc.int/resource/docs/2012/cop18/eng/l12.pdf>

USAID (U.S. Agency for International Development). 2012. *USAID Climate Change and Development Strategy: 2012-2016.* Washington, DC. January. <http://www.cgdev.org/doc/Rethinking%20Aid/Climate_Change_&_Dev_Strategy.pdf>

USDA/FS (U.S. Department of Agriculture, Forest Service). 2012. *Future of America's Forest and Rangelands: Forest Service 2010 Resources Planning Act Assessment.* Gen. Tech. Rep. WO-87. Washington, DC. 198 p. <http://www.fs.fed.us/research/rpa/assessment/>

U.S. DOE/EIA (U.S. Department of Energy, Energy Information Administration). 2013. *Monthly Energy Review.* DOE/EIA-0035(2013/05). Washington, DC. May. <http://www.eia.gov/mer>

U.S. DOS (U.S. Department of State). 2007 *U.S. Climate Action Report 2006: Fourth National Communication of the United States of America Under the United Nations Framework Convention on Climate Change.* Washington, DC. July. <http://www.state.gov/g/oes/rls/rpts/car/>

———. 2012. *Meeting the Fast Start Commitment: U.S. Climate Finance in Fiscal Year 2012.* Washington, DC. <http://www.state.gov/documents/organization/201130.pdf>

U.S. EPA/OAP (U.S. Environmental Protection Agency, Office of Atmospheric Programs). 2013. *Inventory of U.S. Greenhouse Gas Emissions and Sinks: 1990–2011.* EPA 430-R-13-001. Washington, DC. April. <http://www.epa.gov/climatechange/emissions/usinventoryreport.html>

Sixth National Communication
of the United States of America

Table of Contents

Acronyms & Abbreviations

3D	three-dimensional
AAAS	American Association for the Advancement of Science
AASHE	Association for the Advancement of Sustainability in Higher Education
AB	Assembly Bill
ac	acre
ACCRI	Aviation Climate Change Research Initiative
ACEF	Africa Clean Energy Finance
ACRF	Atmospheric Radiation Measurement Climate Research Facility
ACUPCC	American College & University Presidents' Climate Commitment
AEO2013	*Annual Energy Outlook 2013*
AERONET	AErosol RObotic NETwork
AFVs	alternative fuel vehicles
AGAGE	Advanced Global Atmospheric Gases Experiment
AIP	Africa Infrastructure Program
AIRS	Atmospheric InfraRed Sounder
AISL	Advancing Informal STEM Learning
AISS	Antarctic Integrated System Science
AMF	Atmospheric Radiation Measurement Mobile Facility
AMO	Advanced Manufacturing Office
AMS	American Meteorological Society
ANL	Argonne National Laboratory
AON	Arctic Observing Network
AR5	IPCC Fifth Assessment Report
ARB	Air Resources Board
ARCSS	Arctic System Science Program
ARM	Atmospheric Radiation Measurement
ARPA-E	Advanced Research Projects Agency–Energy
ARRA	American Recovery and Reinvestment Act of 2009

ARS	Agricultural Research Service
ASR	Atmospheric Systems Research Program
ASSP	Arctic Social Sciences Program
ATE	Advanced Technical Education Program
A-Train	afternoon train
ATVM	Advanced Technology Vehicles Manufacturing Loan Program
AZA	Association of Zoos and Aquariums
BAP	Biorefinery Assistance Program
BCF	billion cubic feet
BEA	Bureau of Economic Analysis
BECP	Building Energy Codes Program
BER	Office of Biological and Environmental Research
BIA	Bureau of Indian Affairs
BLM	Bureau of Land Management
BMT	billion metric tons
BOEM	Bureau of Ocean Energy Management
BOR	Bureau of Reclamation
bpd	barrels per day
BRACE	Building Resilience Against Climate Effects
BRC	Bioenergy Research Center
BTO	Building Technologies Office
BTS	Bureau of Transportation Statistics
Btu	British thermal unit
C	carbon
°C	degree Centigrade
C$_2$ES	Center for Climate and Energy Solutions
C$_2$F$_6$	hexafluoroethane
C$_4$F$_{10}$	perfluorobutane
C$_6$F$_{14}$	perfluorohexane
CAA	Clean Air Act
CAAFI	Commercial Aviation Alternative Fuels Initiative
CAFE	Corporate Average Fuel Economy
CAFEC	Central Africa Forest Ecosystems Conservation Project
CALIPSO	Cloud-Aerosol Lidar and Infrared Pathfinder Satellite Observation
CAP	2013 *Climate Action Plan*
CAR	*U.S. Climate Action Report*
CARPE	Central Africa Regional Program for the Environment
CBECS	Commercial Buildings Energy Consumption Survey
C-CAP	Coastal Community Adaptation Program

CCEP	Climate Change Education Partnership
CCRI	Climate Change Research Initiative
CCS	carbon capture and storage
CDC	Centers for Disease Control and Prevention
CEAC	Clean Energy Assistance Center
CEM	Clean Energy Ministerial
CENR	Committee on Environment and Natural Resources
CEOS	Committee on Earth Observation Satellites
CEQ	Council on Environmental Quality
CERC	U.S.–China Clean Energy Research Center
CERES	Clouds and the Earth's Radiant Energy System
CESC	Clean Energy Solutions Center
CESN	Climate Effects Science Network
CfA	Center for Astrophysics
CFP	Climate Friendly Parks
CH$_4$	methane
CHP	combined heat and power
CHPP	Combined Heat and Power Partnership
CI	Conservation International
CIFs	Climate Investment Funds
CLD	Climate and Large-scale Dynamics
CLEAN	Climate Literacy and Energy Awareness Network
CLIP	Climate Leadership In Parks
CLiZEN	Climate Literacy Zoo Education Network
CMIP5	Phase 5 of the Coupled Model Intercomparison Project
CMM	coal mine methane
CMOP	Coalbed Methane Outreach Program
CNH	Dynamics of Coupled and Natural Human Systems
CNRS	Centre Nationale de Recherche Scientifique
CO	carbon monoxide
CO$_2$	carbon dioxide
CO$_2$e	carbon dioxide equivalents
COEA	President's Council of Economic Advisors
COMcheck	commercial compliance software
COOP	Cooperative Observer Program
COP	Conference of the Parties
COSMIC	Constellation Observing System for Meteorology, Ionosphere, and Climate
CPRAL	Coastal Protection and Restoration Authority of Louisiana
CRE	Climate Ready Estuaries program

CrIS	Cross-track Infrared Sounder
CRN	Climate Reference Network
CRP	Conservation Reserve Program
CRSCI	Climate-Ready States and Cities Initiative
CSC	climate science center
CSLF	Carbon Sequestration Leadership Forum
CSP	Climate Services Partnership
CSP	Conservation Stewardship Program
CSREES	Cooperative State Research, Education, and Extension Service
CTA	Conservation Technical Assistance Program
CTF	Clean Technology Fund
CTI	Climate Technology Initiative
CTIC	Conservation Technology Information Center
cu ft	cubic feet
DAAC	Distributed Active Archive Center
DERA	Diesel Emissions Reduction Act
DHS	U.S. Department of Homeland Security
DLR	Deutsches Zentrum für Luft- und Raumfahrt
DMSP	Defense Meteorological Satellites Program
DOC	U.S. Department of Commerce
DoD	U.S. Department of Defense
DOE	U.S. Department of Energy
DOI	U.S. Department of the Interior
DOS	U.S. Department of State
DOT	U.S. Department of Transportation
DR	Discovery Research
DRMS	Decision, Risk, and Management Sciences
EaSM	Earth system model
ECHO	EOS ClearingHOuse
EC-LEDS	Enhancing Capacity for Low-Emission Development Strategies
ECO-Asia	Environmental Cooperation–Asia
ECV	essential climate variable
EERE	Office of Energy Efficiency and Renewable Energy
EERPAT	Energy and Emissions Reduction Policy Analysis Tool
EERS	efficiency resource standard
EESE	Ethics Education in Science and Engineering
EFRC	Energy Frontier Research Center
EESI	Environmental and Energy Study Institute
EIA	U.S. Energy Information Administration

EISA	Energy Independence and Security Act of 2007
ELTI	Environmental Leadership Training Initiative
EMAPS	Environmental Monitoring and Policy Support Project
EMS	environmental management systems
E.O.	executive order
EO-1	Earth Observing-1
EOP	Executive Office of the President
EOR	enhanced oil recovery
EOS	Earth Observing System
EOSDIS	Earth Observing System Data and Information System
EPA	U.S. Environmental Protection Agency
EPAct	Energy Policy Act of 2005
EQIP	Environmental Quality Incentives Program
ESIP	Federation of Earth Science Information Partners
ESM	Earth system model
ESRL	Earth System Research Laboratory
ESSEA	Earth System Science Education Alliance
ETE	Evolution of Terrestrial Ecosystems Program
EUGENE	Ecological Understanding as a Guideline for Evaluation of Nonformal Education
EUMETSAT	European Organization for the Exploitation of Meteorological Satellites
EUPP	Energy Utility Partnership Program
Ex-Im	Export-Import Bank of the United States
°F	degree Fahrenheit
FAA	Federal Aviation Administration
FAO	Food and Agriculture Organization of the United Nations
FCPF	Forest Carbon Partnership Facility
FEMA	Federal Emergency Management Agency
FEMP	Federal Energy Management Program
FESD	Frontiers in Earth System Dynamics
FEWS NET	Famine Early Warning Systems Network
FGHG	fluorinated greenhouse gas
FHA	Federal Housing Administration
FHWA	Federal Highway Administration
FIP	Forest Investment Program
FLIGHT	Facility Level Information on Greenhouse gases Tool
FSA	Farm Service Agency
FSF	fast start finance
ft^2	square feet
ft^3	cubic feet

FTA	Federal Transit Administration
FWS	U.S. Fish and Wildlife Service
FY	fiscal year
G8	Group of Eight
GATE	Graduate Automotive Technology Education
GAW	Global Atmospheric Watch
GCCI	Global Climate Change Initiative
GCEP	Global Change Education Program
GCF	Green Climate Fund
GCIS	Global Climate Information System
GCOS	Global Climate Observing System
GCRA	Global Change Research Act of 1990
GDA	Global Development Alliance
GDP	gross domestic product
GEF	Global Environment Facility
GEO	Group on Earth Observations
GEOSS	Global Earth Observation System of Systems
GFCS	Global Framework for Climate Services
GFW	Global Forest Watch
Gg	gigagram
GHG	greenhouse gas
GIF	Generation IV International Forum
GIS	geographic information system
GLOBE	Global Learning and Observations to Benefit the Environment
GLOSS	Global Sea Level Observing System
g/mi	grams per mile
GMI	Global Methane Initiative
GOES	Geostationary Operational Environmental Satellite
GOES-R	Geostationary Operational Environmental Satellite R-Series
GOOS	Global Ocean Observing System
GPP	*Green Parks Plan*
GPP	Green Power Partnership
GPS	global positioning system
GRACE	Gravity Recovery and Climate Experiment
GREF	Graduate Research Environmental Fellowship
GRP	Grassland Reserve Program
GRUAN	GCOS Reference Upper-Air Network
GTOS	Global Terrestrial Observing System
GWP	global warming potential

H_2CO_3	carbonic acid
ha	hectare
HCFC	hydrochlorofluorocarbon
HDV	heavy-duty vehicle
HFC	hydrofluorocarbon
HFC-23	fluoroform
HHS	U.S. Department of Health and Human Services
HIA	health impact assessment
HPwES	Home Performance with ENERGY STAR
HUD	U.S. Department of Housing and Urban Development
IAC	Industrial Assessment Center
IAM	integrated assessment model
IASI	Infrared Atmospheric Sounding Interferometer
IAV	impact, adaptation, and vulnerability
ICCAGRA	Interagency Coordinating Committee for Airborne Geosciences Research and Applications
ICCATF	Interagency Climate Change Adaptation Task Force
ICCTF	Interagency Climate Change Task Force
ICESat	Ice, Cloud, and land Elevation Satellite
ICSU	International Council for Science
IDEA	Integrated Data and Environmental Applications Center
IEA	International Energy Agency
IECC	International Energy Conservation Code
IES	Integrated Earth Systems
IGERT	Interactive Graduate Education and Research Trainee Program
IGFA	International Group of Funding Agencies for Global Change Research
in	inch
IOC	Intergovernmental Oceanographic Commission
IOOS	Integrated Ocean Observing System
IPCC	Intergovernmental Panel on Climate Change
IPL	Interfaith Power and Light
ISO	International Organization for Standardization
ISS	International Space Station
ISSC	International Social Science Council
ISSE	*Integrative Science for Society and Environment*
ITEP	Institute for Tribal Environmental Professionals
IUFRO	International Union of Forest Research Organizations
IWGCCH	Interagency Working Group on Climate Change and Health
JPSS-1	Joint Polar Satellite System-1

K-12	kindergarten through grade 12
KDFWR	Kentucky Department of Fish and Wildlife Resources
kg	kilogram
km	kilometer
km²	square kilometers
kWh	kilowatt-hour
lb	pound
LCCs	landscape conservation cooperatives
LDCs	least developed countries
LDCF	Least Developed Countries Fund
LDCM	Landsat Data Continuity Mission
LDV	light-duty vehicle
LEAP	Lighting and Energy Access Partnership
LED	light-emitting diode
LEDS	low-emission development strategies
LEDS GP	Low Emission Development Strategies Global Partnership
LFG	landfill gas
LIDAR	light detection and ranging
LLNL	Lawrence Livermore National Laboratory
LMOP	Landfill Methane Outreach Program
LPO	Loan Program Office
LTAR	Long-Term Agro-Ecosystem Research Network
LTER	Long-Term Ecological Research program
LULUCF	land use, land-use change, and forestry
m²	square meters
m³	cubic meters
M2M	Methane to Markets program
MAP-21	Moving Ahead for Progress in the 21st Century Act
MATCH	Metadata Access Tool for Climate and Health
MCC	Millennium Challenge Corporation
MECS	Manufacturing Energy Consumption Survey
MEDUCA	Panama's Ministry of Education
MEF	Major Economies Forum on Energy and Climate
mi	mile
mi²	square miles
MM	million
MMBtu	million British thermal units
MMt	million metric tons
MMtCO₂	million metric tons of carbon dioxide

MODIS	Moderate Resolution Imaging Spectroradiometer
MOVES	Motor Vehicle Emissions Simulator
MPL	micro-pulse lidar
MPLNET	Micro-Pulse Lidar Network
mpg	miles per gallon
MPO	metropolitan planning organization
MSM	multi-scale modeling
MSW	municipal solid waste
MW	megawatts
MWh	megawatt-hour
MY	model year
N$_2$O	nitrous oxide
NACP	North American Carbon Program
NAS	National Academies of Science
NASA	National Aeronautics and Space Administration
NASM	National Air and Space Museum
NASS	National Agricultural Statistical Service
NAST	National Assessment Synthesis Team
NCA	National Climate Assessment
NCDC	National Clean Diesel Campaign
NCDC	National Climatic Data Center
NEI	National Emission Inventory
NEMS	National Energy Modeling System
NEON	National Ecological Observatory Network
NextGen	Next Generation Air Transportation System Plan
NF$_3$	nitrogen trifluoride
NFWPCAP	National Fish, Wildlife, and Plants Climate Adaptation Partnership
NGO	nongovernmental organization
NGSS	Next-Generation Science Standards
NHTSA	National Highway Traffic Safety Administration
NICE	NASA Minority University Research and Education Innovations in Climate Education
NIDIS	National Integrated Drought Information System
NIEHS	National Institute of Environmental Health Sciences
NIFA	National Institute of Food and Agriculture
NMME	National Multi-Model Ensemble
NMNH	National Museum of Natural History
NMVOC	nonmethane volatile organic compound
NOAA	National Oceanic and Atmospheric Administration
NOC	National Ocean Council

NO$_x$	oxides of nitrogen
NPCC2	New York City Panel on Climate Change
NPOESS	National Polar-orbiting Operational Environmental Satellite System
NPP	National Polar-orbiting Partnership
NPS	National Park Service
NRC	National Research Council
NRCS	Natural Resources Conservation Service
NSF	National Science Foundation
NSIDC	National Snow and Ice Data Center
NSTA	National Science Teachers Association
NSTC	National Science and Technology Council
NTER	National Training & Education Resource
NTI	National Transit Institute
NWF	National Wildlife Federation
NWS	National Weather Service
NZP	National Zoological Park
O$_3$	ozone
OAP	Office of Atmospheric Programs
OAQPS	Office of Air Quality Planning and Standards
OAS	Organization of American States
OCO	Orbiting Carbon Observatory
OCS	Outer Continental Shelf
ODS	ozone-depleting substance
OECD	Organisation for Economic Co-operation and Development
OEd	Office of Education
OIG	Office of Inspector General
ONR	Office of Naval Research
OOI	Ocean Observatories Initiative
OPIC	Overseas Private Investment Corporation
OST	White House Office of Science and Technology
OSTP	Office of Science and Technology Policy
OSW	Office of Solid Waste
P2C2	Paleo Perspectives on Climate Change
PAA	Price-Anderson Act of 1957
PACE	Partnership to Advance Clean Energy
PaCIS	Pacific Climate Information System
PARTNER	Partnership for AiR Transportation Noise and Emissions Reduction
PCMDI	Project for Climate Model Diagnosis and Intercomparison
PCNC	Punta Culebra Nature Center

PES	payment for ecosystem services
PFAN	Private Financing Advisory Network
PFC	perfluorocarbon
PM	particulate matter
POES	Polar Operational Environmental Satellite
PPCR	Pilot Program for Climate Resilience
PPD	Presidential Policy Directive on Global Development
PV	photovoltaic
QTR	Quadrennial Technology Review
QuickSCAT	Quick Scatterometer
R&D	research and development
RAMA	Research Moored Array for African-Asian-Australian Monsoon Analysis and Prediction
RAP	Repowering Assistance Program
RCN	Research Coordination Networks
RD&D	research, development, and demonstration
REAP	Rural Energy for America Program
RECS	Residential Energy Consumption Survey
REDD+	reducing emissions from deforestation and forest degradation
REScheck	residential compliance software
RFS	renewable fuel standard
RGGI	Regional Greenhouse Gas Initiative
RISA	Regional Integrated Science and Assessments program
RITA	Research and Innovative Technology Administration
RPA	Regional Plan Association of New York City
RPA	Resources Planning Act of 2010
RPS	renewable energy portfolio standard
RRIF	Railroad Rehabilitation and Improvement Financing
RSL	Arctic Research Support and Logistics
SAGE	Stratospheric Aerosol and Gas Experiment
SAR	Second Assessment Report
SB	Senate Bill
SCAN	Soil Climate Analysis Network
SCBI	Smithsonian Conservation Biology Institute
SCC	social cost of carbon
SCCF	Special Climate Change Fund
SCIPP	Southern Climate Impacts Planning Program
S'COOL	Students' Cloud Observations On-Line
SEAD	Super-Efficient Equipment and Appliances Deployment
SEDAC	Socioeconomic Data and Applications Center

SEES	science, engineering, and education for sustainability
SEP	Sustainable Energy Pathways
SERC	Smithsonian Environmental Research Center
SERVIR	Regional Visualization and Monitoring System
SF$_6$	sulfur hexafluoride
SIDS	small-island developing states
SI-GEO	Smithsonian Institution Global Earth Observatory
SITES	Smithsonian Institution Traveling Exhibition Service
SITN	Solar Instructor Training Network
SLCP	short-lived climate pollutant
SLED	Solar and LED Energy Access initiative
SMM	sustainable materials management
SMR	small modular reactor
SNAP	Significant New Alternatives Policy Program
SNOTEL	SNOpack TELemetry
SO$_2$	sulfur dioxide
SOARS	Significant Opportunities in Atmospheric Research and Science
SORCE	Solar Radiation and Climate Experiment
SOS	Science On a Sphere®
SPURS	Salinity Processes in the Upper Ocean Regional Study
SREP	Scaling-up Renewable Energy Program
SRN	Sustainability Research Network
SSEC	Smithsonian Science Education Center
STEM	science, technology, engineering, and math
STRI	Smithsonian Tropical Research Institute
SURE	Summer Undergraduate Research Experience
SURFRAD	Surface Radiation Budget Network
SUV	sport utility vehicle
t	metric ton
T&D	transmission and distribution
TAO	Tropical–Atmosphere–Ocean
TAP	Technical Assistance Partnership
TCI	Transportation Climate Initiative
TCP	Terrestrial Carbon Processes
TEAM	Travel Efficiency Assessment Method
TFA	Tropical Forest Alliance
TFCA	Tropical Forest Conservation Act
TfL	Transport for London
Tg	teragram

TIGER	Transportation Investment Generating Economic Recovery
TRMM	Tropical Rainfall Measuring Mission
TSI	total solar irradiance
TUES	Transforming Undergraduate Education in STEM
TVA	Tennessee Valley Authority
UCAR	University Corporation for Atmospheric Research
UK	United Kingdom
ULTRA	Urban Long-Term Research Area
UN	United Nations
UNDP	United Nations Development Programme
UNEP	United Nations Environment Programme
UNFCCC	United Nations Framework Convention on Climate Change
UNIFEM	United Nations Development Fund for Women
UNOLS	University-National Oceanographic Laboratory System
USACE	U.S. Army Corps of Engineers
U.S.-ACEF	U.S.-Africa Clean Energy Finance
USAID	U.S. Agency for International Development
USCRN	U.S. Climate Reference Network
USDA	U.S. Department of Agriculture
USFS	U.S. Forest Service
USGCRP	U.S. Global Change Research Program
USGEO	U.S. Group on Earth Observations
USGS	U.S. Geological Survey
VALE	Voluntary Airport Low Emission Program
VCOP	Voluntary Code of Practice for the Reduction of Emissions of HFC & PFC Fire Protection Agents
VMT	vehicle miles traveled
VOC	volatile organic compound
VOSClim	Voluntary Observing Ship Climate Program
VOCTEC	Vocational Training and Education for Climate Energy
VTO	Vehicle Technologies Office
WAP	Weatherization Assistance Program
WCRP	World Climate Research Programme
WMO	World Meteorological Organization
WRI	World Resources Institute
WSC	Water Sustainability and Climate

1

Executive Summary

Climate change represents one of the greatest challenges of our time, with profound and wide-ranging implications for development, economic growth, the environment, and international security. The United States is committed to continuing enhanced action, together with the global community, to lead the global effort to achieve a low-emission, climate-resilient future. This 2014 *U.S. National Communication* describes actions the United States is taking to confront climate change and prepare for its impacts. The report highlights major federal, state, and local initiatives and outlines U.S. efforts to assist other countries in addressing climate change.

NATIONAL CIRCUMSTANCES

Chapter 2 of this report outlines the national circumstances of the United States and how they affect U.S. greenhouse gas (GHG) emissions. The United States is a large country with a diverse geography. The nation stretches across seven time zones, from the Atlantic Seaboard to the Hawaiian Islands, and encompasses a full range of tropical, temperate, and Arctic ecosystems. The total U.S. land area is 3,548,112 square miles (9,192,000 square kilometers); about 28 percent of this land is owned and managed by the federal government in a system of parks, forests, wilderness areas, wildlife refuges, and other public lands.

The United States is a federal republic, whose government is divided into three distinct branches: executive, legislative, and judicial. Each branch plays a separate, significant role in the processes that shape laws and policies related to climate change. In addition, the governments of U.S. states and localities promulgate energy regulations and land-use policies, and their laws and policies together have a substantial influence on the U.S. response to climate change.

As of 2013, the United States is the third most populous country in the world, with an estimated population of 316 million. From 1990 to 2008, the U.S. population grew by 54.5 million, at an average annual rate of just over 1 percent, for a total growth of approximately 22 percent since 1990. However, that growth has slowed somewhat since the global recession in 2008, with an average annual population growth rate of less than 1 percent in 2009, 2010, 2011, and 2012. Nevertheless, the growth rate of the U.S. population was still among the highest in the world among advanced economies during the last five years.

The U.S. economy is the largest national economy in the world, with a nominal gross domestic product (GDP) of $15.7 trillion in 2012, slightly smaller than the GDP of the European Union. The U.S. per capita GDP in 2012 was just over $49,600. Between 1990 and 2008, the U.S. economy grew by more than 60 percent (in constant 2005 dollars), one of the highest growth rates among advanced economies in this time frame. Between 2008 and 2013, however, the U.S. economy averaged only 0.6 percent in real annual GDP growth.

The United States is the world's second-largest producer and consumer of energy. The nation has large reserves of energy sources currently available for production, including fossil fuels, uranium ore, renewable biomass, and hydropower. Other renewable energy sources, such as solar and wind power, currently represent approximately 2 percent of the total energy resources used in the United States.

Several of the long-term trends identified in the 2010 *U.S. Climate Action Report* (CAR)—such as the historical pairing of economic growth and increased energy use—have slowed or reversed because of U.S. national circumstances. As economic growth has slowed since 2008, GHG emissions have also declined. Recent U.S. investments in energy efficiency have also been a factor in the continued decline in U.S. energy intensity. In the coming decades, U.S. energy intensity is projected to decline significantly, allowing the economy to grow while GHG emissions decline. Investments in renewable energy have led to rapid growth of wind, solar, and geothermal power in the energy mix. Solar power capacity grew by approximately 100 percent from 2008 through 2011, and wind power capacity grew by approximately 116 percent during that same period (U.S.DOE/EIA 2012).[1]

A major contributor to the decline in U.S. GHG emissions has been the displacement of coal with natural gas that is extracted from shale rock formations through hydraulic fracturing and horizontal drilling. The production of "shale gas" has grown rapidly in recent years. In 1996, U.S. shale gas wells produced 8.5 billion cubic meters (m^3) (0.3 trillion cubic feet [ft^3]) of natural gas, representing 1.6 percent of U.S. gas production. By 2011, production of shale gas had increased to 241 billion m^3 (8.5 trillion ft^3) of natural gas, or 30 percent of U.S. gas production. The extraction and use of shale gas are projected to continue to grow during the next several years.

The U.S. transportation system has evolved to meet the needs of a highly mobile, dispersed population and a large economy. Automobiles and light trucks still dominate the passenger transportation system, and the highway share of passenger miles traveled. In 2013, the most recent year of available data, automobiles and light trucks constituted about 87 percent of the passenger miles traveled, down 2 percent from the highway share listed in the 2010 CAR. Air travel accounted for slightly more than 11 percent (up 1.5 percent from the 2010 CAR), and mass transit and rail travel combined accounted for only about 1 percent of passenger miles traveled.

GREENHOUSE GAS INVENTORY

Chapter 3 summarizes U.S. anthropogenic GHG emission trends from 1990 through 2011. The estimates presented in the report were calculated using methodologies consistent with those recommended by the Intergovernmental Panel on Climate Change. A complete accounting of GHGs in the United States is referenced in Chapter 3 of this report in Figure 3-1 and Table 3-1. In 2011, total U.S. GHG emissions were 6,702.3 teragrams (Tg) of carbon dioxide equivalents (CO_2e). Overall, total U.S. emissions rose by 8 percent from 1990 through 2011. Over that same period, U.S. GDP increased by 66 percent, and population increased by 25 percent. CO_2 emissions accounted for approximately 84 percent of total U.S. GHG emissions in 2011.

As the largest source of U.S. GHG emissions, CO_2 from fossil fuel combustion has accounted for approximately 78 percent of global warming potential-weighted emissions since 1990. Emissions of CO_2 from fossil fuel combustion increased at an average annual rate of 0.5 percent from 1990 through 2011. The fundamental factors influencing this trend include (1) a generally growing domestic economy over the last 22 years, and (2) an overall growth in emissions from electricity generation and transportation activities. Between 1990 and the end of 2011, CO_2 emissions from fossil fuel combustion increased from 4,748.5 Tg CO_2e to 5,277.2 Tg CO_2e, an 11 percent total increase over the 22-year period. Historically, changes in emissions from fossil fuel combustion have been the dominant factor affecting U.S. emission trends.

Methane (CH_4) accounted for approximately 9 percent of total U.S. GHG emissions in 2011, with natural gas systems being the largest source of CH_4 emissions. U.S. emissions of CH_4 declined by 8 percent from 1990 through 2011. This decline was mostly due to both a decrease in emissions from natural gas transmission and storage resulting from increased voluntary reductions, and a decrease in natural gas distribution emissions resulting from a reduction in cast iron and unprotected steel pipelines, as well as an increase in the collection and combustion of landfill gas.

Nitrous oxide (N_2O) accounted for approximately 5 percent of total U.S. GHG emissions in 2011. The main U.S. human activities producing N_2O are agricultural soil management and

[1] *See* U.S. DOE/EIA 2012, Table 8.2a.

stationary fuel combustion. Overall, U.S. emissions of N_2O increased by 4 percent from 1990 through 2011, largely due to the overall increase in N_2O emissions from agricultural soils. However, annual N_2O emissions from agricultural soils fluctuated between 1990 and 2011, largely as a reflection of annual variation in weather patterns, synthetic fertilizer use, and crop production.

Fluorinated substances—hydrofluorocarbons (HFCs), perfluorocarbons (PFCs), and sulfur hexafluoride (SF_6)—accounted for 2 percent of total U.S. GHG emissions in 2011. The increasing use of these compounds since 1995 as substitutes for ozone-depleting substances (ODS) has been largely responsible for their upward emission trends.

Net CO_2 sequestration from land use, land-use change, and forestry (LULUCF) increased by 110.5 Tg CO_2e (14 percent) from 1990 through 2011. This increase was primarily due to growth in the rate of net carbon accumulation in forest carbon stocks, particularly in above-ground and below-ground tree biomass.

POLICIES AND MEASURES

Chapter 4 of this report outlines approximately 100 near-term policies and measures under-taken by the U.S. government to mitigate GHG emissions. These policies and measures promote increased investment in end-use efficiency, clean energy development, and reductions in agricultural GHG emissions. The U.S. government is also working to reduce emissions from the most potent GHGs, with more than a dozen initiatives across five executive agencies targeting CH_4, N_2O, HFCs, PFCs, SF_6, and other fluorinated gases.

A large number of U.S. states and localities are implementing clean energy incentives and clean energy targets as well. These actions range from voluntary emission goals and green building standards to mandatory cap-and-trade laws.

In May 2010, the U.S. Environmental Protection Agency (EPA) issued a regulation establishing a common-sense approach to permitting GHG emissions. As of April 2013, EPA and states have issued nearly 90 permits to large industrial sources that cover GHG emissions.

On April 17, 2012, EPA issued cost-effective regulations to reduce harmful air pollution from the oil and natural gas industry, while allowing continued, responsible growth in U.S. oil and natural gas production. These regulatory standards achieve a significant co-benefit of CH_4 emission reductions, estimated at 32.6 Tg CO_2e in 2015 and 39.9 Tg CO_2e in 2020.

The National Program for Light-Duty Vehicle GHG Emission Standards and Corporate Average Fuel Economy Standards for combined model years (MYs) 2012–2025 are projected to cut in half the GHG emissions of the average MY 2025 vehicle when compared with the average MY 2010 vehicle, reducing 6,000 Tg CO_2e over the lifetimes of the vehicles sold in MYs 2012–2025. Similarly, the National Program for Heavy-Duty Vehicle GHG Emission Standards and Fuel Efficiency Standards for MYs 2014–2018 will significantly reduce GHG emissions and fuel consumption from heavy-duty vehicles. The heavy-duty vehicle program will cut GHGs by 270 Tg CO_2e during the lifetimes of the vehicles sold in MYs 2014–2018.

New lighting energy efficiency standards will phase out the 130-year-old incandescent light bulb by the middle of the next decade and phases out less efficient fluorescent tubes. The new standards are estimated to have a GHG mitigation potential of 36.3 Tg CO_2e in 2015 and 37.7 Tg CO_2e in 2020.

PROJECTED GREENHOUSE GAS EMISSIONS

Chapter 5 provides projections of U.S. GHG emissions through 2030, including the effects of policies and measures in force as of September 2012, the cutoff date for the *Annual Energy Outlook*'s baseline projections of energy-related CO_2 emissions. The 2012 policy baseline scenario presented in this 2014 *U.S. National Communication* does not include the impacts of more recent policies, including the President's June 2013 *Climate Action Plan*, which are presented in the *Biennial Report* that comprises a part of the 2014 *U.S. Climate Action Report* (EOP 2013a).

The projections of U.S. GHG emissions described here reflect national estimates considering population growth, long-term economic growth potential, historic rates of technology improvement, and normal weather patterns. They are based on anticipated trends in technology deployment and adoption, demand-side efficiency gains, fuel switching, and many of the implemented policies and measures discussed in Chapter 4.

Projections are provided in total, by gas and by sector. Gases included in this report are CO_2, CH_4, N_2O, HFCs, PFCs, and SF_6. Sectors reported include energy (subdivided into electric power, residential, commercial, and industrial); transportation; industrial processes; agriculture; waste; and LULUCF. For the LULUCF sector, projections through 2030 are presented as a range based on two alternative scenarios, while a text box describes longer-term trends in the sector.

Given implementation of programs and measures in place as of September 2012 and current economic projections, total gross U.S. GHG emissions are projected to be 4.6 percent lower than 2005 levels in 2020. Between 2005 and 2011 total gross U.S. GHG emissions have declined significantly due a combination of factors, including the economic downturn and fuel switching from coal to natural gas (U.S. EPA/OAP 2013). Emissions are projected to rise gradually between 2011 and 2020. Emissions are projected to remain below the 2005 level through 2030, despite significant increases in population (26 percent) and GDP (69 percent) during that period. More rapid improvements in technologies that emit fewer GHGs, new GHG mitigation requirements, or more rapid adoption of voluntary GHG emission reduction programs could result in lower gross GHG emission levels than in the 2012 policy baseline scenario projection.

Between 2005 and 2020, CO_2 emissions in the 2012 policy baseline scenario projection (measures in place as of 2012) are estimated to decline by 7.5 percent. In contrast, in the 2010 CAR, CO_2 emissions were expected to *increase* by 1.5 percent between 2005 and 2020 (U.S. DOS 2010), a change of about 9 percent, and in the 2006 CAR, emissions were expected to increase by 14 percent between 2004 and 2020 (U.S. DOS 2006). During the same period, CH_4 and N_2O emissions are expected to grow by 3.5 percent and 6.1 percent, respectively. The most rapid growth is expected in fluorinated GHGs (HFCs, PFCs, and SF_6), which are expected to increase by more than 60 percent between 2005 and 2020, driven by increasing use of HFCs as substitutes for ODS.

VULNERABILITY, ASSESSMENT, CLIMATE CHANGE IMPACTS, AND ADAPTATION

Chapter 6 of this report highlights actions taken in the United States to better understand and respond to vulnerabilities and impacts associated with climate change. All levels of government are working together on an array of climate assessments, research, and other activities to understand the potential impacts of climate change on the environment and the economy and to develop methods and tools to enhance adaptation options.

Notably during this reporting period, the United States undertook development of the *Third National Climate Assessment* (NCA), as mandated by the Global Change Research Act of 1990 (GCRA) (NCADAC 2013). The NCA brings together the best peer-reviewed science on climate change and its impacts on the United States, leveraging research across regions and sectors and providing a basis for future assessment and action. The draft Third NCA was developed through a transparent process that included more than 1,000 direct contributors and 240 chapter authors from academia, resource management agencies, and nongovernmental organizations (NGOs), in addition to government scientists. The U.S. government also sponsors some of the world's most advanced scientific research on climate change. The U.S. Global Change Research Program (USGCRP), also established by the GCRA, is designed to coordinate the federal government's $2.6 billion annual investment in global change research.

Chapter 6 describes:

- Climate and global change impacts on the United States;

- Observed and projected regional, sectoral, and cross-cutting vulnerabilities, such as the potential for water scarcity, interruptions in energy production and transmission, and disruption of multimodal transportation systems;

- Continuing and planned research and sustained assessments to improve the understanding of impacts, vulnerabilities, and options for response over time; and

- Ongoing adaptation measures, including examples of adaptation actions taking place at multiple scales throughout the nation.

Through the creation of special programs related to climate adaptation, the U.S. government is working to address its vulnerabilities to both abrupt and more gradual changes in U.S. climate. At the direction of the previous Interagency Climate Change Adaptation Task Force, federal agencies have begun integrating adaptation planning into their operations, missions, and programs, with the first set of agency-specific adaptation plans publicly released in February 2013.[2] Upon the recommendation of the Adaptation Task Force, Congress, and other interagency bodies, federal agencies also developed a series of cross-cutting strategies to reduce the impacts of climate change on the nation's freshwater and ocean resources, and fish, wildlife, and plants. Chapter 6 includes examples of these efforts, discusses these strategies in detail, and describes President Obama's June 2013 *Climate Action Plan* (EOP 2013a) and November 2013 Executive Order 13653, *Preparing the United States for the Impacts of Climate Change* (EOP 2013b), which directs federal agencies to take a series of steps to enhance their efforts to build national climate preparedness and resilience and ensure the safety, health, and well-being of communities in the face of extreme weather and other impacts of climate change.

States, tribes, and localities also have major roles to play in vulnerability assessment and adaptation, given that many decisions are made at the local level. Chapter 6 contains several examples of this work, such as New York City's development of customized heat-warning systems and California's implementation of building standards mandating energy and water efficiency savings.

Finally, the United States is committed to establishing and maintaining climate adaptation assistance for both domestic and international communities through the 2010 Presidential Policy Directive on Global Development, the Global Climate Change Initiative (GCCI),[3] the Climate Services Partnership,[4] and other efforts of the U.S. Agency for International Development (USAID) and the U.S. Department of State.

FINANCIAL RESOURCES AND TRANSFER OF TECHNOLOGY

Chapter 7 outlines U.S. government initiatives and partnerships and U.S. agency roles in climate-related international assistance and technology transfer. This chapter of the 2014 *U.S. National Communication* provides details on U.S. climate finance by channels and instruments, thematic pillar, and region. It also describes U.S. efforts to mobilize private climate finance, and provides examples of U.S. contributions to capacity building and transfer of technology.

Since the period covered by the 2010 CAR, climate change has become a major thrust of U.S. diplomatic and development assistance efforts. The 2010 Presidential Policy Directive on Global Development identified the GCCI as one of three priority U.S. development initiatives. The GCCI provides a platform upon which the United States builds climate change considerations into its foreign assistance operations. The 2010 *U.S. Quadrennial Diplomacy and Development Review* also identified climate change as one of the main pillars of U.S. diplomacy and international development (U.S. DOS and USAID 2010).

Through the GCCI and other enhanced climate-related investments, the United States has significantly ramped up its provision of climate finance and is assisting dozens of developing countries to mitigate and adapt to climate change.

The United States is using the full range of mechanisms—bilateral, multilateral, and private finance—to invest strategically in building lasting resilience to unavoidable climate impacts; reducing emissions from deforestation and land degradation; and supporting low-carbon development strategies and the transition to a sustainable, clean energy economy. The nation is working hard to ensure that U.S. support is efficient, effective, innovative, based on country-owned plans, and focused on achieving measurable results, with a long-term view of economic and environmental sustainability.

As noted in the Doha Agreements, developed country Parties successfully achieved the "fast start" finance goal. The United States provided $7.5 billion during fiscal years 2010, 2011, and 2012 to more than 120 countries through bilateral and multilateral channels, meeting the

[2] *See* http://www.epa.gov/climatechange/impacts-adaptation/fed-programs.html.

[3] *See* http://www.whitehouse.gov/the-press-office/2010/09/22/fact-sheet-us-global-development-policy.

[4] *See* http://www.climate-services.org/. The Climate Services Partnership, which emerged from the GCCI, helps build the capacity of developing country climate and weather services to provide useful information to decision makers.

President's commitment to provide America's fair share of the collective pledge. This $7.5 billion consists of more than $4.7 billion of congressionally appropriated assistance and more than $2.7 from U.S. development finance and export credit agencies. The $4.7 billion in appropriated assistance levels represents a fourfold increase in annual climate assistance since 2009, with a ninefold increase in adaptation assistance.

Maintaining a strong core of public climate finance is essential, and the United States intends to maintain its commitment to climate change as an important component in the U.S. assistance budget. Private investment will inevitably play an increasingly important role as developing countries put mitigation and adaptation policies and actions into place. The nation is working to combine its significant, but finite, public resources with targeted, smart policies to mobilize maximum private investment into climate-friendly activities. The United States is actively pursuing strategies to encourage private investment in low-carbon, climate-resilient activities in developing countries, and to support countries in their efforts to create a policy framework that will attract investment in clean energy and other climate-supportive activities. Continuing to execute this vision will be important, as developed countries, including the United States, work toward a collective goal of mobilizing $100 billion per year in climate change finance for developing countries by 2020, in the context of meaningful mitigation actions and transparency on implementation.

The United States has also been working with its developed country partners to collectively develop and coordinate strategies for scaling up climate-friendly investment in developing countries. In April 2013, the United States held an inaugural meeting of climate ministers and senior officials from development and finance ministries to explore ways to coordinate more closely on the issue of how to use public resources and policies to mobilize the maximum amount of total investment in climate action. The developed countries in attendance agreed to focus on strengthening and augmenting key tools that are provided through existing public finance institutions that operate at the nexus with the private sector: development finance institutions, multilateral development banks, key multilateral climate change funds, and export credit agencies. The United States has played and will continue to play an active role internationally to help coordinate this work going forward.

RESEARCH AND SYSTEMATIC OBSERVATION

Chapter 8 describes how the United States is providing the fundamental scientific and technological foundation for understanding the causes and consequences of climate and global change, reducing scientific uncertainties, and supporting adaptation and mitigation actions aimed at managing risks and producing benefits at local, regional, and global scales. The chapter covers three broad areas: research on global change, systematic observations, and research and development of technologies to address climate change.

The United States has always placed a high priority on research to understand global change. U.S. federal agencies have put forward a coordinated set of investments in global change science to gain new theoretical knowledge of Earth system processes; to maintain and enhance a mix of atmospheric, oceanic, land-, and space-based observing systems; to advance predictive capabilities through the next generation of numerical modeling; to promote advances in computational capabilities, data management, and information sharing; and to further develop an expert scientific workforce in the United States and worldwide. These include major investments under the American Recovery and Reinvestment Act (Recovery Act) to enhance research infrastructure, build next-generation cyberinfrastructure assets, and award many new research grants and graduate fellowships.

Over the past three years, the United States has enhanced coordination with other nations and international organizations on global change research activities, promoted increased international access to scientific data and information, and fostered increased participation in international global change research by developing nations.

All of these research and assessment activities depend on the existence of a comprehensive, continuous, integrated, and sustained set of physical, chemical, biological, and societal observations of global change and its impacts. These observations are essential for improving the understanding of the components and processes of the Earth system and the causes and consequences of global change. The United States supports a large number of remote-sensing satellite platforms, as well as a broad network of Earth-based global atmospheric, oceanic, and terrestrial observation systems that are essential to climate monitoring globally. These systems are a baseline Earth-observing system and include Earth-observing satellites and extensive nonsatellite observational capabilities across multiple federal agencies that participate in the USGCRP.

Over the last three years, the United States achieved new milestones with the launch of critical new satellite-observing systems, including the Suomi National Polar-orbiting Partnership, the Landsat Data Continuity Mission/Landsat-8, and Aquarius (in partnership with the Space Agency of Argentina). In addition, new surface-based networks, such as the National Ecological Observatory Network and the Ocean Observatories Initiative are well on their way to operation, creating a next generation of *in situ* observing capabilities. And the Atmospheric Radiation Measurement Climate Research Facility, through the U.S. Department of Energy (DOE) Office of Science, received $60 million in Recovery Act funding to enhance its climate change research capabilities, by deploying an expansive array of new instruments.

Finally, this chapter details how the U.S. government is supporting clean energy and climate change mitigation technologies. The technology research and innovation activities within all of these areas, which span multiple federal agencies, are organized around the goals of reducing emissions from energy supply, energy end use, and infrastructure; capturing and sequestering CO_2 emissions; reducing emissions of other GHGs; and measuring and monitoring emissions. They also include bolstering the contributions of basic science to the development of new technologies and monitoring systems. These efforts build on such initiatives as the creation of the DOE Advanced Research Projects Agency-Energy, to spur a revolution in clean energy technologies.

EDUCATION, TRAINING, AND OUTREACH

Chapter 9 outlines the expansion of U.S. climate change education, training, and outreach efforts since the 2010 CAR. Climate change communication faces many challenges, and federal agencies, civil society, and individuals have invested in numerous initiatives to establish a climate-literate citizenry. In the U.S. National Research Council report *America's Climate Choices*, the authors find that "climate change is difficult to communicate by its very nature." However, "education and communication are among the most powerful tools the nation has to bring hidden hazards to public attention, understanding, and action" (NRC 2011).

The United States is working to focus and evolve the use of these tools. Numerous federal, NGO, and individual efforts have supported sustained and robust educational and communications initiatives to develop a climate-literate citizenry and skilled workforce. These include initiatives in schools, online (e.g., Climate.gov), and in the workplace, among many others. When citizens have knowledge of the causes, likelihood, and severity of climate impacts, as well as of the range, cost, and efficacy of options to adapt to impacts, they are more prepared to effectively address the risks and opportunities of climate change. Furthermore, since 2010, more Americans than ever before experienced the impacts of climate change first-hand in the form of extreme events, such as prolonged droughts and stronger and more frequent wildfires, resulting in increased public interest and an opportunity for engagement on climate literacy issues.

U.S. federal agencies—including USAID; the Departments of Agriculture, Commerce, Energy, the Interior, and Transportation; EPA; the National Aeronautics and Space Administration; National Oceanic and Atmospheric Administration; and the National Science Foundation—work on a wide range of climate change education, training, and outreach programs. A USGCRP Communication and Education Interagency Working Group was formed in 2008 to coordinate these efforts and develop an integrated national approach to climate change. This group expanded the work of the Communications Interagency Working Group established in 2004. Efforts by industry, states, local governments, universities, schools, and NGOs are essential complements to more than 100 federal programs that educate industry and the public regarding climate change. The combined efforts of the U.S. federal, state, and local governments and private entities are ensuring that the American public is able to understand and address climate change, in terms of both stabilizing and reducing emissions of GHGs, and also increasing capacity to adapt to the consequences of climate change.

2

National Circumstances

Greenhouse gas (GHG) emissions in the United States are influenced by a multitude of factors. These include population and density trends, economic growth, energy production and consumption, technological development, use of land and natural resources, as well as climate and geographic conditions. This chapter focuses on both current national circumstances and departures from historical trends since the 2010 *U.S. Climate Action Report* (2010 CAR) was submitted to the United Nations Framework Convention on Climate Change. This chapter also discusses the impact of the changes in national circumstances on GHG emissions and removals (U.S. DOS 2010).

KEY DEVELOPMENTS

Several aspects of U.S. national circumstances have changed in the past four years. Some of the most important changes follow.

Challenging Economic Environment

The U.S. economy is still emerging from the aftermath of the economic downturn that followed the financial crisis of 2007-2008. The U.S. unemployment rate in May 2013 (7.5 percent) was more than 3 percent higher than its pre-crisis level (4.4 percent in May 2007), and national output is still below its potential, according to the Congressional Budget Office (CBO 2013). U.S. gross domestic product (GDP) has grown every quarter since the third quarter of 2009, and private nonfarm payroll employment has grown every month since March 2010. During the period of this 2014 U.S. *U.S. Climate Action Report* (2010-2013), the United States has produced fewer GHG emissions annually than it did before the financial crisis. Even as U.S. economic output increased, GHG emissions largely were steady or declining. In 2011, GHG emissions declined by 108 teragrams of carbon dioxide equivalents (Tg CO_2e) (-1.6 percent) from 2010, despite the 1.7 percent growth in the U.S. economy that year.

Economic Policies

The United States has adopted several policies to mitigate the economic effects of the downturn, while making the U.S. economy more energy efficient and less carbon intensive. The 2009 American Recovery and Reinvestment Act (Recovery Act) and subsequent actions by the Federal Reserve of the United States have stimulated U.S. economic activity. The Recovery Act invested in more energy-efficient homes and appliances, as well as provided funds that helped decarbonize U.S. transportation and electricity generation.

Energy Mix

The discovery and exploitation of vast reserves of U.S. natural gas have reduced the domestic price per British thermal unit (Btu) of natural gas and sparked demand for natural gas as both a baseload fuel for electricity generation and a heating fuel for U.S. households. In 2012, natural gas generated 30.4 percent of the nation's electric power, up from 17.8 percent of total electricity generation in 2004.[1] As a result, the use of coal for electric power has declined. Coal now represents 37.4 percent of the energy mix, down from a 50 percent share in 2005. Wind power, solar power, biomass, and geothermal energy generated 5.4 percent of total U.S. electricity in 2012 and represent a significant share of new U.S. electrical generation. During

[1] U.S. Department of Energy (DOE)/ Energy Information Administration (EIA) 2013g, Table 7.2a. Net Electricity Generation: Total (All Sectors). *See* http://www.eia.gov/totalenergy/data/ monthly/pdf/sec7_5.pdf.

the first quarter of 2013, 82 percent of new U.S. electrical capacity was from renewable energy sources. In 2012, conventional hydroelectric power generated 6.8 percent of total electricity generation, and nuclear energy generated 19.0 percent.

Transportation Patterns

Since a peak in 2008, Americans drove 2.6 percent fewer passenger miles annually as of July 2013 than they did before the financial crisis. Generational preferences, the effects of the recession, the high cost of oil, and new urban development patterns increasingly move Americans to mass transit and other modes of transport. Fewer passenger miles translate to fewer GHG emissions from mobile sources. Cars are also becoming more fuel efficient, due to both a shift in consumer demand and federal and state policies.

Legal Framework for Acting on Climate Change

In April 2007, the Supreme Court of the United States ruled that GHGs are air pollutants covered by the Clean Air Act and must be regulated by the U.S. Environmental Protection Agency (EPA) if they may reasonably be anticipated to endanger public health or welfare. In December 2009, EPA issued its Endangerment Finding, which found that current and projected levels of six GHGs threaten the health and human welfare of current and future generations (U.S. EPA 2009). Since this finding, EPA has set in place rules and regulations to limit GHG emissions from motor vehicles and has proposed national limits on the amount of GHG emissions future power plants will be allowed to emit.

Extreme Weather Events

The United States has experienced several extreme weather events since 2010, which have inflicted major damage and raised awareness of the rising costs of climate change.

Evolving Public Attitudes Toward Climate Change

Though U.S. public opinion on climate change remains polarized, the public's concern about climate change is on the rise nationwide. Although numbers vary depending on the polling questions, in several 2013 surveys more than 60 percent of Americans said that climate is changing and that it is important to address this issue for the sake of current and future generations. A majority of Americans support the increased deployment of clean and renewable energy and regulation of power plant emissions (ESSI 2013).

GOVERNMENT STRUCTURE

The United States is a federal republic. As such, local, state, and federal governments share responsibility for the nation's economic development, energy, natural resources, and many other issues affecting the welfare of Americans. At the national level, a number of federal agencies, commissions, and advisory offices to the President are involved in developing, coordinating, and implementing nationwide policies to act on climate change.

The U.S. government is divided into three distinct branches: executive, legislative, and judicial. Each branch possesses distinct powers, but each is also not completely independent of the other. This creates a system of "checks and balances" and separates the powers to create, implement, and adjudicate laws.

Executive Branch

The executive branch is charged with implementing and enforcing the laws of the United States. The President of the United States is the U.S. Head of State and oversees the executive branch. The President is advised by a Cabinet that includes the Vice President and the heads of 15 executive agencies—the Departments of State, Treasury, Defense, Justice, Interior, Agriculture, Commerce, Labor, Health and Human Services, Housing and Urban Development, Transportation, Energy, Education, Veterans Affairs, and Homeland Security. Other positions with Cabinet rank include the President's Chief of Staff, the EPA Administrator, the Director of the Office of Management and Budget, the U.S. Trade Representative, the Chair of the Council of Economic Advisers, the U.S. Ambassador to the United Nations, and the Administrator of the Small Business Administration.

The Executive Office of the President, overseen by the President's Chief of Staff, includes a number of offices that play important roles in U.S. climate policy, such as the Office of Energy and Climate Change, the Office of Science and Technology Policy, the Council on Environmental Quality, and the National Security Council. The executive branch also includes a number of independent commissions, boards, and agencies that play a role in domestic climate policy, such as the Federal Energy Regulatory Commission and the Export-Import Bank. Collectively, executive branch institutions cover a wide range of responsibilities, such as implementing environmental and energy regulations passed by the legislative branch through the rulemaking process, serving America's interests overseas, developing and maintaining the federal highway and air transit systems, researching the next generation of energy technologies, and managing the nation's abundant public lands.

Legislative Branch

The legislative branch consists of the two bodies in the U.S. Congress—the House of Representatives (House) and the Senate—which are the primary lawmaking bodies of the U.S. government. This branch represents the U.S. citizenry through a bicameral system intended to balance power between representation based on population and representation based on statehood. The Senate is composed of 100 members, two from each of the 50 U.S. states. The House is composed of 435 members; each member represents a single congressional district of approximately 650,000 people.

Each of the two bodies of Congress has the authority to develop legislation. A completed bill must receive a majority of votes in both the House and the Senate, and any differences between the House and the Senate versions must be reconciled before that bill can be sent to the President for consideration to be signed into law. The legislation becomes effective upon the President's signature.

In Congress, climate change is addressed by individual members and committees that are charged with developing legislation on energy and other relevant issues relevant to climate change. In the House, the Committees on Appropriations; Agriculture; Science, Space, and Technology; Ways and Means; Natural Resources; and Energy and Commerce, among others, play vital roles in developing legislation related to climate change. In the Senate, the Committees on Environment and Public Works; Finance; Foreign Relations; Agriculture; Commerce, Science, and Transportation; and Energy and Natural Resources develop legislation and are critical venues for debate.

Because the legislative process requires the support of both chambers of Congress and also involves the executive branch, a strong base of support is necessary to enact new legislation. As climate legislation is developed, this high threshold will remain very relevant.

Judicial Branch

The judicial branch is the federal court system responsible for, among other things, interpreting the U.S. Constitution. It includes the Supreme Court, which is the highest court in the United States. The judicial branch in particular plays a significant role in defining the jurisdiction of the executive departments and, in the case of climate change, interpreting laws related to the conduct of climate and energy policies.

Governance of Energy and Climate Change Policy

Jurisdiction for addressing climate change within the federal government cuts across each of the three branches. Within the executive branch alone, some two dozen federal agencies and executive offices work together to advise, develop, and implement policies that help the U.S. government understand the workings of the Earth's climate system, reduce GHG emissions and U.S. dependence on oil, promote a clean energy economy, and assess and respond to the adverse effects of climate change. Chapters 4, 6, 7, 8, and 9 of this report describe the activities of these agencies related to these policies.

As with many other policy areas, jurisdiction for energy policy is shared by federal and state governments. Economic regulation of the energy distribution segment is a state responsibility, with the Federal Energy Regulatory Commission regulating wholesale sales and transportation of natural gas and electricity. In the absence of comprehensive federal climate change

[2] Full projections of the U.S. Census Bureau are free of charge and accessible to the global community. *See* http://www.census.gov.

legislation, U.S. states have increasingly enacted climate change legislation or other policies designed to promote clean energy. Examples of these policies are described in Chapter 4 of this report. Similarly, land-use oversight is subject to mixed jurisdiction, with localities playing strong roles as well. Many activities related to adaptation policy are being initiated by state and local entities. Examples of these activities are provided in Chapter 6.

POPULATION PROFILE

Population changes and growth patterns are fundamental drivers of trends in energy consumption, land use, housing density, and transportation, all of which have a significant effect on U.S. GHG emissions. The United States is the third most populous country in the world, with an estimated population of 316 million. From 1990 to 2012, the U.S. population grew by 64.3 million, at an average annual rate of just over 1 percent, for a total growth of approximately 25 percent since 1990. However, that growth has slowed somewhat since the global recession. Average annual population growth in the United States was less than 1 percent in 2010, 2011, and 2012. Even so, the growth rate of the U.S. population was among the highest in the world among advanced economies over the last four years. The U.S. Census Bureau projects that the annual growth rate will shrink slowly from about 0.77 percent in 2015 to 0.5 percent in 2050, when the U.S. population is projected to be almost 400 million.

The U.S. is ranked 149th worldwide in population density and 161st in emissions per capita per square kilometer (km^2).[2] Population density trends show that more Americans are moving into cities and metropolitan areas. In 2012, urban areas—defined as densely developed residential, commercial, and other nonresidential areas—accounted for 80.7 percent of the U.S. population, up from 79.0 percent in 2000. In general, increasing urbanization changes commuter patterns and reduces GHG emissions from the transportation sector. However, compared with cities in many other industrialized countries, major U.S. cities have relatively low population densities, and U.S. urban commuters use more energy for transportation and generate higher GHG emissions per person. In addition, within any metropolitan region, the population density, walkability of neighborhoods, and access to public transit vary substantially. As a result, the average GHG emissions from household transportation differ significantly.

GEOGRAPHIC PROFILE

The United States is one of the largest countries in the world, with a total area of 9,192,000 km^2 (3,548,112 square miles [mi^2]) stretching over seven time zones. The topography is diverse, featuring deserts, lakes, mountains, plains, and forests. The federal government owns and manages the natural resources on about 28 percent of U.S. land, most of which is managed as part of the national systems of parks, forests, wilderness areas, wildlife refuges, and other public lands. More than 60 percent of land area is privately owned, 9 percent is owned by state and local governments, and 2 percent is held in trust by the United States for the benefit of various Native American tribes.

CLIMATE PROFILE

The climate of the United States is highly diverse, ranging from tropical conditions in south Florida and Hawaii to arctic and alpine conditions in Alaska and across the Rocky Mountains. Temperatures for the continental United States show a strong gradient across regions and seasons, from very high temperatures in southern coastal states where the annual average temperatures exceed 21°C (70°F), to much cooler conditions in the northern parts of the country along the Canadian border, with seasonal differences as great as 50°C (90°F) and 10°C (50°F), respectively, between summer and winter in the northern Great Plains.

Similarly, precipitation varies across the country and by seasons, measuring more than 127 centimeters (cm) (50 inches [in]) per year along the Gulf of Mexico, while annual precipitation can be less than 30 cm (12 in) in the Intermountain West and Southwest. The peak rainfall season also varies by region. Many parts of the Great Plains and Midwest experience late-spring peaks, West Coast states have a distinct rainy season during winter, the Desert Southwest is influenced by summer's North American Monsoon, and many Gulf and Atlantic coastal regions experience summertime peaks.

The United States is subject to almost every kind of weather extreme, including severe thunderstorms, almost 1,500 tornadoes per year, and an average of 17 hurricanes that make landfall along the Gulf and Atlantic coasts each decade. At any given time, approximately 20 percent of the country experiences drought conditions. Differing U.S. climate conditions can be expressed by the number of annual heating and cooling degree-days, which represent the number of degrees that the daily average temperature—the mean of the maximum and the minimum temperatures for a 24-hour period—is below (necessitating heating) or above (necessitating cooling) 18.3°C (65°F). For example, a weather station reporting a mean daily temperature of 4°C (40°F) would report 25 heating degree-days. From 2001 to 2011, the number of heating degree-days averaged 4,324, which was 2.3 percent below the 20th-century average.[3] Over the same period, the annual number of cooling degree-days averaged 1,343, which was 6.0 percent above the long-term average.[4]

ECONOMIC PROFILE

The U.S. economy is currently the largest national economy in the world, with a nominal GDP of $15.7 trillion in 2012. The U.S. per capita GDP in 2012 was just over $49,600. Between 1990 and 2008, the U.S. economy grew by more than 60 percent (in constant 2005 dollars)—one of the highest growth rates among advanced economies in this time frame. However, between 2008 and 2013, the U.S. economy averaged only 0.6 percent in real GDP growth per year. As economic growth has slowed, GHG emissions have declined slightly since 2008. Recent U.S. investments in energy efficiency have also been a factor in the continued decline in U.S. energy intensity, which is projected to decline significantly over the coming decades (Figure 2-1).

Figure 2-1 **U.S. Primary Energy Consumption per Real Dollar Gross Domestic Product** (2005$)

Between 2008 and 2013, the U.S. economy averaged only 0.6 percent in real GDP growth per year as emissions continued to decline. U.S. energy intensity is projected to decline significantly over the coming decades.

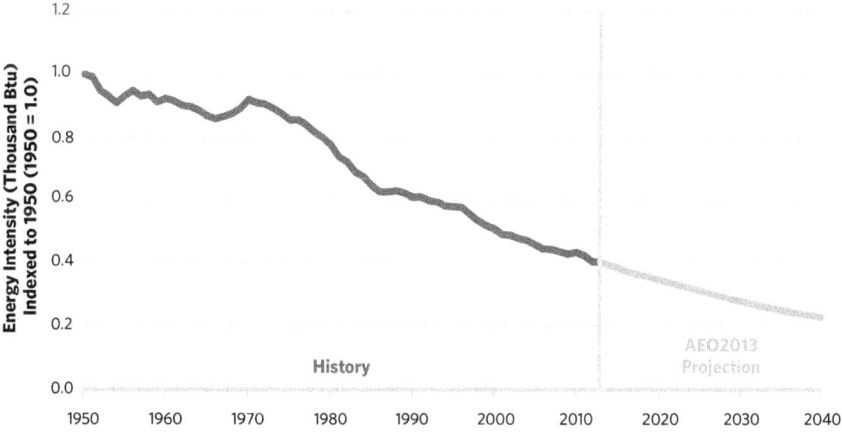

Sources: U.S. DOE/EIA 2012 and 2013b.

ENERGY RESERVES AND PRODUCTION

The United States is the world's second-largest producer and consumer of energy. The major energy sources consumed in the United States are petroleum, natural gas, coal, nuclear, and renewable energy. Renewable energy sources, including solar, wind, hydropower, and geothermal, have rapidly expanded. For example, solar power generation grew by more than 400 percent from 2008 through 2012, and wind power generation grew by more 150 percent during that same period.[5] While the three major fossil fuels—petroleum, natural gas, and coal—have dominated the U.S. fuel mix, recent increases in the domestic production of petroleum

[3] U.S. DOE/EIA 2012, Table 1.7, Heating Degree-Days by Month, 1949–2011. *See* http://www.eia.gov/totalenergy/data/annual/showtext.cfm?t=ptb0107.

[4] U.S. DOE/EIA 2012, p. 19.

[5] U.S. DOE/EIA 2012, Table 8.2a, Electricity Net Generation: Total (All Sectors), 1949–2011. *See* http://www.eia.gov/totalenergy/data/annual/showtext.cfm?t=ptb0802a.

Figure 2-2 **Estimated U.S. Energy Use in 2012: ~95.1 Quads**

The United States is the world's second-largest producer and consumer of energy. This figure shows the sources of energy and source end points within the U.S. economy.

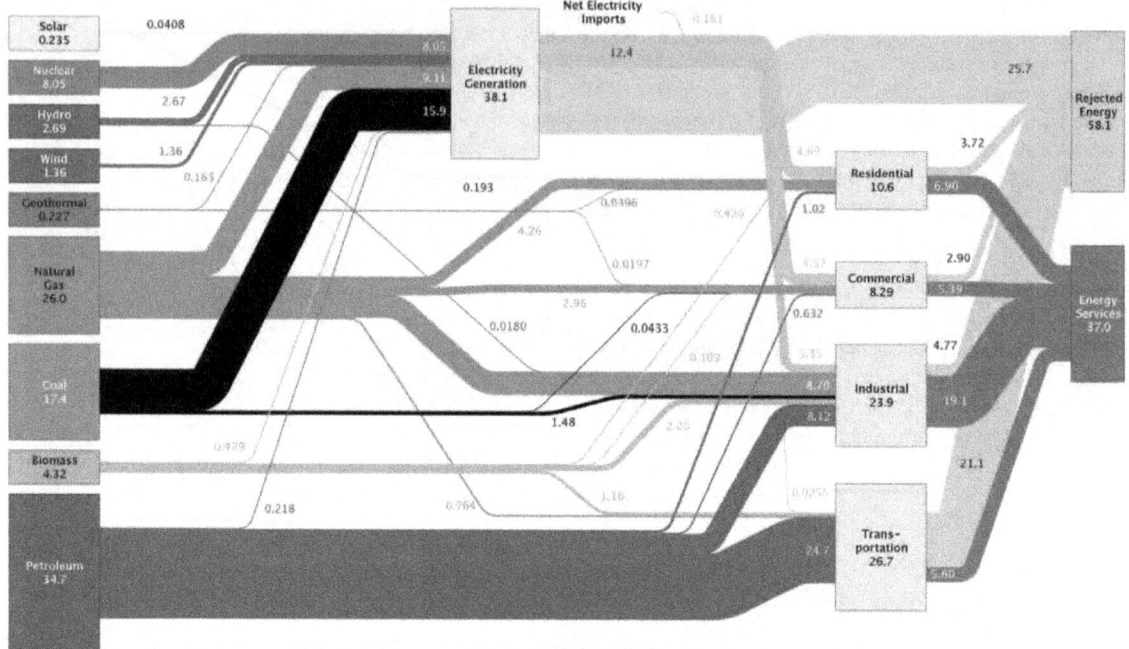

Sources: Lawrence Livermore National Laboratory (LLNL-MI-410527) 2013 and U.S. DOE/EIA 2013g.
Notes:

- Distributed electricity represents only retail electricity sales and does not include self-generation.
- EIA reports consumption of renewable resources (i.e., hydro, wind, geothermal, and solar) for electricity in Btu-equivalent values by assuming a typical fossil fuel plant "heat rate."
- The efficiency of electricity production is calculated as the total retail electricity delivered divided by the primary energy input into electricity generation.
- End-use efficiency is estimated as 65 percent for the residential and commercial sectors, 80 percent for the industrial sector, and 21 percent for the transportation sector.
- Totals may not equal the sum of components due to independent rounding.

liquids and natural gas have prompted shifts between the uses of fossil fuels (largely from coal-fired to natural gas-fired power generation).[6] Figure 2-2 provides an overview of energy flows through the U.S. economy in 2012. This section focuses on changes in U.S. energy supply and demand since the 2010 CAR, which covered changes through 2008.

Fossil Fuels

The current base of U.S. energy resources used is fossil fuels, accounting for approximately 68.4 percent of all U.S. energy consumption in 2012.[7]

Coal

Coal is the fuel most frequently used for power generation and has the highest emissions of CO_2 per unit of energy for conventional fuel sources. The use of coal in electricity generation steadily declined to 37.4 percent in 2012, down from 50 percent of the fuel mix in 2005. The United States uses about 890 million short tons of coal per year.[8] Current estimated recoverable coal reserves would supply the U.S. demand for energy, assuming constant 2011 rates of consumption, for approximately 258 years. As of December 31, 2008, of the estimated world recoverable coal reserves of 948 billion short tons, the United States holds the world's largest share (27 percent), followed by Russia (18 percent), China (13 percent), and Australia (9 percent).

[6] U.S. DOE/EIA 2013d. *See* http:// www.eia.gov/todayinenergy/detail. cfm?id=11951&src=Total-b1.

[7] U.S. DOE/EIA 2013g, Table 7.2a, Electricity Net Generation: Total (All Sectors). *See* http://www.eia.gov/ totalenergy/data/monthly/pdf/sec7_5. pdf.

[8] U.S. DOE/EIA 2013g, Table 6.2, Coal Consumption by Sector. *See* http://www. eia.gov/totalenergy/data/monthly/pdf/ sec6_4.pdf.

Natural Gas

Due to the advent of innovative drilling techniques, such as horizontal drilling and hydraulic fracturing, the United States has experienced a boom in shale gas and oil exploration and extraction (Figure 2-3), and natural gas has recently become an increasingly prominent U.S. fuel source. Electricity generation from natural gas increased from 17.8 percent in 2004 to 30.4 percent in 2012. The rapid increase in natural gas production has also heightened awareness of the possible negative environmental impacts of natural gas production through hydraulic fracturing if responsible production practices are not followed.

Proved U.S. reserves of dry natural gas are rapidly increasing. Between 2007 and 2011, they grew by 28.1 percent—from 6,734 billion cubic meters (m^3) (237,726 billion cubic feet [ft^3]) to 9,464 m^3 (334,067 billion ft^3). In 2012, the United States produced 682 billion m^3 (24,062 billion ft^3) of dry natural gas, a 19.4 percent increase since 2008.[9] Imports totaled 89 billion m^3 (3,135 billion ft^3) in 2012, while exports increased by 8 percent from 2011 to 46 billion m^3 (1,619 billion ft^3) in 2012.[10] This growth has led to greater domestic natural gas supply and relatively low prices in the United States, thus reducing U.S. reliance on foreign natural gas.[11]

Oil

Horizontal drilling and hydraulic fracturing in shale and other very low-permeability formations continue to drive record increases in proved oil. Field production of crude oil increased from an average of 5 million barrels per day (bpd) in 2008 to 7 million bpd by the end of 2012. Proved domestic reserves of crude oil were 19.1 billion barrels at the end of 2008; by the end of 2011, they had risen to 26.5 billion barrels, a 38.8 percent increase.[12] Crude oil imports in 2008 totaled 9.78 million bpd, with an additional 3.14 million bpd of refined products; by 2012, that number had fallen to 8.49 million bpd, with another 2.56 million bpd of petroleum products imported.[13] In 2012, the United States relied on net petroleum imports to meet approximately 40 percent of its petroleum needs, the lowest level since 1991.[14] The countries from which the United States imports the largest shares of crude oil and petroleum products include Canada (28 percent), Saudi Arabia (13 percent), Mexico (10 percent), Venezuela (9 percent), and Russia (5 percent).[15]

Figure 2-3 **Lower 48 States Shale Plays**

Through enhanced technologies, such as hydraulic fracturing and horizontal drilling, producers have been able to exploit previously inaccessible shale deposits throughout the United States.

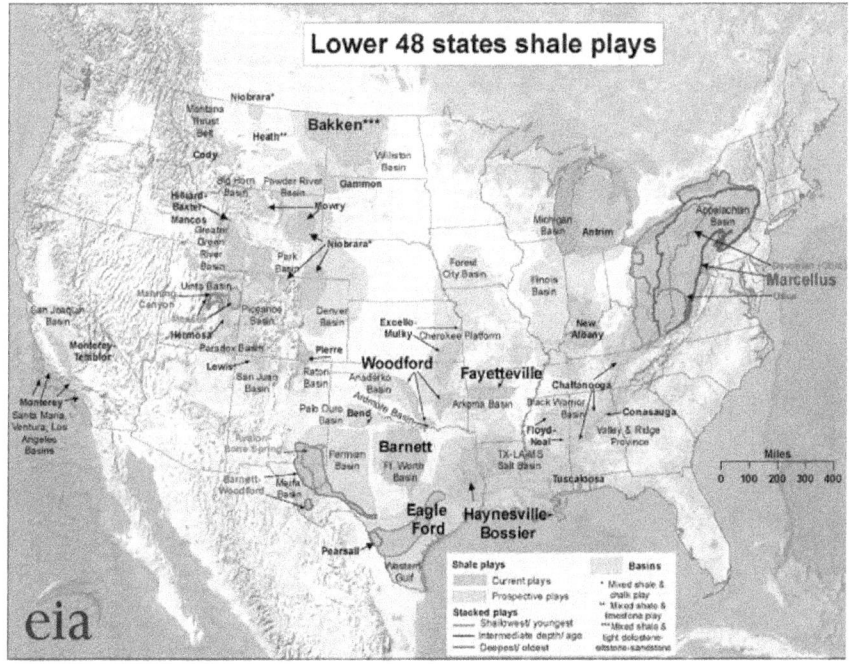

Source: U.S. Energy Information Administration, based on data from various published studies. Updated May 9, 2011.

[9] U.S. DOE/EIA. Natural Gas Summary. *See* http://www.eia.gov/dnav/ng/ng_sum_lsum_dcu_nus_a.htm.

[10] U.S. DOE/EIA 2013i, U.S. Natural Gas Imports & Exports 2012. *See* http://www.eia.gov/naturalgas/importsexports/annual/.

[11] Ibid.

[12] U.S. DOE/EIA. Table 5: Total U.S. Proved Reserves of Crude Oil and Lease Condensate, Crude Oil, and Lease Condensate, 2002–2011. *See* http://www.eia.gov/naturalgas/crudeoilreserves/pdf/table_5.pdf.

[13] U.S. DOE/EIA. Imports by Area of Entry. *See* http://www.eia.gov/dnav/pet/pet_move_imp_dc_nus-z00_mbblpd_a.htm.

[14] U.S. DOE/EIA 2013f. *See* http://www.eia.gov/tools/faqs/faq.cfm?id=32&t=6.

[15] U.S. DOE/EIA 2013e. *See* http://www.eia.gov/energy_in_brief/article/foreign_oil_dependence.cfm.

Figure 2-3 shows current and prospective shale gas and oil plays in the contiguous United States. Horizontal drilling and hydraulic fracturing have opened up previously inaccessible deposits of natural gas and oil.

Nuclear Energy

In 2012, nuclear energy from 104 operating reactor units accounted for 19 percent of all electricity generated in the United States. The U.S. supply of uranium, the fuel used for nuclear fission, is mostly imported from other countries, with about 17.4 percent of the uranium purchased in 2012 being supplied by the United States.[16] Most of these reserves can be found in Wyoming, Texas, New Mexico, Arizona, Colorado, and Utah. The average yearly U.S. uranium concentrate production in 2010–2012 was 1.9 million kilograms (kg) (4.1 million pounds [lb]), up from an average yearly production of 1 million kg (2.3 million lb) during 2003–2005.[17]

Renewable Energy

Renewable energy represents a rapidly growing source of U.S. energy production. In 2012, renewable energy accounted for 5.4 percent of U.S. electric generation excluding conventional hydropower, or 12.2 percent including conventional hydropower.[18] Though there is currently no federally mandated standard for the use of renewable energy sources for electric generation, as of 2013, 29 states have legislatively mandated a renewable energy portfolio standard (RPS). The RPS requirements vary by state, though many states have mandated that 15–25 percent of electricity sales come from renewable sources by 2020 or 2025.

The Energy Policy Act of 2005 established federally mandated investment tax credits for those investing in residential, commercial, and industrial renewable energy, and extended the production tax credit for renewable energy electricity generation through 2012. Similarly, the Energy Improvement and Extension Act of 2008 extended the investment tax credit until 2016, and the American Taxpayer Relief Act of 2012 extended the production tax credit for one year.

These policies have played a primary role in the rapid expansion of electricity generated from renewable resources, such as solar energy (Figure 2-4) and wind. Conventional hydropower remains the largest renewable source of electricity generation, producing 277 billion kilowatt-hours (kWh) in 2012.[19] Electricity production from renewable sources, excluding conventional hydropower, totaled 219 billion kWh in 2012, which represents a 73.5 percent increase in production from 2008. Major growth is visible in the wind power industry alone, with electricity generation from wind increasing by 153 percent from 2008 levels to reach more than 140 billion kWh in 2012.[20] In 2012, wind energy was the number one source of new U.S. electricity generation capacity for the first time—representing 43 percent of all new electric additions.[21]

Figure 2-4 **U.S. Solar Electricity Generation**

Investment tax credit policies resulting from the the Energy Improvement and Extension Act of 2008 and American Recovery and Reinvestment Act of 2009 have played a primary role in the rapid expansion of electricity generated from renewable resources, such as solar energy.

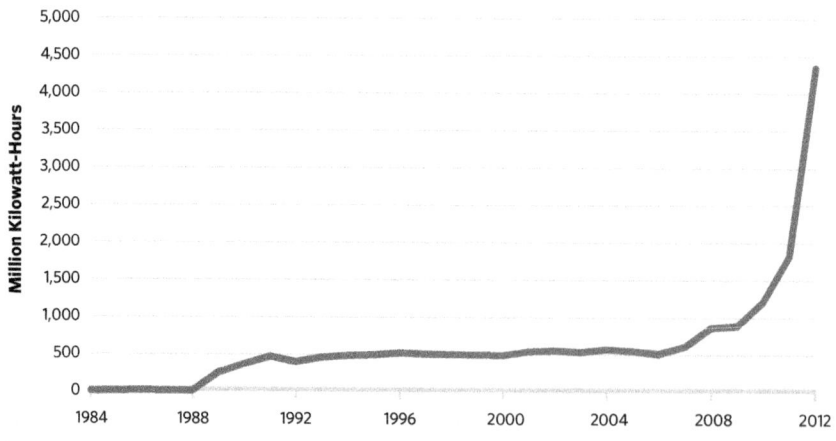

Sources: U.S. DOE/EIA 2012, Table 8.2a, and U.S. DOE/EIA 2013c, Table 1.20.B.

[16] U.S. DOE/EIA, Uranium Marketing Annual Report. *See* http://www.eia.gov/uranium/marketing/html/summarytable1a.cfm.

[17] U.S. DOE/EIA. 2012, Domestic Uranium Production Report. *See* http://www.eia.gov/uranium/production/annual/pdf/dupr.pdf.

[18] U.S. DOE/EIA 2012, Table 8.2a. Electricity Net Generation: Total (All Sectors), 1949–2011. *See* http://www.eia.gov/totalenergy/data/annual/showtext.cfm?t=ptb0802a.

[19] U.S. DOE/EIA 2013g, Table 7.2a, Electricity Net Generation: Total (All Sectors). *See* http://www.eia.gov/totalenergy/data/monthly/pdf/sec7_5.pdf.

[20] Ibid.

[21] U.S. DOE 2013. *See* http://energy.gov/articles/energy-dept-reports-us-wind-energy-production-and-manufacturing-reaches-record-highs.

Electricity

Total U.S. electricity generation was at 4,054 billion kWh in 2012, down 1.58 percent from 2008, and down by 0.02 percent compared with 2005 levels, but up by 8.5 percent compared with generation levels in 2001. The U.S. Department of Energy's (DOE's) Energy Information Administration (EIA) projects that U.S. electricity demand will continue to rise by 17.6 percent between 2012 and 2030.[22]

In 2012, U.S. electricity generation was largely powered by coal-fired power plants, at 37 percent of total generation. However, compared with previous years, the share of electricity generated from coal is declining, down from 51.2 percent in 2000 and 44 percent in 2009. This declining trend is due to rapid growth in natural gas-fired generation, which has risen from 16 percent of total electric generation in 2000 to more than 30 percent in 2012.

Federal Energy Subsidies

The U.S. federal government provides a number of subsidies and interventions in the energy market, including direct expenditures to producers or consumers, tax expenditures, research and development, and loans and loan guarantees. Between 2007 and 2010 (the latest data available), the value of direct financial interventions and subsidies in energy markets doubled, growing from $17.9 billion to $37.2 billion. In broad categories, the largest increase was for conservation and end-use subsidies (particularly for renewables), followed to a lesser degree by increases in electricity-related subsidies and subsidies for fuels used outside the electricity sector. A key factor in the increased support for conservation programs, end-use technologies, and renewables was the passage of several pieces of legislation responding to the recent economic downturn, particularly the Recovery Act and the Energy Improvement and Extension Act. This growth in energy-specific subsidies between fiscal years 2007 and 2010 does not closely correspond with changes in energy consumption and production over the same period. In fact, overall energy consumption actually fell from 101 quadrillion Btu to 98 quadrillion Btu between 2007 and 2010 due to increasing domestic production of shale gas, crude oil, and renewable energy.

ENERGY CONSUMPTION[23]

The United States currently consumes energy from petroleum, natural gas, coal, nuclear, conventional hydropower, and renewables. While fossil fuels remain predominant, 2005–2012 trends show swift—and ongoing—evolution of the fuel mix toward cleaner sources, with natural gas and renewables increasingly displacing coal and petroleum (Figure 2-5). Petroleum, the single largest source, accounted for 36.5 percent of total primary energy consumption in 2012, down from 37.5 percent in 2008 and 40.3 percent in 2005. Coal declined from 22.8 percent in 2005 to 18.3 percent in 2012, a level surpassed by natural gas at 27.3 percent, representing an approximately equal increase of 5 percentage points. Over the same time frame, conventional hydropower stayed level at about 3 percent, nonhydro renewables expanded from 3.5 to 6.5 percent, and nuclear grew moderately, from 8.1 to 8.5 percent.

Total U.S. energy consumption continues a recent trend of overall decline, falling 2.4 percent between 2011 and 2012.[24] This follows recession-driven drops of 2.0 percent between 2007 and 2008 and 4.7 percent between 2008 and 2009, resulting in a 6.1 percent decline in energy consumption between 2007 and 2012.[25] These shifts reflect both fluctuating economic growth, from recession into early recovery, and generally increasing sectoral and economy-wide energy efficiency. EIA's *Annual Energy Outlook 2013 (AEO2013)* Reference case forecasts growing primary energy consumption from 2013 onward, mainly supplied by natural gas and renewables (U.S. DOE/EIA 2013b).

The rates of U.S. energy consumption, per capita and per unit of economic output, are on descending long-term trajectories (Figure 2-6). In 2011, per capita energy use fell by 1.3 percent from 2010, to 312 million Btu per person, comparable to levels last seen in the 1980s.[26] Energy consumption per unit of GDP fell by 2.1 percent from 2010 to 2011, to 7,310 Btu per dollar (2005 dollars).[27] EIA projects per capita consumption to fall below 270 million Btu per person by 2034, largely from mandated efficiency gains in appliances and vehicles.[28]

[22] U.S. DOE/EIA 2013b, Electricity Supply, Disposition, Prices and Emissions, Total Electricity Use 2012=3837 and 2030=4513.

[23] U.S. DOE/EIA 2013b and U.S. DOE/EIA 2013g, Table 1.1, Total Energy Flow, 2011. *See* http://www.eia.gov/totalenergy/data/monthly/pdf/sec1_3.pdf.

[24] Ibid.

[25] Ibid.

[26] U.S. DOE/EIA 2012. Table 1.5, Energy Consumption, Expenditures, and Emissions Indicators Estimates, Selected Years, 1949–2011. *See* http://www.eia.gov/totalenergy/data/annual/pdf/sec1_13.pdf.

[27] Ibid.

[28] U.S. DOE/EIA 2013b, p. 59.

Figure 2-5   **U.S. Primary Energy Consumption Highlights: 2005-2012**

While fossil fuels remain predominant in the U.S. energy profile, 2005-2012 trends show the swift—and ongoing—evolution of the U.S. fuel mix toward cleaner sources, with natural gas and renewables increasingly displacing coal and petroleum.

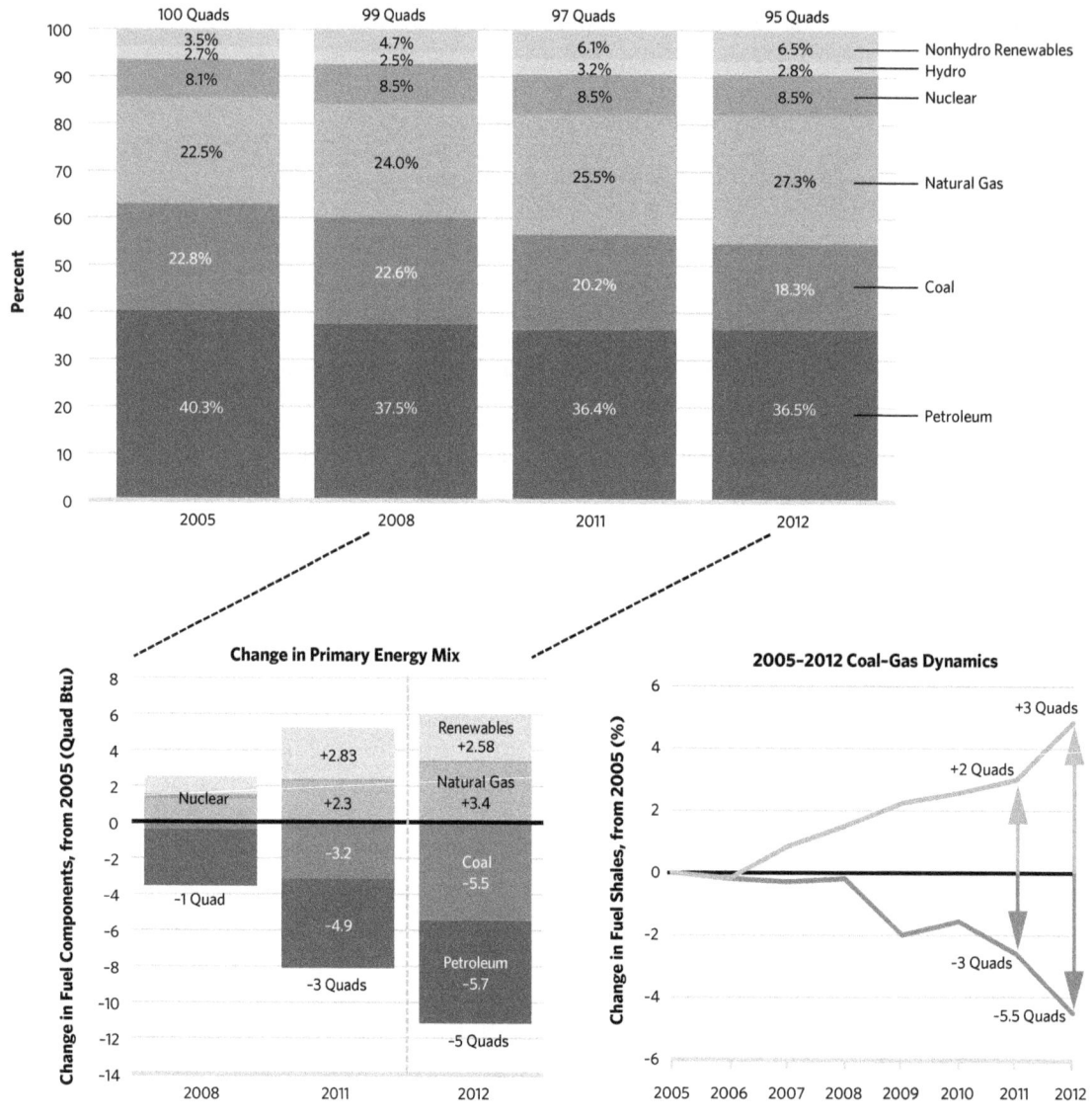

Source: U.S. DOE/EIA 2013a.

Decreasing energy intensity and increasing source decarbonization directly drive steadily declining U.S. carbon intensity. The ratio of CO_2 emissions to real GDP (2005 dollars) fell by 7 percent between 2008 and 2011, from 456 to 413 metric tons of CO_2 per million dollars. This ratio fell by 13.1 percent between 2005 and 2011.[29]

Residential Sector

The residential sector's energy base fluctuates according to season, region, year, and prevailing economic conditions. Although petroleum and natural gas use typically varies more elastically than electricity consumption, demand for all three decreased from 2008 through 2012. Consumption of petroleum, as fuel oil or liquefied petroleum gas, has been in decline since a peak of 861,000 bpd in 2003, dropping to 758,000 bpd in 2008 and 602,000 bpd in 2012.

[29] U.S. DOE/EIA, International Energy Statistics. See http://www.eia.gov/cfapps/ipdbproject/iedindex3.cfm?tid=91&pid=46&aid=31&cid=US,&syid=2005&eyid=2011&unit=MTCDPUSD.

Figure 2-6 **U.S. Energy and Carbon Intensity Trends**

The rates of U.S. energy consumption, per capita and per unit of economic output, are on descending long-term trajectories. In 2011, per capita energy use fell by 1.26 percent from 2010, to 312 million Btu per person, comparable to levels last seen in the 1980s.

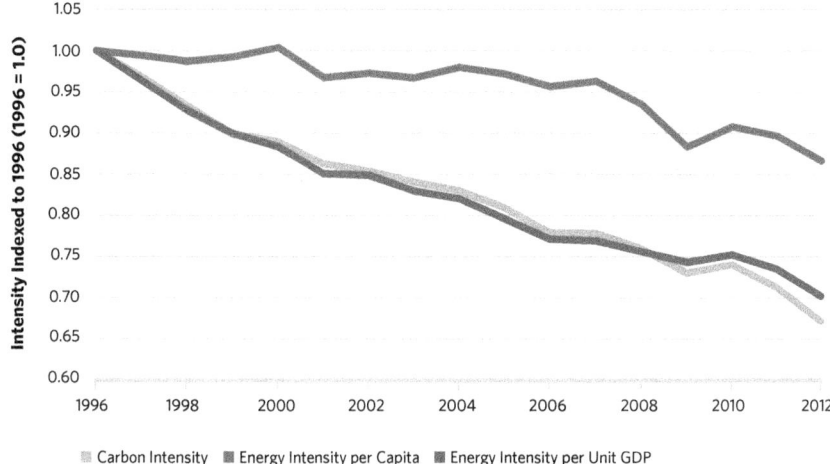

Carbon Intensity   Energy Intensity per Capita   Energy Intensity per Unit GDP

Source: U.S. DOE/EIA 2012. Table 1.5, Energy Consumption, Expenditures, and Emissions Indicators Estimates, Selected Years, 1949-2011. See http://www.eia.gov/totalenergy/data/annual/pdf/sec1_13.pdf.

Consumption of natural gas has fluctuated as well in recent years, declining after a 2003 peak of 144 billion m^3 (5,079 billion ft^3) to 124 billion m^3 (4,368 billion ft^3) in 2006—a level not seen since 1987—then increasing again to 139 billion m^3 (4,892 billion ft^3) in 2008, before trending down to 118 billion m^3 (4,180 billion ft^3) in 2012.

The residential sector, made up of living quarters for private households, uses energy for various applications: space heating, water heating, air conditioning, lighting, refrigeration, cooking, appliances, and electronics. In 2012, residential energy consumption, including electricity losses, totaled 20.2 quadrillion Btu (21.2 percent of total consumption), down from 21.7 quadrillion Btu (21.9 percent) in 2008, representing a 6.9 percent decline.[30] Residential fossil CO_2 emissions (including the emissions associated with electricity consumed in the residential sector), representing 20.0 percent of total energy CO_2 (equaling the sector's 2000 share) also fell by 14.5 percent, from 1.2 Tg CO_2e in 2008 to 1.1 Tg CO_2e in 2012, a 16 percent reduction from 2005.

Commercial Sector

The commercial sector is made up of service facilities and equipment used by businesses, federal and local governments, private and public organizations, institutional living quarters, and sewage treatment plants. The most common uses of energy in this sector include space ventilation and air conditioning, water heating, lighting, refrigeration, cooking, and operation of office and other equipment. Less common uses of energy include transportation.

As of 2012, electricity accounted for 78.5 percent of the commercial sector's energy use, followed by natural gas at just under 17 percent.[31] Demand responds largely to a combination of prices, among other market factors, and weather, although the impact of weather is less significant in commercial than in residential buildings. Since the period covered by the 2010 CAR, demand for electricity has declined gradually, falling by 0.7 percent between 2009 and 2012.[32] After notable increases of 6.3 percent in 2007 and 4.7 percent in 2008, demand for natural gas fell for three out of four years (all but 2011), declining most steeply by 7.8 percent in 2012.[33] In 2012, total commercial energy consumption was 4.9 percent lower than in 2008, more than offsetting a 3 percent increase between 2005 and 2008.[34] At 17.5 quadrillion Btu, the commercial sector's energy use represented 18.4 percent of total U.S. energy demand in 2012.[35]

[30] U.S. DOE/EIA 2013g, Table 2.1, Energy Consumption by Sector. See http://www.eia.gov/totalenergy/data/monthly/pdf/sec2_3.pdf.

[31] U.S. DOE/EIA 2013g, Table 2.3, Commercial Energy Sector Consumption. See http://www.eia.gov/totalenergy/data/monthly/pdf/sec2_7.pdf.

[32] Ibid.

[33] Ibid.

[34] Ibid.

[35] U.S. DOE/EIA 2013g, Table 2.1. Energy Consumption by Sector. See http://www.eia.gov/totalenergy/data/monthly/pdf/sec2_3.pdf.

Industrial Sector

The U.S. industrial sector consists of all facilities and equipment used for producing, processing, or assembling goods, including manufacturing, mining, agriculture, and construction. The sector depends largely on coal, natural gas, and petroleum for its energy use. In 2012, electricity use, including system losses, represented around one-third of all energy consumed in the industrial sector.[36]

Since 2008, natural gas has narrowly displaced petroleum as the primary energy source.[37] In 2012, natural gas and petroleum accounted for 28.3 percent and 26.4 percent of energy consumption, respectively. Renewable energy use—primarily biomass—surpassed coal in 2007.[38] Between 2008 and 2012, industrial renewable energy consumption expanded from 6.5 to 7.2 percent, while coal dropped from 5.9 to 4.8 percent.[39]

Industrial sector energy use fell from 43 percent of total energy consumption in 1973 to 32.0 percent in 2007, and grew to 32.3 percent in 2012.[40] Industry consumed 2.1 percent less energy between 2008 and 2012, largely from coal and petroleum, because of a decline in electrical system energy losses.[41] Within the industrial sector, energy consumption decreased by 2.0 percent in 2008 and 10.2 percent in 2009, then increased by 5.5 percent in 2010 and 1.4 percent in 2011, and again decreased by 2.0 percent in 2012.[42] At 30.7 quadrillion Btu, the industrial sector's energy use represented 32.3 percent of total U.S. energy demand in 2012.[43]

Approximately three-fifths of the total energy used in the industrial sector is for manufacturing, with chemicals and allied products, petroleum and coal products, paper and nonmetallic minerals, and primary metals accounting for most of this share. The top five energy-consuming industries—bulk chemicals, refining, paper, steel, and food—account for around 60 percent of industrial energy use, but comprise only 26 percent of shipments. Projected slow growth in these energy-intensive industries is likely to result from increased foreign competition, reduced domestic demand for raw materials and the basic goods they produce, and movement of investment capital to more profitable areas.[44] EIA's *AEO2013* Reference case projects that, despite a 76 percent increase in industrial shipments, industrial energy consumption will grow by only 19 percent between 2011 and 2040, primarily because of a shift in the share of shipments from energy-intensive manufacturing to plastics, computers, transportation equipment, and other less energy-intensive industries.[45]

TRANSPORTATION

The U.S. transportation system has evolved to meet the needs of a highly mobile, dispersed population and a large, dynamic economy. While the transportation system supports the movement of people and goods and the economic vitality of the country, efforts are underway to ensure that it is also as sustainable as possible.

Over the years, the United States has developed an extensive multimodal system that includes road, air, rail, and water transport capable of moving large volumes of people and goods long distances. Automobiles and light trucks still dominate the passenger transportation system, and the highway share of passenger miles traveled in 2013 was about 87 percent of the total, down 2 percent from the 2010 CAR. Air travel accounted for slightly more than 11 percent of passenger miles traveled (up 1.5 percent from the 2010 CAR), and mass transit and rail travel combined accounted for only about 1 percent. For-hire transport services, as a portion of GDP, have barely changed since the 2010 CAR, accounting for 3.0 percent of GDP in 2011.[46]

Highway Vehicles

The trends in highway vehicles have not changed appreciably in the past decade. Between 2008 and 2011, the number of passenger vehicles declined by 1.1 percent, reaching 253.1 million in 2011.[47] This degree of vehicle ownership is a result of population distribution, land-use patterns, location of work and shopping, and public preferences for personal mobility. Single-occupant passenger automobiles dominated daily trips between home and workplace in 2009, with more than three-quarters of the nation's workforce individually driving to and from work (McKenzie and Rapino 2011). Just more than 10 percent of workers commuted in carpools of two or more people, around 5 percent used public transportation, and the rest of the workforce used other means (biking, walking, taxis, etc.) (McKenzie 2010).

[36] U.S. DOE/EIA 2013g, Industrial Sector Energy Consumption Estimates. *See* http://www.eia.gov/totalenergy/data/monthly/pdf/sec2_9.pdf.

[37] Ibid.

[38] Ibid.

[39] Ibid.

[40] U.S. DOE/EIA 2013g, Table 2.1. Energy Consumption by Sector. *See* http://www.eia.gov/totalenergy/data/monthly/pdf/sec2_3.pdf.

[41] U.S. DOE/EIA 2013g, Table 2.4. Industrial Sector Energy Consumption Estimates. *See* http://www.eia.gov/totalenergy/data/monthly/pdf/sec2_9.pdf.

[42] U.S. DOE/EIA 2013g, Table 2.1. Energy Consumption by Sector. *See* http://www.eia.gov/totalenergy/data/monthly/pdf/sec2_3.pdf.

[43] Ibid.

[44] U.S. DOE/EIA 2013b. *See* http://www.eia.gov/forecasts/aeo/pdf/0383(2013).pdf.

[45] U.S. DOE/EIA 2013b, Industrial. *See* http://www.eia.gov/forecasts/aeo/sector_industrial_all.cfm.

[46] U.S. Department of Transportation (DOT)/ Bureau of Transportation Statistics (BTS), Table 3-2: U.S. Gross Domestic Product (GDP) Attributed to For-Hire Transportation Services (see http://www.rita.dot.gov/bts/sites/rita.dot.gov.bts/files/publications/national_transportation_statistics/html/table_03_02.html); and U.S. DOT/ Research and Innovative Technology Administration (RITA)/BTSa, National Transportation Statistics 2013 (see http://www.rita.dot.gov/bts/sites/rita.dot.gov.bts/files/NTS_Entire_Q1.pdf).

[47] U.S. DOT/BTS, Table 4-9: Motor Vehicle Fuel Consumption and Travel. *See* http://www.rita.dot.gov/bts/sites/rita.dot.gov.bts/files/publications/national_transportation_statistics/html/table_04_09.html.

Private vehicles, which include automobiles, light trucks, vans, and motorcycles, are used for 84 percent of all trips nationwide. Most (55 percent) of these trips are made by car or van, 18 percent by sport utility vehicle (SUV), and 10 percent by pickup truck.[48] The largest sources of transportation GHGs in 2011 were passenger cars (41.2 percent); light-duty trucks, which include SUVs, pickup trucks, and minivans (17.4 percent); freight trucks (21.0 percent); rail (6.5 percent); and commercial aircraft (6.1 percent). These figures include direct emissions from fossil fuel combustion, as well as hydrofluorocarbon emissions from mobile air conditioners and refrigerated transport allocated to these vehicle types (U.S. EPA/OAP 2013).

The number of miles driven is another major factor affecting energy use in the highway sector. The number of vehicle miles traveled by passenger cars and light-duty trucks increased by 34 percent from 1990 through 2011.[49] From 2006 through 2008, the total number of vehicle miles driven each year reached around 4.8 trillion km (3 trillion mi), but in 2012 dropped to 4.7 trillion km (3.0 trillion mi), a decline of almost 2 percent.[50]

The fuel economy of passenger cars, light trucks, SUVs, and vans plays a large role in determining energy consumption and GHG emissions from the highway transport sector. The average fuel economy of passenger cars in use in the United States reached an average 14.0 km/liter (32.8 mpg) in 2008-2010.[51] The average fuel economy of new model year 2012 passenger cars and light trucks sold in the United States was 15.1 km/liter (35.6 mpg) and 10.6 km/liter (25.0 mpg),[52] respectively.[53]

Fuel economy standards, known as Corporate Average Fuel Economy (CAFE) standards, and GHG emission standards for new vehicles, play an integral role in determining the fuel economy of passenger cars and light trucks in the United States. New laws and policies outlined in Chapter 4 of this report will result in substantial increases in fuel economy over the next 11 years, and are projected to require the overall fleet to reach an average CO_2 emissions level of 163 grams per mile in 2025, while nearly doubling new vehicle fuel economy.

Air Carriers

U.S. airlines carried 0.6 percent more domestic passengers in 2012 and 2 percent more international passengers than in 2011, for a system-wide increase of 0.8 percent.[54] Collectively, in 2011, the 728 million passengers traveling on U.S.-based airlines traveled 1,302 billion km (809 billion mi). On average, a passenger traveling domestically traveled 1,341 km (883 mi).[55] Since the low of 704 million passengers in 2009, airline ridership has risen, but has yet to reach the high levels experienced prior to the economic recession of 2008.

The impact of the economic recession, coupled with the high price of fuel and lower demand for travel, led airlines to cut back on available capacity by reducing the number of flights—especially those involving smaller aircraft. For example, airlines reduced the number of domestic scheduled passenger flights by 13.9 percent between June 2007 and June 2012 (U.S. DOT/OIG 2012).

Freight

Between 2007 and 2009 (the latest year for which freight data are available), U.S. freight transportation declined by 8.3 percent to 4.3 trillion ton-miles, representing an average decline of 2.8 percent per year, compared with a 1.3 percent average annual growth between 2003 and 2007.[56] Rail accounts for the largest share of total freight ton-miles (36.8 percent), followed by trucks (30.8 percent), pipelines (21.1 percent), waterways (11.1 percent), and air (less than 1 percent).[57]

In recent years, increases in fuel costs, a slight decrease in the number of trucks on the road, and improved energy efficiency have affected the number of gallons of fuel burned by commercial trucks. From 2007 through 2010, truck fuel consumption declined by nearly 5 percent. Fuel use in Class I freight railroads declined by 14 percent, from 15.5 billion liters (4.1 billion gallons) in 2007 to 13.2 billion liters (3.5 billion gallons)in 2010.[58] In terms of energy consumption per ton-mile in 2010, trucking accounted for the largest share, followed by water, which was a distant second.[59]

[48] U.S. DOT/Federal Highway Administration (FHWA). Our Nation's Highways, 2011: 2. Highway Travel. See http://www.fhwa.dot.gov/policyinformation/pubs/hf/pl11028/chapter2.cfm#fig22.

[49] U.S. EPA, Sources of Greenhouse Gas Emissions: Transportation Sector Emissions. See http://www.epa.gov/climatechange/ghgemissions/sources/transportation.html.

[50] U.S. DOE/Energy Efficiency and Renewable Energy (EERE), Alternative Fuels Data Center, Annual Vehicle Miles Traveled in the U.S. See http://www.afdc.energy.gov/data/#tab/all.

[51] "U.S. DOT/RITA/BTS, Table 4-23: Average Fuel Efficiency of U.S. Light Duty Vehicles. See http://www.rita.dot.gov/bts/sites/rita.dot.gov.bts/files/publications/national_transportation_statistics/html/table_04_23.html.

[52] Ibid. See also Light Truck Fuel Economy Standard Rulemaking, MY 2008-2011, Final Rule, p. 12. See http://www.nhtsa.gov/fuel-economy.

[53] See Average Fuel Economy Standards, Passenger Cars and Light Trucks, MY 2011, See Updated CAFE Final Rule, p. 3, at http://www.nhtsa.gov/fuel-economy

[54] U.S. DOT/RITA/BTS 2013b, "Total Passengers on U.S Airlines and Foreign Airlines U.S. Flights Increased 1.3% in 2012 from 2011." See http://www.rita.dot.gov/bts/press_releases/bts016_13.

[55] U.S. DOT/RITA/BTS, "U.S. Airline Revenue Passenger-Miles and Load Factor," Feb. 2013. See http://www.rita.dot.gov/bts/publications/multimodal_transportation_indicators/2013_02/passenger_usage/us_airline_revenue

[56] U.S. DOT/RITA/BTS, Table 1-50: U.S. Ton-Miles of Freight. See http://www.rita.dot.gov/bts/sites/rita.dot.gov.bts/files/publications/national_transportation_statistics/html/table_01_50.html.

[57] Ibid.

[58] U.S. DOT/FHWA, Freight Facts and Figures 2012, Tables 5-7 and 5-7M: Fuel Consumption by Transportation Mode: 2007-2010. See http://www.ops.fhwa.dot.gov/freight/freight_analysis/nat_freight_stats/docs/12factsfigures/table5_7.htm.

[59] Ibid., Table 5-8: Energy Consumption by Selected Freight Transportation Mode: 2007-2010." See http://www.ops.fhwa.dot.gov/freight/freight_analysis/nat_freight_stats/docs/12factsfigures/table5_8.htm.

In 2011, 17.6 billion tons of freight moved throughout the U.S. transportation system (U.S. Congress 2013). Trucks led in both tonnage and dollar value, carrying more than 70 percent of all freight in 2009.[60]

INDUSTRY

The U.S. industrial sector boasts a wide array of light and heavy industries in manufacturing and nonmanufacturing subsectors, the latter of which include mining, agriculture, and construction. Private goods-producing industries accounted for slightly more than 18 percent of total GDP in 2012, and utilities accounted for another 1.9 percent of GDP.

The industrial sector as a whole represents 20 percent of total U.S. GHG emissions (2011 data). Compared with the period covered under the 2010 CAR, the portion of GHG emissions produced by industry has shrunk dramatically (from 28 percent of 2007 emissions to 20 percent of 2011 emissions).

WASTE

In 2011, the United States generated approximately 250 million metric tons of municipal solid waste (MSW), about 3 million metric tons less than 2005.[61] Paper and paperboard products made up the largest component of MSW generated by weight (28 percent), and food waste comprised the second-largest material component (14.5 percent). Glass, metals, plastics, wood, and food each constituted between 5 and 13 percent of the total MSW generated, while rubber, leather, and textiles combined made up about 8 percent of the MSW (U.S. EPA/OSW 2013a).

Recycling and composting have been the most significant change in waste management from a GHG perspective. In 2011, Americans composted or recycled 86.9 million metric tons of MSW, which saved more than 1.1 quadrillion Btu of energy and provides an annual benefit of more than 183 million metric tons of CO_2e emissions reduced, comparable with removing the emissions from more than 34 million passenger vehicles (U.S. EPA/OSW 2013b). On average, Americans recycled and composted 0.7 million kilograms (kg) (1.53 million pounds [lb]) of waste, or 2.0 kg (4.4 lb) per person per day (U.S. EPA/OSW 2013a).

From 1990 to 2011, the recycling rate increased from slightly more than 16 percent to 34.7 percent. Of the remaining MSW generated, about 12 percent was combusted, and less than 54 percent was disposed of in landfills. The number of operating MSW landfills in the United States has decreased substantially over the past 20 years, from about 8,000 in 1988 to about 1,908 in 2009, while the average landfill size has increased (U.S. EPA/OSW 2013b).

The United States is working to reduce methane emissions from landfills by encouraging the recovery and beneficial use of landfill gas (LFG) as an energy source. EPA operates a Landfill Methane Outreach Program, a voluntary assistance program that forms partnerships with communities, landfill owners, utilities, power marketers, states, project developers, tribes, and nonprofit organizations to overcome barriers to project development by helping them assess project feasibility, find financing, and market the benefits of project development to the community. As of June 2012, there were 594 operational LFG energy projects in the United States and approximately 540 landfills that are good candidates for projects.[62]

BUILDING STOCK AND URBAN STRUCTURE

Buildings are large users of energy. Their number, size, and distribution and the appliances and heating and cooling systems that go into them influence energy consumption and GHG emissions. As of 2012, buildings accounted for about 39.7 percent (37.7 quadrillion Btu) of total U.S. energy consumption, 41.2 percent (7 quadrillion Btu) more than the transportation sector and 22.8 percent (11 quadrillion Btu) more than the industrial sector.[63]

Residential Buildings

The U.S. housing market is gradually strengthening since the U.S. economic slowdown in 2007–2009, with home prices continuing to rise and existing home sales increasing. Between 2010 and 2012, the number of privately owned housing units under construction increased by nearly 30 percent.[64] In 2011, there were an estimated 132 million housing units in the United

[60] U.S. DOT/FHWA, Our Nation's Highways, 2011: 2. Highway Travel. See http://www.fhwa.dot.gov/policyinformation/pubs/hf/pl11028/chapter2.cfm.

[61] U.S. EPA, Municipal Solid Waste. See http://www.epa.gov/epawaste/nonhaz/municipal/index.htm.

[62] U.S. EPA, Landfill Methane Outreach Program. See http://www.epa.gov/lmop/basic-info/index.html.

[63] U.S. DOE, "How Much Energy is Consumed in Residential and Commercial Buildings in the United States?" See http://www.eia.gov/tools/faqs/faq.cfm?id=86&t=1.

[64] U.S. Census Bureau, New Privately Owned Housing Units Under Construction, Annual Data: 1969–2012. See http://www.census.gov/construction/nrc/pdf/underann.pdf.

States, 61.6 percent of which were single, detached dwellings and 25.9 percent of which were housing units in multi-unit structures.[65]

While new U.S. homes are larger and more plentiful, their energy efficiency has increased significantly. In 2012, more than 100,000 new homes earned the ENERGY STAR® certification, implying at least a 30 percent energy savings for heating and cooling relative to comparable homes built to current code and bringing the total number of certified homes to more than 1.4 million (U.S. EPA 2013a). On average, homes built between 2000 and 2005 used 14 percent less energy per square foot than homes built in the 1980s and 40 percent less energy per square foot than homes built before 1950. However, there has been a trend toward larger homes. Specifically, single-family homes built between 2000 and 2005 are 29 percent larger on average than those built in the 1980s, and thus have greater requirements for heating, cooling, and lighting.[66]

Commercial Buildings

Between 2000 and 2010, commercial floor space rose approximately 1.8 percent per year.[67] EIA estimates that commercial floor space will grow by 28 percent between 2009 and 2035. In 2003 (the most recent data available), there were nearly 4.9 million commercial buildings with more than 6.7 billion square meters (71.7 billion square feet) of floor space. Much of this growth has been related to the rapidly expanding information, financial, and health services sectors.

Commercial primary energy consumption grew by 65.5 percent between 1980 and 2009.[68] Electricity (78.5 percent) and natural gas (16.9 percent) are the two largest sources of energy used in commercial buildings. In aggregate, commercial buildings represented 46.4 percent of building energy consumption and 18.4 percent of U.S. energy consumption in 2012.[69] The top three end uses in the commercial buildings sector are space heating, lighting, and cooling, which represent close to half of commercial site energy consumption.[70]

AGRICULTURE AND GRAZING

Agriculture in the United States is highly productive. U.S. croplands produce a wide variety of food and fiber crops, feed grains, oil seeds, fruits and vegetables, and other agricultural commodities for both domestic and international markets. Although the United States harvests roughly the same area as it did in 1910, U.S. agriculture feeds a population three times larger, with crops still available for export. Technological changes account for most of the increased productivity. In 2007, there were 1,685,339 farms with cropland in the United States.[71] U.S. cropland was 164 million hectares (ha) (406 million acres[ac]), about 9 percent lower than in 1997.[72]

Soils vary across the landscape in response to the effects of climate, topography, vegetation, and other organisms (including humans) on the rate and direction of soil development processes acting on parent materials over time. In the United States, the wide range and endless combinations of these factors have resulted in a great range of soils with widely varying properties. Soils provide an effective natural filter that protects groundwater and surface water by removing potential contaminants applied on or in the soil. Soils across the United States have the potential to sequester substantial amounts of organic and inorganic carbon, and through this sequestration have the potential to help reduce atmospheric CO_2 levels. Although soils vary in their resistance and resilience, all are subject to degradation through erosion, salinization, and other effects without proper management.

Sources of GHG emissions from U.S. croplands include nitrous oxide (N_2O) from nitrogen fertilizer use and methane from farm animals' enteric fermentation and manure management. Agricultural soil management activities, such as fertilizer application and other cropping practices, were the largest source of U.S. N_2O emissions in 2011, accounting for 69.3 percent.

Conservation is an important objective of U.S. farm policy. The U.S. Department of Agriculture administers conservation programs that have been highly successful at removing environmentally sensitive lands from commodity production and encouraging farmers to adopt conservation practices on working agricultural lands. In terms of GHG mitigation, the largest of these programs, the Conservation Reserve Program (CRP), seeks to reduce soil erosion,

[65] U.S. Census Bureau, American FactFinder (see http://factfinder2.census.gov/faces/tableservices/jsf/pages/productview.xhtml?pid=ACS_12_1YR_B25001&prodType=table); and U.S. Census Bureau, Statistical Abstract of the United States: 2012, Table 989, Housing Units by Units in Structure and State: 2009 (see http://www.census.gov/compendia/statab/2012/tables/12s0989.pdf).

[66] U.S. DOE, Chapter 2, Residential Sector. See http://buildingsdatabook.eren.doe.gov/ChapterIntro2.aspx.

[67] U.S. DOE, Table 3.2.1, Total Commercial Floorspace and Number of Buildings, by Year. See http://buildingsdatabook.eren.doe.gov/TableView.aspx?table=3.2.1.

[68] U.S. DOE, 2009 Annual Energy Review. See http://www.eia.gov/totalenergy/data/annual/index.cfm.

[69] Ibid.

[70] Ibid.

[71] U.S. Department of Agriculture (USDA)/National Agricultural Statistics Service (NASS), 2007 Census of Agriculture. See http://www.agcensus.usda.gov/Publications/2007/Full_Report/usv1.pdf.

[72] Ibid.

improve water quality, and enhance wildlife habitat by retiring environmentally sensitive lands from crop production. As of June 2013, about 11 million ha (27 million ac) are under contract in CRP on 389,722 farms.[73]

FORESTS

U.S. forests are predominately natural stands of native species, and vary from the complex hardwood forests in the East to the highly productive conifer forests of the Pacific Coast. Forests established through planting of tree species comprise more than 26 million ha (63 million ac), or 8 percent of all forests, and nearly all planted stands are established with native species.

In 1907, forests comprised an estimated 34 percent of the total U.S land area (307 million ha, or 758 million ac), which has remained roughly the same, as of 2010 (303 million ha, or 748 million ac).[74] Historically, most of the forestland loss was due to agricultural conversions in the late 19th century, but today most losses are due to intensive uses, such as urban development. Since 1990, net forestland area has increased by approximately 0.6 million ha (1.4 million ac) per year, as marginal agriculture and pasture lands previously converted from forestland in the 19th century have reverted to forestland faster than new losses to urban or other uses.

Of the 305 million ha (751 million ac) of U.S. forestland, nearly 208 million ha (514 million ac) are timberland, most of which is privately owned in the conterminous United States. However, a significant area of forestland is reserved forests, which in 2007 accounted for 10 percent of all forestland, or about 30 million ha (75 million ac) (USDA/FS 2012).

Most timber removals come from private lands, with the South providing nearly two-thirds of all domestic timber. Management inputs over the past several decades have been gradually increasing the production of marketable wood in U.S. forests, especially on private forestland in the South. The United States currently grows more wood than it harvests, with a growth-to-harvest ratio of nearly 2 to 1. As the average age of U.S. forests continues to rise and growth continues to exceed removals, standing volume has increased by 37 percent since 1953 to a level of nearly 33 billion m^3 (1,165 billion ft^3).

Existing U.S. forests are an important net sink for atmospheric carbon. Improved forest management practices, the regeneration of previously cleared forest areas, and timber harvesting and use have resulted in net sequestration of CO_2 every year since 1990. In 2011, the land use, land-use change, and forestry sector absorbed a net of 905.0 Tg of CO_2. This sequestration represents an offset of 17.1 percent of U.S. fossil fuel emissions (5,277 Tg CO_2e) (U.S. EPA/OAP 2013).

[73] USDA/CRP, Status—End of June 2013. See http://www.fsa.usda.gov/Internet/ FSA_File/june2013onepager.pdf.

[74] Trading Economics, Forest Area (% of Land Area) in the United States. See http://www.tradingeconomics.com/ united-states/forest-area-percent-of-land-area-wb-data.html.

3

Greenhouse Gas Inventory

An emissions inventory that identifies and quantifies a country's primary anthropogenic[1] sources and sinks of greenhouse gases (GHGs) is essential for addressing climate change. This inventory adheres to both (1) a comprehensive and detailed set of methodologies for estimating sources and sinks of anthropogenic GHGs, and (2) a common and consistent mechanism that enables Parties to the United Nations Framework Convention on Climate Change (UNFCCC) to compare the relative contributions of different emission sources and GHGs to climate change.

By ratifying the Convention, Parties "shall develop, periodically update, publish and make available ... national inventories of anthropogenic emissions by sources and removals by sinks of all greenhouse gases not controlled by the Montreal Protocol, using comparable methodologies...."[2] The United States views the *Inventory of U.S. Greenhouse Gas Emissions and Sinks: 1990-2011* (1990-2011 Inventory) as an opportunity to fulfill these commitments (U.S. EPA/ OAP 2013).

This chapter summarizes the latest information on U.S. anthropogenic GHG emission trends from 1990 through 2011. To ensure that the U.S. emissions inventory is comparable with those of other UNFCCC Parties, the estimates presented here were calculated using methodologies consistent with those recommended in the *Revised 1996 IPCC Guidelines for National Greenhouse Gas Inventories* (IPCC/UNEP/OECD/IEA 1997), the Intergovernmental Panel on Climate Change (IPCC) *Good Practice Guidance and Uncertainty Management in National Greenhouse Gas Inventories* (IPCC 2000), and the IPCC *Good Practice Guidance for Land Use, Land-Use Change, and Forestry* (IPCC 2003). Additionally, the U.S. emissions inventory has continued to incorporate new methodologies and data from the *2006 IPCC Guidelines for National Greenhouse Gas Inventories* (IPCC 2006). The structure of the 1990-2011 Inventory is consistent with the UNFCCC guidelines for inventory reporting (UNFCCC 2006). For most source categories, the IPCC methodologies were expanded, resulting in a more comprehensive and detailed estimate of emissions (Box 3-1). Consistent with the 1990-2011 Inventory, emissions in this chapter are presented in teragrams[3] of carbon dioxide equivalents (Tg CO_2e).[4]

BACKGROUND INFORMATION

GHGs trap heat and make the planet warmer. The most important GHGs directly emitted as a result of human activities include carbon dioxide (CO_2), methane (CH_4), nitrous oxide (N_2O), and several other fluorine-containing halogenated substances. Although the direct GHGs CO_2, CH_4, and N_2O occur naturally in the atmosphere, human activities have changed their atmospheric concentrations. From the pre-industrial era (i.e., ending about 1750) to 2010, concentrations of CO_2, CH_4, and N_2O have increased globally by 39, 158, and 18 percent, respectively (IPCC 2007 and NOAA/ESRL 2009). The 1990-2011 Inventory estimates the total national GHG emissions and removals associated with human activities across the United States.

[1] The term "anthropogenic," in this context, refers to GHG emissions and removals that are a direct result of human activities or are the result of natural processes that have been affected by human activities (IPCC/ UNEP/OECD/IEA 1997).

[2] Article 4(1)(a) of the UNFCCC (also identified in Article 12). Subsequent decisions by the Conference of the Parties elaborated the role of Annex I Parties in preparing national inventories. *See* http://unfccc.int.

[3] One teragram is equal to 1,012 grams or one million metric tons.

[4] Further information is provided in this chapter's Box 3-2: Global Warming Potentials.

Box 3-1 **Recalculations of Inventory Estimates**

Each year, emission and sink estimates are recalculated and revised for all years in the *Inventory of U.S. Greenhouse Gas Emissions and Sinks,* as attempts are made to improve both the analyses themselves, through the use of better methods or data, and the overall usefulness of the report. In this effort, the United States follows the 2006 IPCC guidelines (IPCC 2006), which state, "Both methodological changes and refinements over time are an essential part of improving inventory quality. It is good practice to change or refine methods "when: available data have changed; the previously used method is not consistent with the IPCC guidelines for that category; a category has become key; the previously used method is insufficient to reflect mitigation activities in a transparent manner; the capacity for inventory preparation has increased; new inventory methods become available; and for correction of errors." In general, recalculations are made to the U.S. GHG emission estimates either to incorporate new methodologies or, most commonly, to update recent historical data.

In each inventory report, the results of all methodology changes and historical data updates are presented in the "Recalculations and Improvements" chapter. If applicable, detailed descriptions of each recalculation are contained within each emission source's description in the report. In general, when methodological changes have been implemented, the entire time series has been recalculated to reflect the change, per the 2006 IPCC guidelines (IPCC 2006). In the case of the most recent inventory report, the time series is 1990 through 2011. Changes in historical data are generally the result of changes in statistical data supplied by other agencies. References for the data are provided for additional information.

More information on the most recent changes is provided in the "Recalculations and Improvements" chapter of the *Inventory of U.S. Greenhouse Gas Emissions and Sinks: 1990–2011* (U.S. EPA/OAP 2013), and previous inventory reports can further describe the changes in calculation methods and data since the *U.S. Climate Action Report 2010* (U.S. DOS 2010).

RECENT TRENDS IN U.S. GREENHOUSE GAS EMISSIONS AND SINKS

In 2011, total U.S. GHG emissions were 6,702.3 Tg CO_2e. Total U.S. emissions have increased by 8.4 percent from 1990 to 2011. Emissions decreased from 2010 to 2011 by 1.6 percent (108.0 Tg CO_2e), due to a decrease in the carbon intensity of fuels consumed to generate electricity resulting from lower coal consumption, higher natural gas consumption, and significantly higher use of hydropower. Additionally, relatively mild winter conditions, especially in the South Atlantic region of the United States where electricity is an important heating fuel, resulted in an overall decrease in electricity demand in most sectors. Since 1990, U.S. emissions have increased at an average annual rate of 0.4 percent.

Figures 3-1 through 3-3 illustrate the overall trends in total U.S. GHG emissions by gas, annual changes, and absolute change since 1990. Table 3-1 provides a detailed summary of U.S. GHG emissions and sinks for 1990 through 2011. These data and trends are further detailed in the 1990–2011 Inventory. In 2011, total net U.S. GHG emissions (i.e., including net sequestration from land use, land-use change, and forestry [LULUCF] activities) were 5,797.3 Tg CO_2e. This represents a 6.5 percent reduction below 2005 levels.

Figure 3-4 illustrates the relative contribution of the direct GHGs to total U.S. emissions in 2011. The primary GHG emitted by human activities in the United States was CO_2, representing approximately 83.7 percent of total GHG emissions. The largest source of CO_2, and of overall GHG emissions, was fossil fuel combustion. CH_4 emissions, which have decreased by 8.2 percent since 1990, resulted primarily from natural gas systems, enteric fermentation associated with domestic livestock, and decomposition of wastes in landfills. Agricultural soil management, mobile source fuel combustion, and stationary source fuel combustion were the major sources of N_2O emissions. Emissions from substitutes for ozone-depleting substances and emissions of hydrofluorocarbon (HFC)-23 (fluoroform) during the production of hydrochlorofluorocarbon (HCFC)-22 were the primary contributors to aggregate HFC emissions. Perfluorocarbon (PFC) emissions resulted from semiconductor manufacturing and as a by-product of primary aluminum production, while electrical transmission and distribution systems accounted for most sulfur hexafluoride (SF_6) emissions.

Overall, from 1990 to 2011, total emissions of CO_2 increased by 504.0 Tg CO_2e (9.9 percent), while total emissions of CH_4 decreased by 52.7 Tg CO_2e (8.2 percent), and N_2O increased by 12.6 Tg CO_2e (3.6 percent). During the same period, aggregate weighted emissions of HFCs,

Figure 3-1 **U.S. Greenhouse Gas Emissions by Gas**

Between 2007 (2010 CAR data) and 2011, U.S. emissions from all GHGs declined by a total of 561 Tg CO$_2$e, or 7.2 percent. Total U.S. emissions increased by 8.4 percent from 1990 to 2011.

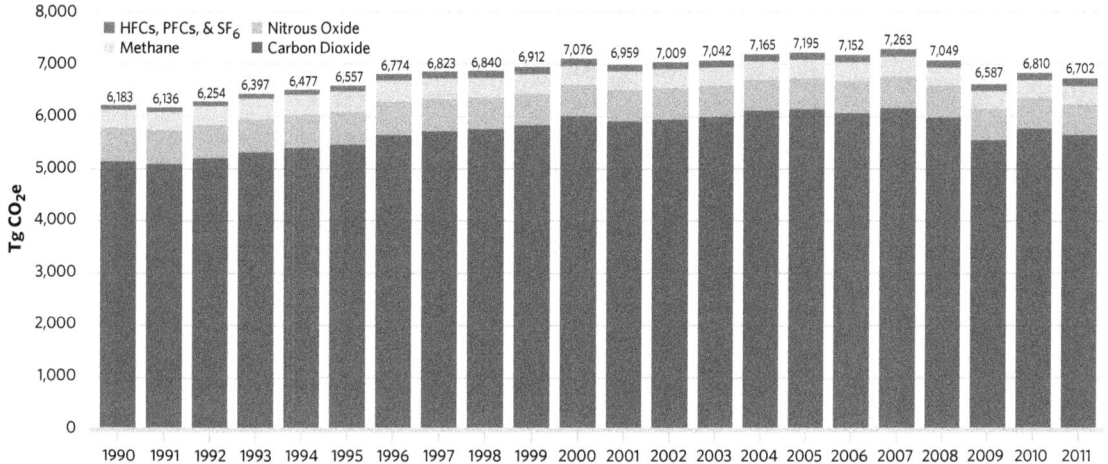

Figure 3-2 **Annual Percentage Change in U.S. Greenhouse Gas Emissions**

Between 2008 and 2011, U.S. GHG emissions fell by 4.9 percent. The average annual rate of increase from 1991 through 2011 was 0.4 percent.

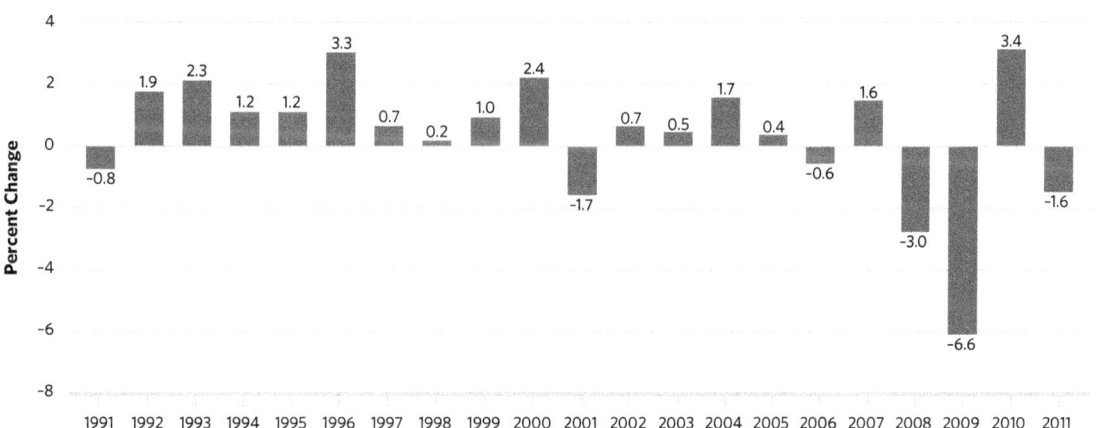

Figure 3-3 **Cumulative Change in Annual U.S. Greenhouse Gas Emissions Relative to 1991**

From 1991 through 2011, total U.S. GHG emissions rose by 159 Tg CO$_2$e, an increase of 9.2 percent. Between 2007 (2010 CAR data) and 2011, U.S. GHG emissions declined by 561 Tg CO$_2$e, or 7.7 percent.

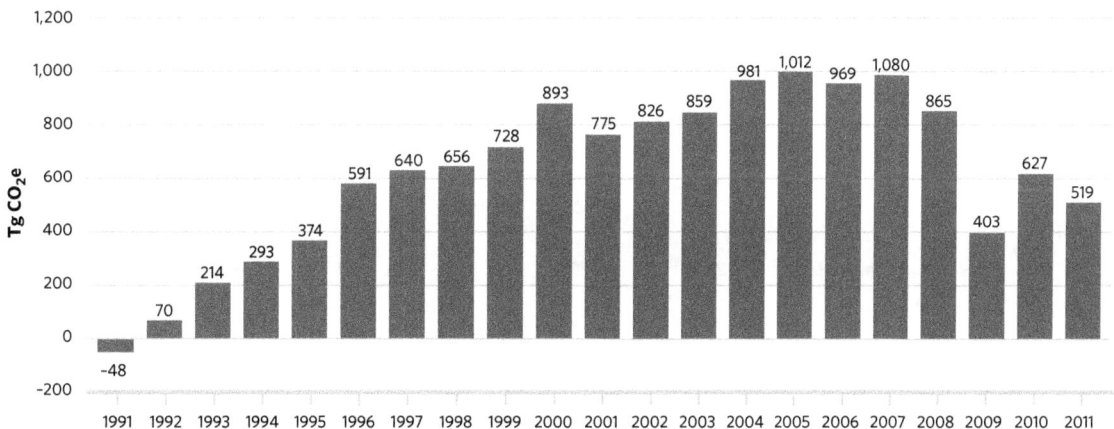

Table 3-1 **Recent Trends in U.S. Greenhouse Gas Emissions and Sinks** (Tg CO₂e)

In 2011, total U.S. GHG emissions were 6,702.3 Tg CO₂e, representing a 8.4 percent increase since 1990, and a 7.7 percent decrease since 2007 (2010 CAR data).

Gas/Source	1990	2005	2007	2008	2009	2010	2011
Carbon Dioxide (CO₂)	**5,108.8**	**6,109.3**	**6,128.6**	**5,944.8**	**5,517.9**	**5,736.4**	**5,612.9**
Fossil Fuel Combustion	4,748.5	5,748.7	5,767.7	5,590.6	5,222.4	5,408.1	5,277.2
Electricity Generation	1,820.8	2,402.1	2,412.8	2,360.9	2,146.4	2,259.2	2,158.5
Transportation	1,494.0	1,891.7	1,904.7	1,816.0	1,749.2	1,763.9	1,745.0
Industrial	848.6	823.4	844.4	802.0	722.6	780.2	773.2
Residential	338.3	357.9	341.6	347.0	337.0	334.6	328.8
Commercial	219.0	223.5	218.9	223.8	223.4	221.8	222.1
U.S. Territories	27.9	50.0	45.2	41.0	43.8	49.6	49.7
Non-Energy Use of Fuels	117.4	142.7	134.9	139.5	124.0	132.8	130.6
Iron & Steel and Metallurgical Coke Production	99.8	66.7	71.3	66.8	43.0	55.7	64.3
Natural Gas Systems	37.7	29.9	30.9	32.6	32.2	32.3	32.3
Cement Production	33.3	45.2	44.5	40.5	29.0	30.9	31.6
Lime Production	11.5	14.3	14.6	14.3	11.2	13.1	13.8
Incineration of Waste	8.0	12.5	12.7	11.9	11.7	12.0	12.0
Other Process Uses of Carbonates	4.9	6.3	7.4	5.9	7.6	9.6	9.2
Ammonia Production	13.0	9.2	9.1	7.9	7.9	8.7	8.8
Cropland Remaining Cropland	7.1	7.9	8.2	8.6	7.2	8.4	8.1
Urea Consumption for Nonagricultural Purposes	3.8	3.7	4.9	4.1	3.4	4.4	4.3
Petrochemical Production	3.4	4.3	4.1	3.6	2.8	3.5	3.5
Aluminum Production	6.8	4.1	4.3	4.5	3.0	2.7	3.3
Soda Ash Production and Consumption	2.8	3.0	2.9	3.0	2.6	2.7	2.7
Titanium Dioxide Production	1.2	1.8	1.9	1.8	1.6	1.8	1.9
Carbon Dioxide Consumption	1.4	1.3	1.9	1.8	1.8	2.2	1.8
Ferroalloy Production	2.2	1.4	1.6	1.6	1.5	1.7	1.7
Glass Production	1.5	1.9	1.5	1.5	1.0	1.5	1.3
Zinc Production	0.6	1.0	1.0	1.2	0.9	1.2	1.3
Phosphoric Acid Production	1.5	1.3	1.2	1.2	1.0	1.1	1.2
Wetlands Remaining Wetlands	1.0	1.1	1.0	1.0	1.1	1.0	0.9
Lead Production	0.5	0.6	0.6	0.5	0.5	0.5	0.5
Petroleum Systems	0.4	0.3	0.3	0.3	0.3	0.3	0.3
Silicon Carbide Production and Consumption	0.4	0.2	0.2	0.2	0.1	0.2	0.2
Land Use, Land-Use Change, and Forestry (Sink)[a]	*(794.5)*	*(997.8)*	*(929.2)*	*(902.6)*	*(882.6)*	*(888.8)*	*(905.0)*
Wood Biomass and Ethanol Consumption[b]	*218.6*	*228.7*	*238.3*	*251.7*	*245.1*	*264.5*	*264.5*
International Bunker Fuels[c]	*103.5*	*113.1*	*115.3*	*114.3*	*106.4*	*117.0*	*111.3*

Table 3-1 (Continued)   **Recent Trends in U.S. Greenhouse Gas Emissions and Sinks** (Tg CO$_2$e)

Gas/Source	1990	2005	2007	2008	2009	2010	2011
Methane (CH$_4$)	**639.9**	**593.6**	**618.6**	**618.8**	**603.8**	**592.7**	**587.2**
Natural Gas Systems	161.2	159.0	168.4	163.4	150.7	143.6	144.7
Enteric Fermentation	132.7	137.0	141.8	141.4	140.6	139.3	137.4
Landfills	147.8	112.5	111.6	113.6	113.3	106.8	103.0
Coal Mining	84.1	56.9	57.9	67.1	70.3	72.4	63.2
Manure Management	31.5	47.6	52.4	51.5	50.5	51.8	52.0
Petroleum Systems	35.2	29.2	29.8	30.0	30.5	30.8	31.5
Wastewater Treatment	15.9	16.5	16.6	16.6	16.5	16.4	16.2
Forestland Remaining Forestland	2.5	8.0	14.4	8.7	5.7	4.7	14.2
Rice Cultivation	7.1	6.8	6.2	7.2	7.3	8.6	6.6
Stationary Combustion	7.5	6.6	6.4	6.6	6.3	6.3	6.3
Abandoned Underground Coal Mines	6.0	5.5	5.3	5.3	5.1	5.0	4.8
Petrochemical Production	2.3	3.1	3.3	2.9	2.9	3.1	3.1
Mobile Combustion	4.6	2.4	2.1	1.9	1.8	1.8	1.7
Composting	0.3	1.6	1.7	1.7	1.6	1.5	1.5
Iron & Steel and Metallurgical Coke Production	1.0	0.7	0.7	0.6	0.4	0.5	0.6
Field Burning of Agricultural Residues	0.2	0.2	0.2	0.2	0.2	0.2	0.2
Ferroalloy Production	+	+	+	+	+	+	+
Silicon Carbide Production and Consumption	+	+	+	+	+	+	+
Incineration of Waste	+	+	+	+	+	+	+
International Bunker Fuels[c]	*0.1*	*0.1*	*0.1*	*0.1*	*0.1*	*0.1*	*0.1*
Nitrous Oxide (N$_2$O)	**344.3**	**356.1**	**376.1**	**349.7**	**338.7**	**343.9**	**356.9**
Agricultural Soil Management	227.9	237.5	252.3	245.4	242.8	244.5	247.2
Stationary Combustion	12.3	20.6	21.2	21.1	20.7	22.6	22.0
Mobile Combustion	44.0	36.9	29.0	25.5	22.7	20.7	18.5
Manure Management	14.4	17.1	18.0	17.8	17.7	17.8	18.0
Nitric Acid Production	18.2	16.9	19.7	16.9	14.0	16.8	15.5
Forestland Remaining Forestland	2.1	6.9	12.1	7.4	5.0	4.2	11.9
Adipic Acid Production	15.8	7.4	10.7	2.6	2.8	4.4	10.6
Wastewater Treatment	3.5	4.7	4.8	4.9	5.0	5.1	5.2
N$_2$O from Product Uses	4.4	4.4	4.4	4.4	4.4	4.4	4.4
Composting	0.4	1.7	1.8	1.9	1.8	1.7	1.7
Settlements Remaining Settlements	1.0	1.5	1.6	1.5	1.4	1.5	1.5
Incineration of Waste	0.5	0.4	0.4	0.4	0.4	0.4	0.4
Field Burning of Agricultural Residues	0.1	0.1	0.1	0.1	0.1	0.1	0.1
Wetlands Remaining Wetlands	+	+	+	+	+	+	+
International Bunker Fuels[c]	*0.9*	*1.0*	*1.0*	*1.0*	*0.9*	*1.0*	*1.0*

Table 3-1 (Continued)   **Recent Trends in U.S. Greenhouse Gas Emissions and Sinks** (Tg CO_2e)

Gas/Source	1990	2005	2007	2008	2009	2010	2011
Hydrofluorocarbons (HFCs)	**36.9**	**115.0**	**120.0**	**117.5**	**112.0**	**121.3**	**129.0**
Substitution of Ozone-Depleting Substances[d]	0.3	99.0	102.7	103.6	106.3	114.6	121.7
HCFC-22 Production	36.4	15.8	17.0	13.6	5.4	6.4	6.9
Semiconductor Manufacture	0.2	0.2	0.3	0.3	0.2	0.4	0.3
Perfluorocarbons (PFCs)	**20.6**	**6.2**	**7.7**	**6.6**	**4.4**	**5.9**	**7.0**
Semiconductor Manufacture	2.2	3.2	3.8	3.9	2.9	4.4	4.1
Aluminum Production	18.4	3.0	3.8	2.7	1.6	1.6	2.9
Sulfur Hexafluoride (SF$_6$)	**32.6**	**15.0**	**12.3**	**11.4**	**9.8**	**10.1**	**9.4**
Electrical Transmission and Distribution	26.7	11.1	8.8	8.6	8.1	7.8	7.0
Magnesium Production and Processing	5.4	2.9	2.6	1.9	1.1	1.3	1.4
Semiconductor Manufacture	0.5	1.0	0.8	0.9	0.7	1.0	0.9
Total	**6,183.3**	**7,195.3**	**7,263.2**	**7,048.8**	**6,586.6**	**6,810.3**	**6,702.3**
Net Emissions (Sources and Sinks)	**5,388.7**	**6,197.4**	**6,334.0**	**6,146.2**	**5,704.0**	**5,921.5**	**5,797.3**

+ Does not exceed 0.05 Tg CO_2e.

[a] Parentheses indicate negative values or sequestration. The net CO_2 flux total includes both emissions and sequestration, and constitutes a net sink in the United States. Sinks are only included in net emissions totals.

[b] Emissions from Wood Biomass and Ethanol Consumption are not included specifically in summing energy sector totals. Net carbon fluxes from changes in biogenic carbon reservoirs are accounted for in the estimates for Land Use, Land-Use Change, and Forestry.

[c] Emissions from International Bunker Fuels are not included in totals.

[d] Small amounts of PFC emissions also result from this source.

Note: Totals may not sum due to independent rounding. Parentheses indicate negative values or sequestration.

Figure 3-4   **2011 Greenhouse Gas Emissions by Gas**

The primary GHG emitted by human activities in the United States was CO_2, representing approximately 83.7 percent of total GHG emissions.

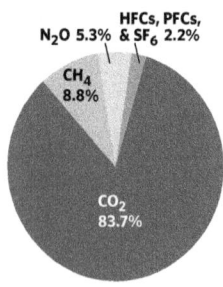

Note: Percentages Based on Tg CO_2e.
Source: U.S. EPA/OAP 2013.

[5] Global CO_2 emissions from fossil fuel combustion were taken from the U.S. Department of Energy, Energy Information Administration, International Energy Statistics 2010. See http://tonto.eia.doe. gov/cfapps/ipdbproject/IEDIndex3.cfm.

PFCs, and SF$_6$ rose by 55.1 Tg CO_2e (61.1 percent). From 1990 to 2011, HFCs increased by 92.0 Tg CO_2e (249.3 percent), PFCs decreased by 13.6 Tg CO_2e (66.1 percent), and SF$_6$ decreased by 23.3 Tg CO_2e (71.3 percent).

Despite being emitted in smaller quantities relative to the other principal GHGs, emissions of HFCs, PFCs, and SF$_6$ are significant because many of these gases have extremely high global warming potentials and, in the cases of PFCs and SF$_6$, long atmospheric lifetimes (Box 3-2). Conversely, U.S. GHG emissions were partly offset by carbon sequestration in forests, trees in urban areas, agricultural soils, and landfilled yard trimmings and food scraps, which, in aggregate, offset 13.5 percent of total emissions in 2011. The following sections describe each gas's contribution to total U.S. GHG emissions in more detail.

Carbon Dioxide Emissions

The global carbon cycle is made up of large carbon flows and reservoirs. Since the Industrial Revolution (i.e., about 1750), global atmospheric concentrations of CO_2 have risen by about 39 percent (IPCC 2007 and NOAA/ESLR 2009), principally due to the combustion of fossil fuels. Within the United States, fossil fuel combustion accounted for 94.0 percent of CO_2 emissions in 2011. Globally, approximately 31,780 Tg of CO_2 were added to the atmosphere through the combustion of fossil fuels in 2010, of which the United States accounted for about 18 percent.[5] Changes in land use and forestry practices can also increase emissions of CO_2 (e.g., through conversion of forestland to agricultural or urban use) or can result in CO_2 removals (or sinks, e.g., through net additions to forest biomass). In addition to fossil fuel combustion, several other sources emit significant quantities of CO_2. These sources include non-energy use of fuels, iron and steel production, and cement production (Figure 3-5).

As the largest source of U.S. GHG emissions, CO_2 from fossil fuel combustion has accounted for approximately 78 percent of GWP-weighted emissions since 1990, and was approximately 79

Box 3-2 **Global Warming Potentials**

Gases in the atmosphere can contribute to the greenhouse effect both directly and indirectly. Direct effects occur when the gas itself absorbs radiation. Indirect radiative forcing occurs when chemical transformations of the substance produce other GHGs, when a gas influences the atmospheric lifetimes of other gases, and/or when a gas affects atmospheric processes that alter Earth's radiative balance (e.g., affect cloud formation or albedo).[6] The IPCC developed the global warming potential (GWP) concept to compare the ability of each GHG to trap heat in the atmosphere relative to another gas.

The GWP of a GHG is defined as the ratio of the time-integrated radiative forcing from the instantaneous release of 1 kilogram (kg) of a trace substance relative to that of 1 kg of a reference gas (IPCC 2001). Direct radiative effects occur when the gas itself is a GHG. The reference gas used is CO_2; therefore, GWP-weighted emissions are measured in teragrams of carbon dioxide equivalents ($TgCO_2e$).[7] All gases in this chapter are presented in units of Tg CO_2e.

The UNFCCC reporting guidelines for national inventories were most recently updated in 2006 (IPCC 2006), but continue to require the use of GWPs from the IPCC Second Assessment Report (SAR) (IPCC 1996). This requirement ensures that current estimates of aggregate GHG emissions for 1990 to 2011 are consistent with estimates developed prior to the publication of the IPCC Third Assessment Report (IPCC 2001) and the IPCC Fourth Assessment Report (IPCC 2007). Therefore, to comply with international reporting standards under the UNFCCC, the United States reports its official emission estimates using the SAR GWP values listed in Table 3-2.

GWPs are not provided for carbon monoxide (CO), oxides of nitrogen (NO_x), nonmethane volatile organic compounds (NMVOCs), sulfur dioxide (SO_2), black carbon, and aerosols because there is no agreed-upon method to estimate the contribution of gases that are short-lived in the atmosphere, are spatially variable, or have only indirect effects on radiative forcing (IPCC 1996).

Table 3-2 **Global Warming Potentials Used in This Report** (100-Year Time Horizon)

Gas	GWP
CO_2	1
CH_4*	21
N_2O	310
HFC-23	11,700
HFC-32	650
HFC-125	2,800
HFC-134a	1,300
HFC-143a	3,800
HFC-152a	140
HFC-227ea	2,900
HFC-236fa	6,300
HFC-4310mee	1,300
CF_4	6,500
C_2F_6	9,200
C_4F_{10}	7,000
C_6F_{14}	7,400
SF_6	23,900

Source: IPCC 1996.

* The CH_4 GWP includes the direct effects and those indirect effects due to the production of tropospheric ozone andstratospheric water vapor. The indirect effect due to the production of CO_2 is not included.

Note: GWP = global warming potential; CO_2 = carbon dioxide; CH_4 = methane; N_2O = nitrous oxide; HFC = hydrofluorocarbon; CF_4 = tetrafluoromethane; C_2F_6 = hexafluoroethane; C_4F_{10} = perfluorobutane; C_6F_{14} = perfluorohexane or tetradecafluorohexane; SF_6 = sulfur hexafluoride.

Figure 3-5 **2011 U.S. Sources of CO_2 Emissions** (Tg CO_2e)

In 2011, CO_2 accounted for 83.7 percent of U.S. GHG emissions, with fossil fuel combustion accounting for 94.0 percent of CO_2 emissions.

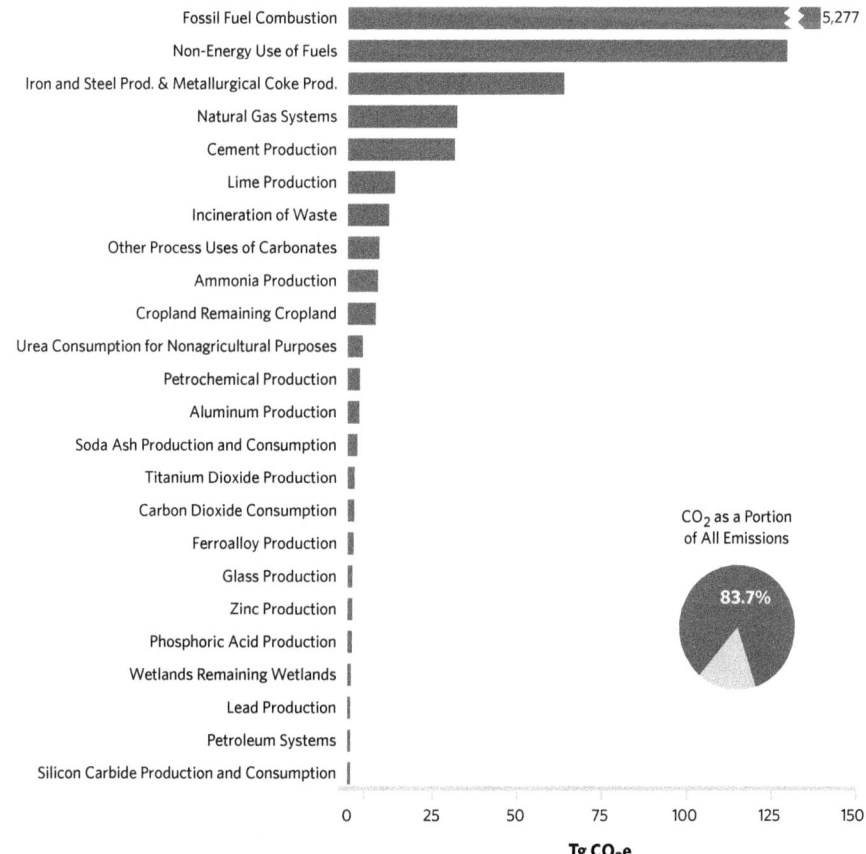

[6] Albedo is a measure of Earth's reflectivity, and is defined as the fraction of the total solar radiation incident on a body that is reflected by it.

[7] Carbon comprises 12/44ths of carbon dioxide by weight.

percent of total GWP-weighted emissions in 2011. Emissions of CO_2 from fossil fuel combustion increased at an average annual rate of 0.5 percent from 1990 to 2011. The fundamental factors influencing this trend include (1) a generally growing domestic economy over the last 22 years, and (2) an overall growth in emissions from electricity generation and transportation activities. Between 1990 and 2011, CO_2 emissions from fossil fuel combustion increased from 4,748.5 Tg CO_2e to 5,277.2 Tg CO_2e—an 11.1 percent total increase over the 22-year period. From 2010 to 2011, these emissions decreased by 130.9 Tg CO_2e (2.4 percent).

Historically, changes in emissions from fossil fuel combustion have been the dominant factor affecting U.S. emission trends. Changes in CO_2 emissions from fossil fuel combustion are influenced by many long-term and short-term factors, including population and economic growth, energy price fluctuations, technological changes, and seasonal temperatures. In the short term, the overall consumption of fossil fuels in the United States fluctuates primarily in response to changes in general economic conditions, energy prices, weather, and the availability of nonfossil alternatives.

For example, a year with increased consumption of goods and services, low fuel prices, severe summer and winter weather conditions, nuclear plant closures, and lower precipitation feeding hydroelectric dams would likely have proportionally greater fossil fuel consumption than a year with poor economic performance, high fuel prices, mild temperatures, and increased output from nuclear and hydroelectric plants. In the long term, energy consumption patterns respond to changes that affect the scale of consumption (e.g., population, number of cars, and size of houses); the efficiency with which energy is used in equipment (e.g., cars, power plants, steel mills, and light bulbs); and behavioral choices (e.g., walking, bicycling, or telecommuting to work instead of driving).

The five major fuel-consuming sectors contributing to CO_2 emissions from fossil fuel combustion are electricity generation, transportation, industrial, residential, and commercial. The electricity generation sector produces CO_2 emissions as it consumes fossil fuel to provide electricity to one of the other four "end-use" sectors. For the discussion that follows, electricity generation emissions have been distributed to each end-use sector on the basis of each sector's share of aggregate electricity consumption. This method of distributing emissions assumes that each end-use sector consumes electricity that is generated from the national average mix of fuels according to their carbon intensity. Emissions from electricity generation are also addressed separately after the end-use sectors have been discussed.

Note that emissions from U.S. territories are calculated separately due to a lack of specific consumption data for the individual end-use sectors. Figures 3-6 and 3-7 and Table 3-3 summarize CO_2 emissions from fossil fuel combustion by end-use sector.

Figure 3-6 **2011 U.S. CO_2 Emissions from Fossil Fuel Combustion by Sector and Fuel Type**

In 2011, U.S. transportation sector emissions were primarily from petroleum consumption, while electricity generation emissions were primarily from coal consumption.

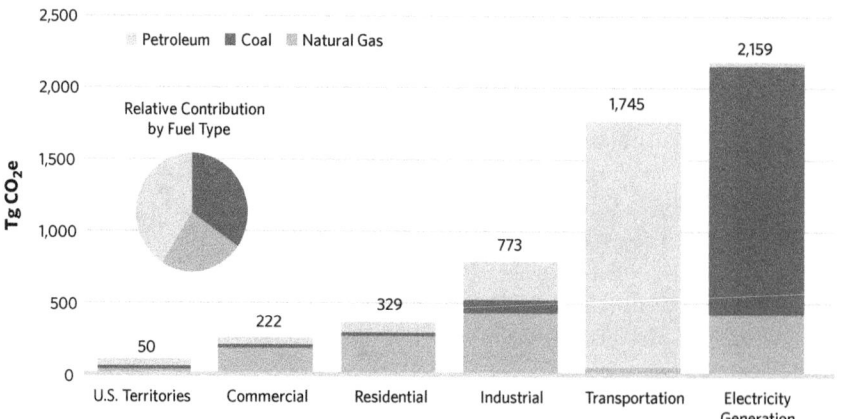

Figure 3-7 **2011 U.S. End-Use Sector Emissions of CO_2, CH_4, and N_2O from Fossil Fuel Combustion**

In 2011, direct fossil fuel combustion accounted for the vast majority of fossil fuel-related CO_2 emissions from the transportation sector (mostly petroleum combustion). Electricity consumption indirectly accounted for most of the fossil fuel-related CO_2 emissions from the commercial, residential, and industrial sectors (mostly coal combustion).

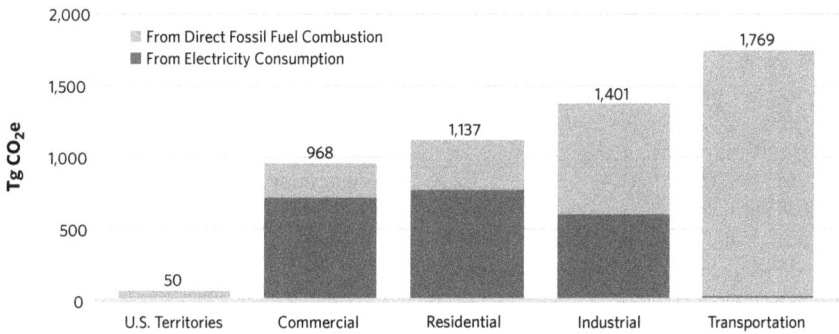

Table 3-3 **CO_2 Emissions from Fossil Fuel Combustion by Fuel-Consuming End-Use Sector** (Tg CO_2e)

The figures below reflect the distribution of electricity generation emissions to each of the four end-use sectors on the basis of each sector's share of aggregate electricity consumption. Between 2007 (2010 CAR data) and 2011, CO_2 emissions decreased by 490.5 Tg CO_2e, or 8.4 percent.

End-Use Sector	1990	2005	2007	2008	2009	2010	2011
Transportation	**1,497.0**	**1,896.5**	**1,909.7**	**1,820.7**	**1,753.7**	**1,768.4**	**1,749.3**
Combustion	1,494.0	1,891.7	1,904.7	1,816.0	1,749.2	1,763.9	1,745.0
Electricity	3.0	4.7	5.1	4.7	4.5	4.5	4.3
Industrial	**1,535.3**	**1,560.4**	**1,559.9**	**1,499.3**	**1,324.6**	**1,421.3**	**1,392.1**
Combustion	848.6	823.4	844.4	802.0	722.6	780.2	773.2
Electricity	686.7	737.0	715.4	697.3	602.0	641.1	618.9
Residential	**931.4**	**1,214.7**	**1,205.2**	**1,189.9**	**1,123.5**	**1,175.0**	**1,125.6**
Combustion	338.3	357.9	341.6	347.0	337.0	334.6	328.8
Electricity	593.0	856.7	863.5	842.9	786.5	840.4	796.9
Commercial	**757.0**	**1,027.2**	**1,047.7**	**1,039.8**	**976.8**	**993.9**	**960.5**
Combustion	219.0	223.5	218.9	223.8	223.4	220.6	222.1
Electricity	538.0	803.7	828.8	816.0	753.5	773.3	738.4
U.S. Territories[a]	**27.9**	**50.0**	**45.2**	**41.0**	**43.8**	**49.6**	**49.7**
Total	**4,748.5**	**5,748.7**	**5,767.7**	**5,590.6**	**5,222.4**	**5,408.1**	**5,277.2**
Electricity Generation	**1,820.8**	**2,402.1**	**2,412.8**	**2,360.9**	**2,146.4**	**2,259.2**	**2,158.5**

[a] Fuel consumption by U.S. territories (i.e., American Samoa, Guam, Puerto Rico, U.S. Virgin Islands, Wake Island, and other U.S. Pacific Islands) is included.

Note: Totals may not sum due to independent rounding.

Transportation End-Use Sector

Transportation activities (excluding international bunker fuels) accounted for 33 percent of CO_2 emissions from fossil fuel combustion in 2011.[8] Virtually all of the energy consumed in this end-use sector came from petroleum products. Nearly 65 percent of the emissions resulted from gasoline consumption for personal vehicle use. The remaining emissions came from other transportation activities, including the combustion of diesel fuel in heavy-duty vehicles and jet fuel in aircraft. From 1990 to 2011, transportation emissions rose by 17 percent, principally because of increased demand for travel and the stagnation of fuel efficiency across the U.S. vehicle fleet.

[8] If emissions from international bunker fuels are included, the transportation end-use sector accounted for 34.5 percent of U.S. emissions from fossil fuel combustion in 2011.

The number of vehicle miles traveled by light-duty motor vehicles (passenger cars and light-duty trucks) increased by 34 percent from 1990 to 2011, as a result of a confluence of factors, including population growth, economic growth, urban sprawl, and low fuel prices over much of this period. However, the more recent trend for transportation has shown a general decline in emissions, due to recent slow growth in economic activity, higher fuel prices, and an associated decrease in the demand for passenger transportation. Additionally, light-duty motor vehicles are also becoming more fuel efficient, due to both a shift in consumer demand and federal and state policies.

Industrial End-Use Sector

Industrial CO_2 emissions, resulting both directly from the combustion of fossil fuels and indirectly from the generation of electricity consumed by industry, accounted for 26 percent of CO_2 from fossil fuel combustion in 2011. Approximately 56 percent of these emissions resulted from direct fossil fuel combustion to produce steam and/or heat for industrial processes. The remaining emissions resulted from consuming electricity for motors, electric furnaces, ovens, lighting, and other applications. In contrast to the other end-use sectors, emissions from industry have steadily declined since 1990. This decline is due to structural changes in the U.S. economy (i.e., shifts from a manufacturing-based to a service-based economy), fuel switching, and efficiency improvements.

Residential and Commercial End-Use Sectors

The residential and commercial end-use sectors accounted for 21 and 18 percent, respectively, of CO_2 emissions from fossil fuel combustion in 2011. Both sectors relied heavily on electricity for meeting energy demands, with 71 and 77 percent, respectively, of their emissions attributable to electricity consumption for lighting, heating, cooling, and operating appliances. The remaining emissions were due to the consumption of natural gas and petroleum for heating and cooking. Emissions from the residential and commercial end-use sectors have increased by 21 percent and 27 percent since 1990, respectively, due to increasing electricity consumption for lighting, heating, air conditioning, and operating appliances.

Electricity Generation

The United States relies on electricity to meet a significant portion of its energy demands. Electricity generators consumed 36 percent of U.S. energy from fossil fuels and emitted 41 percent of the CO_2 from fossil fuel combustion in 2011. The type of fuel combusted by electricity generators has a significant effect on their emissions. For example, some electricity is generated with low-CO_2-emitting energy technologies, particularly nonfossil options, such as nuclear, hydroelectric, or geothermal energy. Electricity generators relied on coal for approximately 42 percent their total energy requirements in 2011, and accounted for 95 percent of all coal consumed for energy in the United States in 2011.

Recently, the carbon intensity of fuels consumed to generate electricity has decreased, due to lower consumption of coal and higher consumption of natural gas and other energy sources. The discovery and exploitation of vast reserves of natural gas in the United States have reduced its domestic price per energy unit and have sparked demand for natural gas as a baseload fuel for electricity generation. Across the time series, changes in electricity demand and the carbon intensity of fuels used for electricity generation have had a significant impact on CO_2 emissions.

Other significant CO_2 trends include:

- CO_2 emissions from non-energy use of fossil fuels increased by 13.1 Tg CO_2e (11.2 percent) from 1990 through 2011. Emissions from non-energy uses of fossil fuels were 130.6 Tg CO_2e in 2011, which constituted 2.3 percent of total national CO_2 emissions, or approximately the same proportion as in 1990.

- CO_2 emissions from iron and steel production and metallurgical coke production increased by 8.5 Tg CO_2e (15.3 percent) from 2010 to 2011, continuing a two-year trend of increasing emissions, primarily due to increased steel production associated with improved economic conditions. Despite this, from 1990 through 2011, emissions declined by 35.5 Tg CO_2e (35.6 percent), as a result of the restructuring of the industry, technological improvements, and increased scrap utilization.

- In 2011, CO_2 emissions from cement production increased by 0.7 Tg CO_2e (2.3 percent) from 2010. After decreasing in 1991 by 2.2 percent from 1990 levels, emissions from cement production grew every year through 2006. From 2006 through 2011, emissions have fluctuated due to the economic recession and associated decrease in demand for construction materials. Overall, from 1990 to 2011, emissions from cement production decreased by 1.6 Tg CO_2e (4.9 percent).

- Net CO_2 uptake from LULUCF grew by 110.5 Tg CO_2e (13.9 percent) from 1990 through 2011. This increase was primarily due to a higher rate of net carbon accumulation in forest carbon stocks, particularly in above-ground and below-ground tree biomass, and harvested wood pools. Annual carbon accumulation in landfilled yard trimmings and food scraps slowed over this period, while the rate of carbon accumulation in urban trees accelerated.

Methane Emissions

CH_4 is more than 20 times as effective as CO_2 at trapping heat in the atmosphere (IPCC 1996). Over the last 250 years, the concentration of CH_4 in the atmosphere increased by 158 percent (IPCC 2007). Anthropogenic sources of CH_4 include natural gas and petroleum systems, agricultural activities, landfills, coal mining, wastewater treatment, stationary and mobile combustion, and certain industrial processes (Figure 3-8).

Some significant trends in U.S. emissions of CH_4 include:

- Natural gas systems were the largest anthropogenic source category of CH_4 emissions in the United States in 2011, with 144.7 Tg CO_2e of CH_4 emitted into the atmosphere. This

Figure 3-8 **2011 U.S. Sources of Methane Emissions**

In 2011, CH_4 accounted for 8.8 percent of U.S. GHG emissions on a global warming potential-weighted basis. Natural gas systems comprised the largest source of CH_4, accounting for 144.7 Tg CO_2e, or 24.6 percent of total CH_4 emissions. Enteric fermentation followed close behind, contributing 137.4 Tg CO_2e, or 23.4 percent.

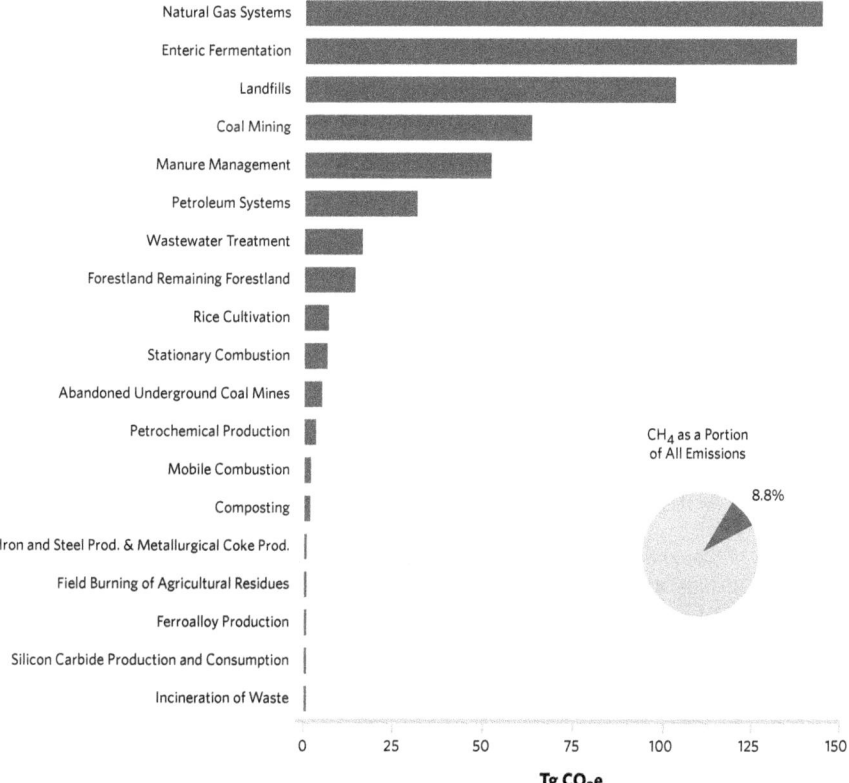

represented a 16.5 Tg CO_2e (10.2 percent) decrease since 1990, largely due to lower emissions from transmission and storage resulting from both increased voluntary reductions and decreased distribution emissions from cast iron and steel pipelines. Emissions from field production accounted for approximately 37 percent of CH_4 emissions from natural gas systems in 2011.

CH_4 emissions from field production decreased by 12 percent from 1990 through 2011. However, the trend was not stable over the time series. Emissions from field production rose by 43 percent from 1990 through 2006, and then declined by 38 percent from 2006 to 2011. The drivers of this trend include increased voluntary reductions and the effects of the recent global economic slowdown.

- Enteric fermentation is the second-largest anthropogenic source of CH_4 emissions in the United States. In 2011, CH_4 emissions from enteric fermentation were 137.4 Tg CO_2e (23.4 percent of total CH_4 emissions), an increase of 4.6 Tg CO_2e (3.5 percent) since 1990. This increase generally follows the trends in cattle populations. From 1990 through 1995, emissions from enteric fermentation rose, but then fell from 1996 through 2001, mainly due to fluctuations in beef cattle populations and improved digestibility of feed for feedlot cattle. Emissions generally increased from 2002 through 2007, though with a slight decrease in 2004, as both dairy and beef cattle populations grew and the literature for dairy cow diets indicated poorer feed digestibility for those years. Emissions decreased again from 2008 through 2011, as beef cattle populations again declined.

- Landfills are the third-largest anthropogenic source of CH_4 emissions in the United States, accounting for 17.5 percent of total CH_4 emissions (103.0 Tg CO_2e) in 2011. From 1990 through 2011, CH_4 emissions from landfills decreased by 44.7 Tg CO_2e (30.3 percent), with small increases occurring in some interim years, despite the higher volume of municipal solid waste (MSW) placed in landfills. This downward trend can be attributed to a 21 percent reduction in decomposable materials (i.e., paper and paperboard, food scraps, and yard trimmings) discarded in MSW landfills over the time series, and an increase in landfill gas collected and combusted.[9]

- In 2011, CH_4 emissions from coal mining were 63.2 Tg CO_2e—a 9.2 Tg CO_2e (12.6 percent) decrease from 2010 emission levels. The overall decline of 20.8 Tg CO_2e (24.8 percent) from 1990 resulted from the mining of less gassy coal from underground mines and the increased use of CH_4 collected from degasification systems.

- Methane emissions from manure management rose by 65.3 percent, from 31.5 Tg CO_2e in 1990 to 52.0 Tg CO_2e in 2011. The majority of this increase was from swine and dairy cow manure, reflecting the general trend in manure management toward greater use of liquid systems, which increases CH_4 emissions. This trend is the combined result of a shift to larger facilities, and to facilities in the West and Southwest, all of which tend to use liquid systems. Also, new regulations limiting the application of manure nutrients have shifted manure management practices at smaller dairies from daily spread to manure managed and stored on site.

Nitrous Oxide Emissions

N_2O is produced by biological processes that occur in soil and water and by a variety of anthropogenic activities in the agricultural, energy-related, industrial, and waste management fields. While total N_2O emissions are much lower than CO_2 emissions, N_2O is approximately 300 times more powerful than CO_2 at trapping heat in the atmosphere (IPCC 1996). Since 1750, the global atmospheric concentration of N_2O has risen by approximately 19 percent (IPCC 2007). The main U.S. anthropogenic activities producing N_2O are agricultural soil management, stationary fuel combustion, fuel combustion in motor vehicles, manure management, and nitric acid production (Figure 3-9).

Some significant trends in U.S. emissions of N_2O include:

- Agricultural soils accounted for approximately 69.3 percent (247.2 Tg CO_2e) of N_2O emissions and 3.7 percent of total emissions in the United States in 2011. Annual N_2O emissions

Figure 3-9   **2011 U.S. Sources of Nitrous Oxide Emissions**

In 2011, N_2O accounted for 5.3 percent of U.S. GHG emissions on a global warming potential-weighted basis. Agricultural soil management was the largest U.S. source of N_2O, producing 247.2 Tg CO_2e, or 69.3 percent of N_2O emissions.

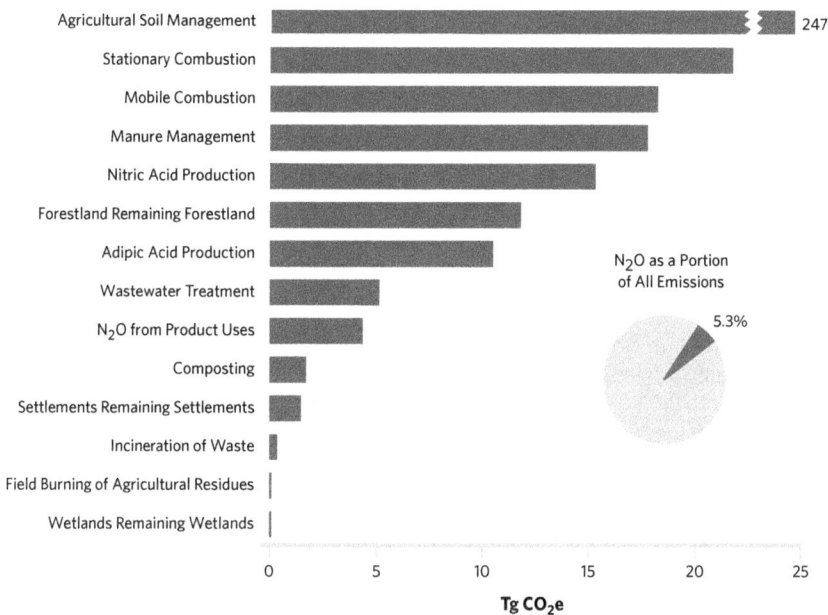

from agricultural soils fluctuated between 1990 and 2011, although overall emissions were 8.5 percent higher in 2011 than in 1990. The annual fluctuation was largely a reflection of annual variation in weather patterns, synthetic fertilizer use, and crop production.

- N_2O emissions from stationary combustion increased by 9.7 Tg CO_2e (79.3 percent) from 1990 through 2011, primarily as a result of the growth of coal fluidized bed boilers in the electric power sector.

- In 2011, mobile combustion produced 18.5 Tg CO_2e (5.2 percent) of U.S. N_2O emissions. Although N_2O emissions from mobile combustion decreased by 58.0 percent from 1990 through 2011, they increased by 25.6 percent from 1990 through 1998, because of control technologies that reduced NO_x emissions but boosted N_2O emissions. Since 1998, newer control technologies have led to an overall decline of 36.8 Tg CO_2e (66.6 percent) in N_2O from this source.

- N_2O emissions from adipic acid production were 10.6 Tg CO_2e in 2011, and have decreased significantly in recent years due to the widespread installation of pollution control measures.

HFC, PFC, and SF_6 Emissions

HFCs are a family of synthetic chemicals that are used as alternatives to ozone-depleting substances (ODS), which are being phased out under the 1987 Montreal Protocol and 1990 Clean Air Act Amendments. Because HFCs and PFCs do not deplete the stratospheric ozone layer, they are acceptable alternatives under the Montreal Protocol.

PFCs are another family of synthetic chemicals that are emitted primarily from the production of semiconductors and as a by-product during the production of primary aluminum. A small amount of PFCs, which like HFCs do not deplete the ozone layer, are also used as alternatives to ODS.

However, these compounds, along with SF_6, are potent GHGs. Besides having high GWPs, SF_6 and PFCs have extremely long atmospheric lifetimes, resulting in their essentially irreversible accumulation in the atmosphere once emitted.

In addition to the use of HFCs and PFCs as alternatives to ODS, other sources of these gases include electrical transmission and distribution systems, HCFC-22 production, semiconductor manufacturing, aluminum production, and magnesium production and processing (Figure 3-10).

Some significant trends in U.S. HFC, PFC, and SF_6 emissions include:

* Emissions resulting from the substitution of ODS (e.g., chlorofluorocarbons [CFCs]) have been consistently increasing, from 0.3 Tg CO_2e in 1990 to 121.7 Tg CO_2e in 2011. Emissions from ODS substitutes are both the largest and the fastest-growing source of HFC, PFC, and SF_6 emissions. These emissions have been increasing since the phase-out of ODS required under the Montreal Protocol came into effect, especially after 1994, when the first generation of new technologies featuring ODS substitutes fully penetrated the market (excluding most aerosols, from which CFCs were banned in 1978).

* HFC emissions from the production of HCFC-22 decreased by 29.5 Tg CO_2e (81.0 percent) from 1990 through 2011. This reduction was due to (1) a steady decline in the emission rate of HFC-23 (i.e., the amount of HFC-23 emitted/kg of HCFC-22 manufactured); (2) the use of thermal oxidation at some plants to reduce HFC-23 emissions; and (3) a decrease in the domestic production of HCFC-22 as Montreal Protocol and Clean Air Act restrictions took effect.

* SF_6 emissions from electric power transmission and distribution systems decreased by 19.6 Tg CO_2e (73.6 percent) from 1990 through 2011, primarily because of higher purchase prices for SF_6 and efforts by industry to reduce emissions.

* PFC emissions from aluminum production decreased by 15.5 Tg CO_2e (84.0 percent) from 1990 through 2011, due to both industry emission reduction efforts and declines in domestic aluminum production.

OVERVIEW OF SECTOR EMISSIONS AND TRENDS

In accordance with the *Revised 1996 IPCC Guidelines* (IPCC/UNEP/OECD/IEA 1997) and the 2003 *UNFCCC Guidelines on Reporting and Review* (UNFCCC 2003), Figure 3-11 and Table 3-4 aggregate emissions and sinks by sectors, as defined by the IPCC. Emissions of all gases can be summed from each source category from IPCC guidance. From 1990 through 2011, total emissions in the energy, industrial processes, and agriculture sectors grew by 478.4 Tg CO_2e (9.1 percent), 10.3 Tg CO_2e (3.3 percent), and 47.6 Tg CO_2e (11.5 percent), respectively.

Figure 3-10 **2011 U.S. Sources of Hydrofluorocarbons, Perfluorocarbons, and Sulfur Hexafluoride Emissions**

In 2011, HFCs, PFCs, and SF_6 accounted for 2.2 percent of U.S. GHG emissions on a global warming potential-weighted basis. Emissions from the substitution of ozone-depleting substances (e.g., chlorofluorocarbons) have been consistently increasing, from 0.3 Tg CO_2e in 1990 to 121.7 Tg CO_2e in 2011.

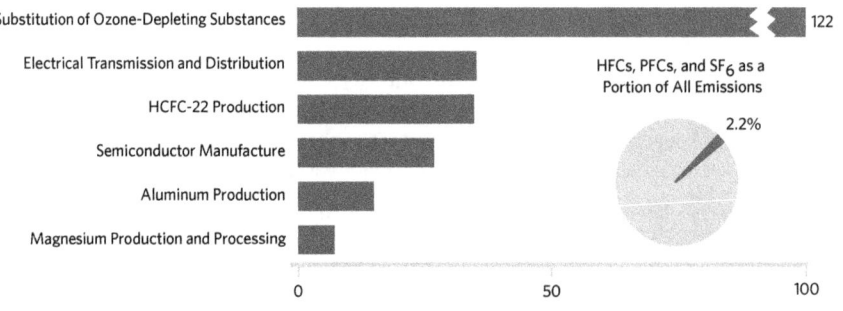

Emissions from the waste and solvent and other product use sectors decreased by 40.2 Tg CO_2e (23.9 percent) and by less than 0.1 Tg CO_2e (0.4 percent), respectively. Over the same period, estimates of net carbon sequestration in the LULUCF sector (magnitude of emissions plus CO_2 flux from all LULUCF source categories) increased by 87.6 Tg CO_2e (11.2 percent).

Energy

The energy sector produces emissions of all GHGs resulting from stationary and mobile energy activities, including fuel combustion and fugitive fuel emissions. Energy-related activities—primarily fossil fuel combustion—accounted for the vast majority of U.S. CO_2 emissions from 1990 through 2011. In 2011, approximately 87 percent of the energy consumed in the

Figure 3-11 **U.S. Greenhouse Gas Emissions and Sinks by IPCC Sector**

Along with Table 3-4, this figure aggregates emissions and sinks by sectors, as defined by the Intergovernmental Panel on Climate Change. Since 2007 (2010 CAR data), GHG emissions in all sectors have decreased, and net sequestration from land use, land-use change, and forestry (LULUCF) have remained relatively stable.

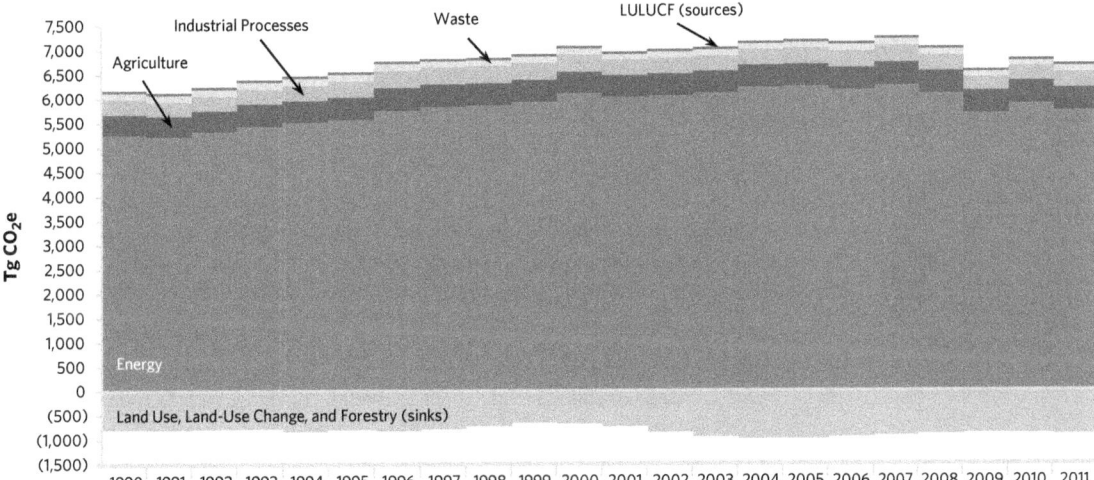

Source: U.S. EPA/OAP 2013.

Table 3-4 **Recent Trends in U.S. Greenhouse Gas Emissions and Sinks by IPCC Sector** (Tg CO_2e)

From 1990 to 2011, total emissions in the energy, industrial processes, and agriculture sectors increased, emissions in the solvent and other product use sector remained unchanged, and emissions in the waste sector decreased. Net sequestration in the land-use change and forestry sector increased by 408.6 Tg CO_2e, or 13.9 percent.

IPCC Sector	1990	2005	2007	2008	2009	2010	2011
Energy	5,267.3	6,251.6	6,266.9	6,096.2	5,699.2	5,889.1	5,745.7
Industrial Processes	316.1	330.8	347.2	318.7	265.3	303.4	326.5
Solvent and Other Product Use	4.4	4.4	4.4	4.4	4.4	4.4	4.4
Agriculture	413.9	446.2	470.9	463.6	459.2	462.3	461.5
Land-Use Change and Forestry	13.7	25.4	37.3	27.2	20.4	19.7	36.6
Waste	167.8	136.9	136.5	138.6	138.1	131.4	127.7
Total Emissions	**6,183.3**	**7,195.3**	**7,263.2**	**7,048.8**	**6,586.6**	**6,810.3**	**6,702.3**
Land-Use Change and Forestry (Sinks)	(794.5)	(997.8)	(929.2)	(902.6)	(882.6)	(888.8)	(905.0)
Net Emissions (Emissions and Sinks)	**5,388.7**	**6,197.4**	**6,334.0**	**6,146.2**	**5,704.0**	**5,921.5**	**5,797.3**

* The net CO_2 flux total includes both emissions and sequestration, and constitutes a sink in the United States. Sinks are only included in net emissions total.

Note: Totals may not sum due to independent rounding. Parentheses indicate negative values or sequestration. IPCC = Intergovernmental Panel on Climate Change.

United States (on a British thermal unit basis) was produced by the combustion of fossil fuels; the remaining 13 percent was produced by other sources, such as hydroelectric, biomass, nuclear, wind, and solar energy (Figure 3-12). Energy-related activities are also responsible for CH_4 and N_2O emissions (43 percent and 11 percent of total U.S. emissions, respectively). Overall, emission sources in the energy sector accounted for a combined 85.7 percent of total U.S. GHG emissions in 2011.

Figure 3-12 **2011 U.S. Energy Consumption by Energy Source**

In 2011, approximately 87 percent of U.S. energy consumed was produced by the combustion of fossil fuels. The remaining 13 percent was produced by other sources, such as hydroelectric, biomass, nuclear, wind, and solar energy.

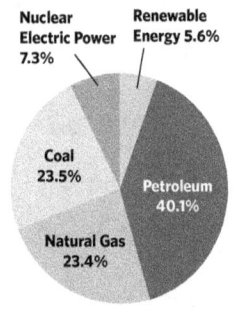

Nuclear Electric Power 7.3%
Renewable Energy 5.6%
Coal 23.5%
Petroleum 40.1%
Natural Gas 23.4%

Industrial Processes

The industrial processes sector contains by-products or fugitive emissions of GHGs from industrial processes not directly related to energy activities, such as fossil fuel combustion. For example, industrial processes can chemically transform raw materials, which often release waste gases, such as CO_2, CH_4, and N_2O. These processes include iron and steel production and metallurgical coke production, cement production, ammonia production and urea consumption, lime production, other process uses of carbonates (e.g., flux stone, flue gas desulfurization, and glass manufacturing), soda ash production and consumption, titanium dioxide production, phosphoric acid production, ferroalloy production, glass production, CO_2 consumption, silicon carbide production and consumption, aluminum production, petrochemical production, nitric acid production, adipic acid production, lead production, and zinc production. Additionally, emissions from industrial processes release HFCs, PFCs, and SF_6. Overall, emission sources in the industrial process sector accounted for 4.9 percent of U.S. GHG emissions in 2011.

Solvent and Other Product Use

The solvent and other product use sector contains GHG emissions that are produced as a by-product of various solvent and other product uses. In the United States, emissions from N_2O from product uses, the only source of GHG emissions from this sector, accounted for about 0.1 percent of total U.S. anthropogenic GHG emissions on a carbon-equivalent basis in 2011.

Agriculture

The agricultural sector contains anthropogenic emissions from agricultural activities (except fuel combustion, which is addressed in the energy sector, and agricultural CO_2 fluxes, which are addressed in the LULUCF sector). Agricultural activities contribute directly to emissions of GHGs through a variety of processes, including the enteric fermentation in domestic livestock, livestock manure management, rice cultivation, agricultural soil management, and field burning of agricultural residues. CH_4 and N_2O were the primary GHGs emitted by agricultural activities. CH_4 emissions from enteric fermentation and manure management represented 23.4 percent and 8.9 percent of total, CH_4 emissions from anthropogenic activities in 2011, respectively. Agricultural soil management activities, such as fertilizer application and other cropping practices, were the largest source of U.S. N_2O emissions in 2011, accounting for 69.3 percent. In 2011, emission sources accounted for in the agricultural sector were responsible for 6.9 percent of total U.S. GHG emissions.

Land Use, Land-Use Change, and Forestry

The LULUCF sector contains emissions of CH_4 and N_2O, and emissions and removals of CO_2 from forest management, other land-use activities, and land-use change. Forest management practices, tree planting in urban areas, the management of agricultural soils, and the landfilling of yard trimmings and food scraps resulted in a net uptake (sequestration) of carbon in the United States. Forests (including vegetation, soils, and harvested wood) accounted for 92 percent of total 2011 net CO_2 flux, urban trees accounted for 8 percent, mineral and organic soil carbon stock changes accounted for 1 percent, and landfilled yard trimmings and food scraps accounted for 1 percent of the total net flux in 2011.

The net forest sequestration is a result of net forest growth and increasing forest area, as well as a net accumulation of carbon stocks in harvested wood pools. The net sequestration in urban forests is a result of net tree growth in these areas. In agricultural soils, mineral and organic soils sequester approximately five times as much carbon as is emitted from these soils through liming and urea fertilization. The mineral soil carbon sequestration is largely due to the conversion of cropland to permanent pastures, grasslands, and hay production, a reduction in summer fallow areas in semi-arid areas, an increase in the adoption of conservation tillage practices, and an increase in the amounts of organic fertilizers (i.e., manure and sewage sludge) applied to agricultural lands. The net sequestration from yard trimmings and food scraps is due to the long-term accumulation of carbon from yard trimmings and food scraps in landfills.

LULUCF activities in 2011 resulted in a net carbon sequestration of 905.0 Tg CO_2e (Table 3-5). This represents an offset of 16.1 percent of total U.S. CO_2 emissions, or 13.5 percent of total GHG emissions in 2011. Between 1990 and 2011, total LULUCF net carbon flux resulted in a 13.9 percent increase in CO_2 sequestration, primarily due to an increase in the rate of net carbon accumulation in forest carbon stocks, particularly in above-ground and below-ground tree biomass, and harvested wood pools. Annual carbon accumulation in landfilled yard trimmings and food scraps slowed over this period, while the rate of annual carbon accumulation increased in urban trees.

Emissions from LULUCF are shown in Table 3-6. Liming of agricultural soils and urea fertilization in 2011 resulted in CO_2 emissions of 8.1 Tg CO_2e (8,117 gigagrams [Gg]). Lands undergoing peat extraction (i.e., peatlands remaining peatlands) resulted in CO_2 emissions of 0.9 Tg CO_2e (918 Gg), and N_2O emissions of less than 0.05 Tg CO_2e. The application of synthetic fertilizers to forest soils in 2011 resulted in direct N_2O emissions of 0.4 Tg CO_2e (1 Gg). Direct N_2O emissions from fertilizer application to forest soils have increased by 455 percent since 1990, but still account for a relatively small portion of overall emissions. Additionally, direct N_2O emissions from fertilizer application to settlement soils in 2011 accounted for 1.5 Tg CO_2e (5 Gg), representing an increase of 51 percent since 1990. Forest fires in 2011 resulted in CH_4 emissions of 14.2 Tg CO_2e (675 Gg), and in N_2O emissions of 11.6 Tg CO_2e (37 Gg).

Waste

The waste sector contains emissions from waste management activities (except incineration of waste, which is addressed in the energy sector). Landfills were the largest source of anthropogenic GHG emissions in the waste sector, accounting for 80.7 percent of this sector's emissions, and 17.5 percent of total U.S. CH_4 emissions.[10] Additionally, wastewater treatment accounts for 16.7 percent of waste emissions, 2.8 percent of U.S. CH_4 emissions, and 1.5 percent of U.S. N_2O emissions. Emissions of CH_4 and N_2O from composting are also accounted for in this sector, generating emissions of 1.5 Tg CO_2e and 1.7 Tg CO_2e, respectively. Overall, emission sources accounted for in the waste sector generated 1.9 percent of total U.S. GHG emissions in 2011.

[10] Landfills also store carbon, due to incomplete degradation of organic materials, such as wood products and yard trimmings, as described in the *Inventory of U.S. Greenhouse Gas Emissions and Sinks: 1990-2011* (U.S. EPA/OAP 2013).

Table 3-5 **Net CO_2 Flux from Land Use, Land-Use Change, and Forestry** (Tg CO_2e)

Between 1990 and 2011, total LULUCF net carbon flux resulted in a 13.9 percent increase in CO_2 sequestration, primarily due to an increase in the rate of net carbon accumulation in forest carbon stocks, particularly in above-ground and below-ground tree biomass, and harvested wood pools.

Sink Category	1990	2005	2007	2008	2009	2010	2011
Forestland Remaining Forestland	(696.8)	(905.0)	(859.3)	(833.3)	(811.3)	(817.6)	(833.5)
Cropland Remaining Cropland	(34.1)	(20.3)	(6.6)	(5.2)	(4.6)	(3.0)	(2.9)
Land Converted to Cropland	21.0	13.5	14.5	14.5	14.5	14.5	14.5
Grassland Remaining Grassland	(5.3)	(1.0)	7.1	7.2	7.3	7.3	7.4
Land Converted to Grassland	(7.7)	(10.2)	(9.0)	(9.0)	(8.9)	(8.8)	(8.8)
Settlements Remaining Settlements	(47.5)	(63.2)	(65.0)	(66.0)	(66.9)	(67.9)	(68.8)
Other (Landfilled Yard Trimmings and Food Scraps)	(24.2)	(11.6)	(10.9)	(10.9)	(12.7)	(13.3)	(13.0)
Total	**(794.5)**	**(997.8)**	**(929.2)**	**(902.6)**	**(882.6)**	**(888.8)**	**(905.0)**

Note: Totals may not sum due to independent rounding. Parentheses indicate net sequestration.

Table 3-6   **Emissions from Land Use, Land-Use Change, and Forestry** (Tg CO_2e)

Between 1990 and 2011, CH_4 emissions from forest fires rose by 407.1 percent, and direct N_2O emissions from fertilizer application to forest soils rose by 455 percent. While these increases are significant, these sources account for a relatively small portion of overall GHG emissions.

Source Category	1990	2005	2007	2008	2009	2010	2011
Carbon Dioxide (CO₂)	**8.1**	**8.9**	**9.2**	**9.6**	**8.3**	**9.4**	**9.0**
Cropland Remaining Cropland: Liming of Agricultural Soils	4.7	4.3	4.5	5.0	3.7	4.7	4.5
Cropland Remaining Cropland: Urea Fertilization	2.4	3.5	3.8	3.6	3.6	3.7	5.3
Wetlands Remaining Wetlands: Peatlands Remaining Peatlands	1.0	1.1	1.0	1.0	1.1	1.0	0.9
Methane (CH₄)	**2.5**	**8.0**	**14.4**	**8.7**	**5.7**	**4.7**	**14.2**
Forestland Remaining Forestland: Forest Fires	2.5	8.0	14.4	8.7	5.7	4.7	14.2
Nitrous Oxide (N₂O)	**3.1**	**8.4**	**13.7**	**8.9**	**6.4**	**5.6**	**13.4**
Forestland Remaining Forestland: Forest Fires	2.0	6.6	11.7	7.1	4.7	3.8	11.6
Forestland Remaining Forestland: Forest Soils	0.1	0.4	0.4	0.4	0.4	0.4	0.4
Settlements Remaining Settlements: Settlement Soils	1.0	1.5	1.6	1.5	1.4	1.5	1.5
Wetlands Remaining Wetlands: Peatlands Remaining Peatlands	+	+	+	+	+	+	+
Total	**13.7**	**25.4**	**37.3**	**27.2**	**20.4**	**19.7**	**36.6**

+ Less than 0.05 Tg CO_2e.

Note: Totals may not sum due to independent rounding.

EMISSIONS BY ECONOMIC SECTOR

Throughout the 1990–2011 Inventory, emission estimates are grouped into six sectors defined by the IPCC: energy, industrial processes, solvent use, agriculture, LULUCF, and waste (U.S. EPA/OAP 2013). While it is important to use this characterization for consistency with UNFCCC reporting guidelines, it is also useful to allocate emissions into more commonly used domestic sectoral categories. This section reports emissions by the following economic sectors: residential, commercial, industry, transportation, electricity generation, agriculture, and U.S. territories. Table 3-7 summarizes emissions from each of these sectors, and Figure 3-13 shows the trend in emissions by sector from 1990 to 2011.

Using this categorization, emissions from electricity generation accounted for the largest portion (33 percent) of U.S. GHG emissions in 2011. Transportation activities, in aggregate, accounted for the second-largest portion (27 percent), while emissions from industry accounted for the third-largest portion (20 percent) of U.S. GHG emissions in 2011. In contrast to electricity generation and transportation, emissions from industry have in general declined over the past decade. The long-term decline in these emissions has been due to structural changes in the U.S. economy (i.e., shifts from a manufacturing-based to a service-based economy), fuel switching, and energy efficiency improvements.

The remaining 20 percent of U.S. GHG emissions were contributed by, in order of importance, the agriculture, commercial, and residential sectors, plus emissions from U.S. territories. Activities related to agriculture accounted for 8 percent of U.S. emissions. Unlike other economic sectors, agricultural sector emissions were dominated by N_2O emissions from agricultural soil management and CH_4 emissions from enteric fermentation. The commercial and residential sectors accounted for 6 percent and 5 percent, respectively, of emissions, and U.S. territories accounted for 1 percent. Emissions from these three sectors primarily consisted of CO_2 emissions from fossil fuel combustion. CO_2 was also emitted and sequestered by a

Table 3-7 **U.S. Greenhouse Gas Emissions Allocated to Economic Sectors** (Tg CO₂e)

Between 2007 (2010 CAR data) and 2011, U.S. GHG emissions from major economic sectors decreased by 560.9 Tg CO₂e, or 7.7 percent. The long-term decline in these emissions has been due to structural changes in the U.S. economy, fuel switching, and energy efficiency improvements.

Implied Sectors	1990	2005	2007	2008	2009	2010	2011
Electric Power Industry	1,866.1	2,445.7	2,455.6	2,402.0	2,187.6	2,303.0	2,200.9
Transportation	1,553.2	2,012.3	2,013.1	1,916.0	1,840.6	1,852.2	1,829.4
Industry	1,538.8	1,416.2	1,456.1	1,398.8	1,244.2	1,331.8	1,332.0
Agriculture	458.0	517.4	555.6	535.3	525.4	528.7	546.6
Commercial	388.1	374.1	372.0	380.9	382.9	376.9	378.0
Residential	345.4	371.3	358.2	366.0	358.1	359.6	357.3
U.S. Territories	33.7	58.2	52.6	49.8	47.9	58.0	58.0
Total Emissions	**6,183.3**	**7,195.3**	**7,263.2**	**7,048.8**	**6,586.6**	**6,810.3**	**6,702.3**
Land Use, Land-Use Change, and Forestry (Sinks)	(794.5)	(997.8)	(929.2)	(902.6)	(882.6)	(888.8)	(905.0)
Net Emissions (Sources and Sinks)	**5,388.7**	**6,197.4**	**6,334.0**	**6,146.2**	**5,704.0**	**5,921.5**	**5,797.3**

Note: Totals may not sum due to independent rounding. Emissions include CO₂, CH₄, N₂O, HFCs, PFCs, and SF₆. Parentheses indicate negative values or sequestration.

Figure 3-13 **U.S. Greenhouse Gas Emissions Allocated to Economic Sectors**

In 2011, electricity generation accounted for the largest portion (33 percent) of U.S. GHG emissions, transportation activities accounted for 27 percent, and industry accounted for 20 percent. In contrast to electricity generation and transportation, emissions from industry have generally declined over the past decade.

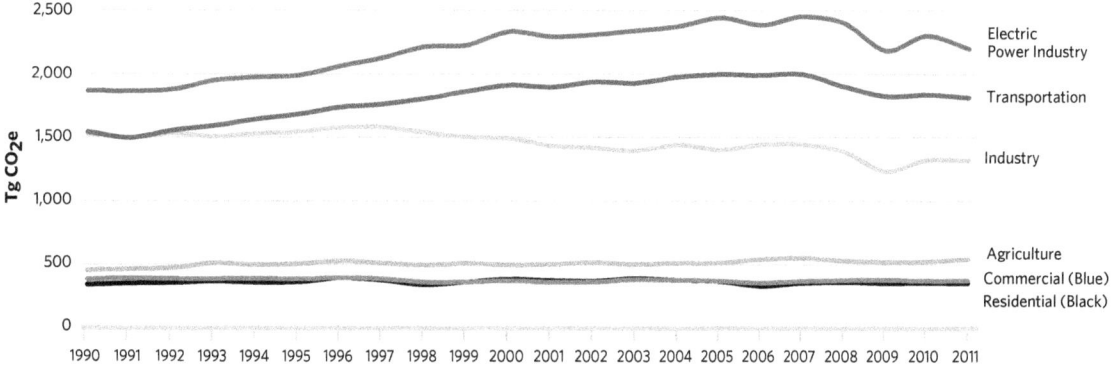

variety of activities related to forest management practices, tree planting in urban areas, the management of agricultural soils, and landfilling of yard trimmings.

Electricity is ultimately consumed in the economic sectors described above. Table 3-8 presents GHG emissions from economic sectors with emissions related to electricity generation distributed into end-use categories (i.e., emissions from electricity generation are allocated to the economic sectors in which the electricity is consumed). To distribute electricity emissions among end-use sectors, emissions from the source categories assigned to electricity generation were allocated to the residential, commercial, industry, transportation, and agriculture economic sectors according to retail sales of electricity.[11] These source categories include

[11] Emissions were not distributed to U.S. territories, since the electricity generation sector only includes emissions related to the generation of electricity in the 50 U.S. states and the District of Columbia.

CO$_2$ from fossil fuel combustion and the use of limestone and dolomite for flue gas desulfurization, CO$_2$ and N$_2$O from incineration of waste, CH$_4$ and N$_2$O from stationary sources, and SF$_6$ from electrical transmission and distribution systems.

When emissions from electricity are distributed among these sectors, industrial activities accounted for the largest share of U.S. GHG emissions (28 percent) in 2011. Transportation is the second-largest contributor to total U.S. GHG emissions (27 percent), and the residential and commercial sectors contributed the next-largest shares in 2011. Emissions from these sectors increase substantially when emissions from electricity are included, due to their relatively large share of electricity consumption (e.g., lighting, appliances). In all sectors except agriculture, CO$_2$ accounts for more than 80 percent of GHG emissions, primarily from the combustion of fossil fuels. Figure 3-14 and Box 3-3 show the trend in these emissions by sector from 1990 to 2011.

Table 3-8 **U.S Greenhouse Gas Emissions by Economic Sector with Electricity-Related Emissions Distributed** (Tg CO$_2$e)

In 2011, after distributing emissions from electricity generation to end-use sectors, industry accounted for 28.3 percent of total U.S. GHG emissions, and the transportation sector accounted for 27.4 percent.

Implied Sectors	1990	2005	2007	2008	2009	2010	2011
Industry	2,181.3	2,102.4	2,113.6	2,036.3	1,789.8	1,916.9	1,897.2
Transportation	1,556.3	2,017.2	2,018.2	1,920.8	1,845.2	1,856.9	1,833.7
Residential	939.5	1,192.4	1,215.6	1,211.1	1,150.8	1,165.2	1,131.0
Commercial	953.1	1,243.6	1,237.1	1,223.6	1,159.6	1,216.3	1,169.8
Agriculture	519.3	581.5	626.2	607.1	593.3	597.1	612.6
U.S. Territories	33.7	58.2	52.6	49.8	47.9	58.0	58.0
Total Emissions	**6,183.3**	**7,195.3**	**7,263.2**	**7,048.8**	**6,586.6**	**6,810.3**	**6,702.3**
Land Use, Land-Use Change, and Forestry (Sinks)	(794.5)	(997.8)	(929.2)	(902.6)	(882.6)	(888.8)	(905.0)
Net Emissions (Sources and Sinks)	**5,388.7**	**6,197.4**	**6,334.0**	**6,146.2**	**5,704.0**	**5,921.5**	**5,797.3**

Note: Parentheses indicate negative values or sequestration.

Figure 3-14 **U.S. Greenhouse Gas Emissions with Electricity Distributed to Economic Sectors**

In 2011, after distributing emissions from electricity the major economic sectors, industrial activities accounted for 28 percent, and transportation accounted for 27 percent. In all sectors, GHG emissions declined from 2007 (2010 CAR data) to 2011.

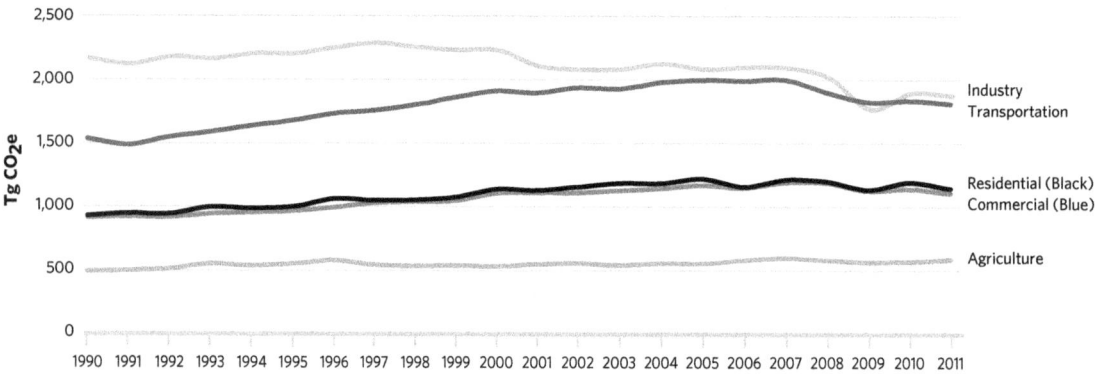

Box 3-3 **Recent Trends in Various U.S. Greenhouse Gas Emissions-Related Data**

Total emissions can be compared with other economic and social indices to highlight changes over time. These comparisons include (1) emissions per unit of aggregate energy consumption, because energy-related activities are the largest sources of emissions; (2) emissions per unit of fossil fuel consumption, because almost all energy-related emissions involve the combustion of fossil fuels; (3) emissions per unit of electricity consumption, because the electric power industry—utilities and nonutilities combined—was the largest source of U.S. GHG emissions in 2011; (4) emissions per unit of total gross domestic product (GDP) as a measure of national economic activity; and (5) emissions per capita.

Table 3-9 provides data on various statistics related to U.S. GHG emissions normalized to 1990 as a baseline year. U.S. GHG emissions have grown at an average annual rate of 0.4 percent since 1990. This rate is slightly faster than that for total energy and fossil fuel consumption, and much slower than that for electricity consumption, overall GDP, and national population (Figure 3-15).

Table 3-9 **Recent Trends in Various U.S. Data** (Index 1990 = 100)

Since 1990, U.S. GHG emissions have grown at an average annual rate of 0.4 percent—slightly faster than the rate for total energy and fossil fuel consumption, and much slower than that for electricity consumption, overall GDP, and national population.

Variable	1990	2005	2007	2008	2009	2010	2011	Growth Rate[a]
Gross Domestic Product[b]	100	157	165	164	159	163	166	2.5%
Electricity Consumption[c]	100	134	137	136	131	137	136	1.5%
Fossil Fuel Consumption[c]	100	119	119	116	109	112	101	0.1%
Energy Consumption[c]	100	119	120	117	111	115	102	0.1%
Population[d]	100	118	121	122	123	124	125	1.1%
Greenhouse Gas Emissions[e]	100	116	117	114	107	110	108	0.4%

[a] Average annual growth rate.
[b] GDP in chained 2005 dollars (BEA 2012).
[c] Energy content-weighted values (EIA 2013).
[d] U.S. Census Bureau (2012).
[e] Global warming potential-weighted values.

Figure 3-15 **U.S. Greenhouse Gas Emissions per Capita and per Dollar of Gross Domestic Product**

Between 1990 and 2011, U.S. GHG emissions per capita and per dollar of GDP declined, despite increases in real GDP and population.

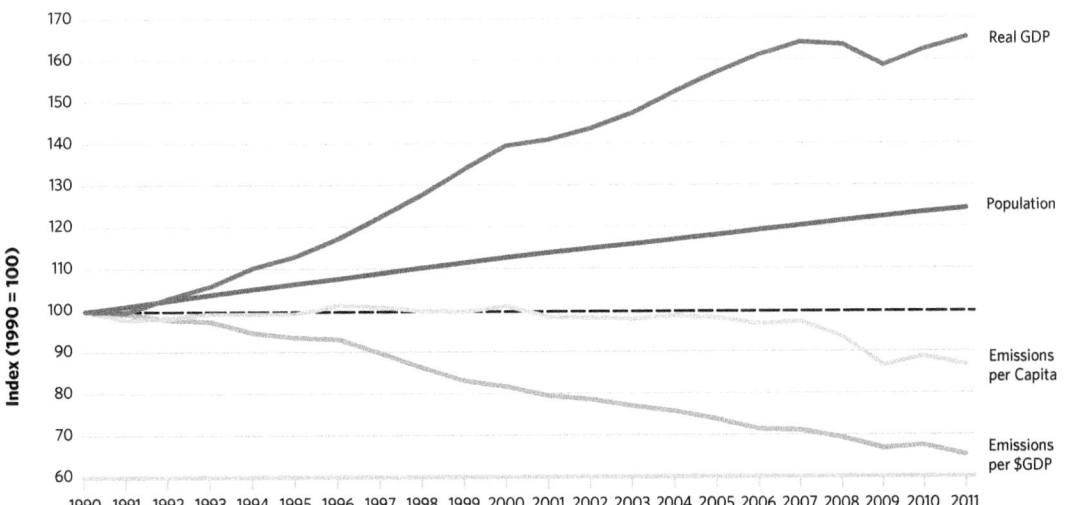

Sources: BEA 2012, U.S. Census Bureau 2012, and emission estimates in U.S. EPA/OAP 2013.

INDIRECT GREENHOUSE GAS EMISSIONS

The reporting requirements of the UNFCCC request that information be provided on indirect GHGs, which include CO, NO_x, NMVOCs, and SO_2 (UNFCCC 2006). These gases do not have a direct global warming effect, but indirectly affect terrestrial radiation absorption by influencing the formation and destruction of tropospheric and stratospheric ozone, or, in the case of SO_2, by affecting the absorptive characteristics of the atmosphere. Additionally, some of these gases may react with other chemical compounds in the atmosphere to form compounds that are GHGs.

CO is produced when carbon-containing fuels are combusted incompletely. NO_x (i.e., NO and NO_2) is created by lightning, fires, fossil fuel combustion, and in the stratosphere from N_2O. NMVOCs—which include hundreds of organic compounds that participate in atmospheric chemical reactions (e.g., propane, butane, xylene, toluene, ethane)—are emitted primarily from transportation, industrial processes, and nonindustrial consumption of organic solvents. In the United States, SO_2 is primarily emitted from coal combustion for electric power generation and the metals industry. Sulfur-containing compounds emitted into the atmosphere tend to exert a negative radiative forcing (i.e., cooling); therefore, they are discussed separately.

One important indirect climate change effect of NMVOCs and NO_x is their role as precursors for tropospheric ozone formation. They can also alter the atmospheric lifetimes of other GHGs. Another example of indirect GHG formation into direct GHGs is CO's interaction with the hydroxyl radical—the major atmospheric sink for CH_4 emissions—to form CO_2. Therefore, increased atmospheric concentrations of CO limit the number of hydroxyl molecules (OH) available to destroy CH_4.

Since 1970, the United States has published estimates of emissions of CO, NO_x, NMVOCs, and SO_2 (U.S. EPA/OAQPS 2009, 2010),[12] which are regulated under the Clean Air Act.[13] Table 3-10 shows that fuel combustion accounts for the majority of emissions of these indirect GHGs. Industrial processes—such as the manufacture of chemical and allied products, metals processing, and industrial uses of solvents—are also significant sources of CO, NO_x, and NMVOCs.

[12] NO_x and CO emission estimates from field burning of agricultural residues were estimated separately. Therefore, they were not taken from U.S. EPA/OAQPS 2009 or U.S. EPA/OAQPS 2010.

[13] Due to redevelopment of the information technology systems for the National Emission Inventory (NEI), publication of the most recent emissions for these pollutants was not available for the *Inventory of U.S. Greenhouse Gas Emissions and Sinks: 1990-2011* (U.S. EPA/OAP 2013). For an overview of the activities and the schedule for developing the 2011 NEI, with the goal of producing Version 1 in the summer of 2013, *see* EPA's NEI Plan of Activities at http://www.epa.gov/ttn/chief/eis/2011nei/2011plan.pdf.

Table 3-10 **Emissions of NO$_x$, CO, NMVOCs, and SO$_2$** (Tg)

Fuel combustion accounts for the majority of emissions of indirect GHGs. Industrial processes and industrial uses of solvents are also significant sources of CO, NO$_x$, and NMVOCs.

Gas/Activity	1990	2005	2007	2008	2009	2010	2011
Nitrogen Oxides (NO$_x$)	**21.7**	**15.9**	**14.4**	**13.5**	**11.5**	**11.5**	**11.5**
Mobile Fossil Fuel Combustion	10.9	9.0	8.0	7.4	6.2	6.2	6.2
Stationary Fossil Fuel Combustion	10.0	5.9	5.4	5.1	4.2	4.2	4.2
Industrial Processes	0.6	0.6	0.5	0.5	0.6	0.6	0.6
Oil and Gas Activities	0.1	+	+	+	+	+	+
Waste Combustion	0.1	+	+	+	+	+	+
Agricultural Burning	+	+	+	+	+	+	+
Solvent Use	+	+	+	+	+	+	+
Waste	+	+	+	+	+	+	+
Carbon Monoxide (CO)	**130.0**	**70.8**	**63.6**	**60.0**	**51.4**	**51.4**	**51.4**
Mobile Fossil Fuel Combustion	119.4	62.7	55.3	51.5	43.4	43.4	43.4
Stationary Fossil Fuel Combustion	5.0	4.6	4.7	4.8	4.5	4.5	4.5
Industrial Processes	4.1	1.6	1.6	1.7	1.5	1.5	1.5
Waste Combustion	1.0	1.4	1.4	1.4	1.4	1.4	1.4
Oil and Gas Activities	0.3	0.3	0.3	0.3	0.3	0.3	0.3
Agricultural Burning	0.2	0.2	0.2	0.2	0.2	0.2	0.2
Waste	+	+	+	+	+	+	+
Solvent Use	+	+	+	+	+	+	+
Non-Methane Volatile Organic Compounds (NMVOCs)	**20.9**	**13.8**	**13.4**	**13.3**	**9.3**	**9.3**	**9.3**
Mobile Fossil Fuel Combustion	10.9	6.3	5.7	5.4	4.2	4.2	4.2
Solvent Use	5.2	3.9	3.8	3.8	2.6	2.6	2.6
Industrial Processes	2.4	2.0	1.9	1.8	1.3	1.3	1.3
Oil and Gas Activities	0.6	0.5	0.5	0.5	0.6	0.6	0.6
Stationary Fossil Fuel Combustion	0.9	0.7	1.1	1.3	0.4	0.4	0.4
Waste Combustion	0.2	0.2	0.2	0.2	0.2	0.2	0.2
Waste	0.7	0.1	0.1	0.1	+	+	+
Agricultural Burning	NA	NA	NA	NA	NA	NA	NA
Sulfur Dioxide (SO$_2$)	**20.9**	**13.5**	**11.8**	**10.4**	**8.6**	**8.6**	**8.6**
Stationary Fossil Fuel Combustion	18.4	11.5	10.2	8.9	7.2	7.2	7.2
Industrial Processes	1.3	0.8	0.8	0.8	0.8	0.8	0.8
Mobile Fossil Fuel Combustion	0.8	0.9	0.6	0.5	0.5	0.5	0.5
Oil and Gas Activities	0.4	0.2	0.2	0.2	0.2	0.2	0.2
Waste Combustion	+	+	+	+	+	+	+
Waste	+	+	+	+	+	+	+
Solvent Use	+	+	+	+	+	+	+
Agricultural Burning	NA	NA	NA	NA	NA	NA	NA

Sources: U.S. EPA 2009 and 2010, except for estimates from field burning of agricultural residues.
+ Does not exceed 0.5 Tg.
Notes: Totals may not sum due to independent rounding. NA = Not Available.

Policies and Measures

INTRODUCTION

Over the past four years, the United States has taken a series of important steps that will reduce the harmful emissions that contribute to climate change, improve public health, and protect the environment. At the federal level, the United States has made significant progress in reducing greenhouse gas (GHG) emissions, including through establishing historic new fuel economy standards for cars and trucks. Other policies and measures have reduced GHG emissions and consumer energy bills through energy efficiency measures, doubling generation of electricity from wind, solar, and geothermal energy, and improving sustainability for federal facilities. These policies and measures reduce pollution and speed the transition to more sustainable sources of energy, industrial processes, and waste management practices.

Within the United States, several regional, state, and local initiatives complement federal efforts to reduce GHG emissions. These include a wide range of policies that affect the energy and transportation sectors, among many others, from direct regulation of GHGs to policies that indirectly reduce emissions. State, tribal, and local governments also have unique authorities to address climate change apart from the federal government, particularly in regulating land-use planning decisions. Taken together, state, local, and federal actions create a broad policy framework to reduce emissions and spur investments in cleaner energy and energy efficiency.

Building on important progress achieved during his first term, in June 2013 President Obama released a *Climate Action Plan* to further reduce the nation's GHG emissions (EOP 2013a). This plan lays out additional executive actions the United States will take—in partnership with states, tribes, communities, and the private sector—to meet the ambitious commitment of reducing U.S. GHG emissions in the range of 17 percent below 2005 levels by 2020. The plan includes such steps as establishing the first-ever carbon pollution standards for both new and existing power plants; setting a new goal to once again double electricity generation from wind, solar, and geothermal power; reducing emissions of highly potent hydrofluorocarbons (HFCs); developing a comprehensive methane (CH_4) emission reduction strategy; and efforts to protect America's forests and critical landscapes. Each of these actions is discussed in further detail in the preceding *Biennial Report*.

This chapter outlines and discusses policies and measures in the following key areas:

* Federal policies and measures, including actions in the transportation, energy, industrial, agricultural, forestry, and waste management sectors, and federal government actions and cross-cutting initiatives.

* Nonfederal policies and measures, including regional, state, and local actions to address climate change.

This *Sixth National Communication* addresses a broader scope of policies and measures than the preceding *Biennial Report*.

FEDERAL POLICIES AND MEASURES

The United States utilizes a combination of near- and long-term regulatory and voluntary activities for climate mitigation. Policies and measures are being implemented across the economy, including in the transportation, energy supply, energy end use, industrial processes, agricultural, waste, and federal facilities sectors. In addition, the United States utilizes cross-cutting policies and measures to encourage cost-effective reductions across multiple sectors. Although significant GHG reductions have been made through existing initiatives, the United States recognizes that opportunities continue to arise to expand and build upon existing regulatory and voluntary programs for further GHG emission reductions. Table 4-1 summarizes the key new initiatives since the *U.S. Climate Action Report 2010* (2010 CAR) (U.S. DOS 2010), including implementation of several new regulatory policies across the transportation, energy, and industrial (non-carbon dioxide [CO_2]) sectors since 2010.

The remainder of this section discusses these and the other new and existing U.S. climate mitigation policies and measures. Policies and measures are organized by sector, listing newly adopted policies and measures, and those with the most significant effect on GHG mitigation first. Table 4-2 at the end of this Federal Policies and Measures section presents estimates of GHG emission reductions. All efforts contribute directly or indirectly to GHG emission reductions, even though many policies and measures are being advanced for other primary purposes, such as to reduce other harmful pollutants; to improve sustainability, economic

Table 4-1   **Significant New Policies Adopted and Key Implementation Progress Made Since CAR 2010**

Since 2010, the United States has been implementing several new regulatory policies that are reducing GHG emissions across the transportation, energy, and industrial (non-carbon dioxide) sectors.

Sector	Policy/Measure	Description of Activity Since CAR 2010
Transport	National Program for Light-Duty Vehicle GHG Emissions and Corporate Average Fuel Economy Standards	The combined model year (MY) 2012–2025 standards are expected to effectively cut in half vehicle GHG emissions, reducing 6,000 teragrams (Tg) of GHGs over the lifetimes of the vehicles sold in MYs 2012–2025—more than the total amount of carbon dioxide (CO_2) emitted by the United States in 2010.
Transport	National Program for Heavy-Duty Vehicle GHG Emissions and Fuel Efficiency Standards	The MY 2014–2018 standards are expected to significantly reduce GHG emissions and fuel consumption from heavy-duty vehicles. The national program will cut 270 Tg of carbon dioxide equivalents (CO_2e) of GHG emissions during the lifetimes of the vehicles sold in MYs 2014–2018.
Energy (Supply)	Carbon Pollution Standard for Future Power Plants	In September 2013, EPA proposed a carbon pollution standard for future fossil-fuel power plants. Power plants account for approximately 40 percent of all U.S. CO_2 emissions, and represent the single-largest source of industrial GHG emissions in the nation.
Energy (Residential, Commercial, Industrial)	Appliance and Equipment Energy Efficiency Standards	Since 2009, 17 new or updated federal standards have been issued, which will help increase annual energy savings by more than 50 percent over the next decade. Products covered by the standards represent about 90 percent of home energy use, 60 percent of commercial building use, and 29 percent of industrial energy use.
Energy (Residential, Commercial, Industrial)	Lighting Energy Efficiency Standards	These standards will phase out the 130-year-old incandescent light bulb by the middle of the next decade and phase out less efficient fluorescent tubes. They are estimated to have a GHG mitigation potential of 36.3 Tg CO_2e in 2015 and 37.7 Tg CO_2e in 2020.
Industrial Processes (Non-CO_2)	Federal Air Standards for the Oil and Natural Gas Industry	On April 17, 2012, EPA issued cost-effective regulations to reduce harmful air pollution from the oil and natural gas industry, while allowing continued, responsible growth in U.S. oil and natural gas production. These regulatory standards are projected to achieve a significant co-benefit of methane emission reductions, estimated at 32.6 Tg CO_2e in 2015 and 39.9 Tg CO_2e in 2020.
Cross-Cutting	Best Available Control Technology for GHG Emissions	In May 2010, EPA issued a regulation establishing a common-sense approach to permitting GHG emissions. As of April 2013, EPA and states have issued nearly 90 permits to large industrial sources that cover GHG emissions.

growth, and rural development; and to spur the development and deployment of new technologies. Similar policies and measures are addressed together in some instances, to convey the comprehensive approach being deployed at the federal level. Please refer to Table 4-2 for descriptions and mitigation estimates of individual measures.

Transportation Sector

U.S. federal policies and measures to reduce GHGs from the transportation sector leverage a mix of regulatory, voluntary, and informational approaches with the greatest estimated mitigation impact from regulatory instruments. Programs are being implemented across multiple federal agencies to improve vehicle efficiency, increase the use of renewable fuels, and encourage the adoption of new technologies and practices. See Table 4-2 for estimates of GHG emission reductions in the transportation sector.

National Program for Light-Duty Vehicle GHG Emissions and CAFE Standards[1]

Responding to the country's critical need to address global climate change and reduce oil consumption, President Obama directed the U.S. Environmental Protection Agency (EPA) and the U.S. Department of Transportation's (DOT's) National Highway Traffic Safety Administration (NHTSA) to work closely together to establish a harmonized National Program for Corporate Average Fuel Economy (CAFE) Standards and GHG standards for light-duty vehicles (LDVs).[2] In April 2010, NHTSA and EPA issued a joint final rule establishing the first phase of standards for model year (MY) 2012–2016 cars and light trucks. At the time of the final rule, the MY 2012–2016 standards were projected to result in an average industry fleetwide tailpipe CO_2 level of 250 grams/mile (g/mi) by MY 2016, including expected reductions in hydrofluorocarbon (HFC) emissions from air conditioners. This would be equivalent to 57.1 kilometers (km) (35.5 mi) per gallon if achieved exclusively through fuel economy improvements. The standards represent the first time EPA promulgated federal emission standards for GHGs using its authority under the Clean Air Act (CAA), and also represent one of the largest increases in stringency since the inception of the CAFE program in the 1970s.

Building on the success of the first phase of the National Program, on July 29, 2011, President Obama announced the second phase of standards for MY 2017–2025 cars and light trucks. Thirteen automobile manufacturers representing more than 90 percent of U.S. vehicle sales announced their support for the program, as well as the United Auto Workers Union. NHTSA and EPA issued a joint final rule establishing these new standards in August 2012. At the time of the final rule, EPA projected the MY 2025 standards would result in an average industry fleetwide level of 163 g/mi of CO_2 in MY 2025, again, including expected reductions in HFC emissions. This would be equivalent to 87.7 km (54.5 mi)/gallon if achieved exclusively through fuel economy improvements.

In total, the standards issued by NHTSA and EPA under the Obama administration are projected to effectively cut in half vehicle GHG emissions and double average vehicle fuel efficiency compared with MY 2011 cars and light trucks. The National Program is projected to save American families more than $1.7 trillion in fuel costs and reduce America's dependence on foreign oil by more than 2 million barrels per day (bpd) in 2025. In addition, the program is expected to cut 6 billion metric tons of GHGs during the lifetimes of the cars and light trucks sold in MYs 2012–2025—more than the total amount of U.S. CO_2 emissions in 2010.

Because California harmonized its state requirements with the federal program, the National Program also ensures that automobile manufacturers can build a single fleet of U.S. vehicles that satisfy the requirements of both the federal and California emission control programs.[3] This will help to reduce costs and regulatory complexity, while providing significant energy security and environmental benefits to the nation as a whole. The program design also ensures that consumers still have a full range of vehicle choices (Box 4-1).

Because the standards for MYs 2022–2025 were set so far in advance, and because NHTSA is legally required to issue CAFE standards in no more than five years at a time, EPA committed in the MY 2017–2025 rule to conduct a mid-term evaluation of the MY 2022–2025 standards, which will be undertaken in a few years concurrent with NHTSA's rulemaking to set final standards for those model years. The evaluation, which will be based on the most

[1] See http://www.nhtsa.gov/ fueleconomy; and http://epa.gov/otaq/ climate/regs-light-duty.htm.

[2] See http://www.whitehouse.gov/ the_press_office/President-Obama-Announces-National-Fuel-Efficiency-Policy/.

[3] California approved an emission control program that reduces GHGs under a single package of standards called Advanced Clean Cars.

Box 4-1 **Assessing the Economic Benefits of Policy Measures**

The U.S. government analyzes the anticipated economic effects of its proposed standards and policies. A key element of these analyses has been the estimation of the potential economic and human welfare benefits of reduced GHGs. Specifically, federal agencies use a metric known as the social cost of carbon (SCC) to estimate the dollar value of the benefits of regulatory actions that affect CO_2 emissions.

The SCC is a present-value calculation of the avoided worldwide damages—e.g., the benefits associated with a 1-ton decrease in CO_2 emissions in a particular year—and thus the value of the benefits from a commensurate reduction in emissions. It is meant to be a comprehensive measure, including losses due to changes in net agricultural productivity, human health risks, property damages from increased flood risk, and the loss of ecosystem services.

In 2010, in an effort to promote consistency in how federal agencies calculate the social benefits of reducing CO_2 emissions, the U.S. government selected four SCC values for use in regulatory analyses. These values were first used in DOE's energy conservation standards for small motors in 2010. The U.S. government updated its SCC estimates in 2013 to reflect how climate change impacts are represented in the latest peer-reviewed versions of the three academic models from which the SCC is estimated.[4] The four SCC estimates for 2020 are: $12, $43, and $64 per metric ton (average SCC at discount rates of 5, 3, and 2.5 percent, respectively) and $128 per metric ton (95th percentile SCC at a 3 percent discount rate) in 2007 dollars.

up-to-date information available on technology availability, cost, and all other relevant factors, could lead federal agencies to make the final standards for MYs 2022–2025 more stringent, less stringent, or unchanged from their current levels.

National Program for Heavy-Duty Vehicle GHG Emissions and Fuel Efficiency Standards[5]

Heavy-duty vehicles (HDVs) are a significant contributor to GHG emissions and fuel consumption from the U.S. transportation sector. The contribution from these vehicles is second only to LDVs in this sector. In May 2010, President Obama directed EPA and NHTSA to develop a joint rulemaking to establish fuel efficiency and GHG emission standards for commercial medium- and heavy-duty on-highway vehicles and work trucks beginning in MY 2014.[6] In September 2011, EPA and NHTSA issued the joint final rule establishing the first phase of standards for MY 2014–2018 HDVs.

The MY 2014–2018 standards are expected to achieve up to a 23 percent reduction in GHG emissions and fuel consumption for semis (combination trucks), and up to a 9 percent reduction for buses; special-purpose trucks, such as garbage trucks; and other vocational vehicles. The Heavy-Duty National Program is estimated to save truck owners more than $50 billion in fuel costs and reduce America's dependence on oil by more than 530 million barrels. In addition, the program will cut a projected 270 million metric tons (MMt) of GHG emissions over the lifetimes of the vehicles sold in MYs 2014–2018.

Renewable Fuel Standard Program[7]

The Energy Independence and Security Act of 2007 (EISA) made several changes to the Renewable Fuel Standard (RFS) program, which was originally implemented under the Energy Policy Act (EPAct) of 2005. These changes included a significant increase in the volume of renewable fuel that must be used in transportation fuel each year. By 2022, 36 billion gallons of renewable fuel are required annually—a fivefold increase over the volumes included in EPAct. The statute also includes volume requirements for biomass-based diesel and other advanced biofuels, including 16 billion gallons of cellulosic biofuel annually by 2022. The revised requirements also include new definitions and criteria for both renewable fuels and the feedstocks used to produce them, including new life-cycle GHG emission thresholds for renewable fuels.

EPA, which issued a final rule in February 2010, is currently implementing the RFS program, including continually increasing the number of pathways (combinations of biofuel feedstock, production technologies, and fuels produced) as the program matures and opportunities for biofuel production expand. The RFS program is anticipated to achieve significant reductions in both petroleum use and GHGs.

[4] *See* http://www.whitehouse.gov/sites/default/files/omb/inforeg/social_cost_of_carbon_for_ria_2013_update.pdf.

[5] *See* http://www.nhtsa.gov/About+NHTSA/Press+Releases/2011/White+House+Announces+First+Ever+Oil+Savings+Standards+for+Heavy+Duty+Trucks,+Buses; and http://www.epa.gov/otaq/climate/regs-heavy-duty.htm.

[6] *See* http://www.whitehouse.gov/the-press-office/presidential-memorandum-regarding-fuel-efficiency-standards.

[7] *See* www.epa.gov/otaq/alternative-renewablefuels.

SmartWay Transport Partnership[8]

The SmartWay Transport Partnership is an innovative collaboration with businesses and other stakeholders to decrease climate and other emissions from the movement of goods, by increasing energy efficiency while significantly reducing GHGs and air pollution. EPA provides tools and models to help SmartWay Transport Partners—including shippers and the trucking, rail, and marine shipping companies that deliver their goods—to adopt cost-effective strategies to save fuel and reduce GHG emissions.

To date, more than 3,000 companies and organizations have joined the SmartWay partnership. Freight shippers meet their goals by selecting the greenest carriers and modes to fit their shipping needs, while trucking and rail companies meet their goals by improving freight transport efficiency.

The SmartWay program is also working with other governments and organizations around the world to establish international benchmarks for clean, efficient freight transportation. In 2012, EPA and Natural Resources Canada announced the expansion of SmartWay into Canada. EPA estimates that SmartWay could help the shipping industry reduce up to 43 teragrans of carbon dioxide equivalent (Tg CO_2e) emissions by 2020.

Other SmartWay initiatives include the evaluation of fuel-saving technologies and SmartWay designation of efficient heavy-duty trucks and trailers. SmartWay-designated tractor-trailers can save 10–20 percent annually in fuel and CO_2 emissions compared with typical long-haul trucks. SmartWay also promotes idle-reduction programs for trucks and locomotives and has developed guidance on idle reduction policies and programs for states. SmartWay's Supply Chain initiative is developing new tools to help companies quantify and track freight transport environmental performance across all modes, including truck, marine, rail, and aviation.

Light-Duty Vehicle Fuel Economy and Environment Label[9]

Building on EPA's 35-year history of labeling vehicles, EPA and DOT redesigned the Fuel Economy and Environment labels found on all new vehicles. The labels have historically provided information on fuel economy and annual fuel costs that can be compared across all vehicles. The redesigned labels continue this tradition, and additionally provide information on energy use, relative cost of refueling, and environmental ratings for GHGs and smog. The labels can be compared across all new vehicles, including advanced technologies, such as plug-in hybrids and electric vehicles. This information allows the car-buying public to take into account fuel and environmental considerations, including GHG emissions and relative refueling costs.

National Clean Diesel Campaign[10]

EPA's National Clean Diesel Campaign (NCDC) works aggressively to reduce diesel emissions across the country, through the implementation of proven emission control technologies and innovative strategies with the involvement of national, state, and local partners. Many of the clean diesel strategies that NCDC promotes to mitigate nitrogen oxides (NO_x) and particulate matter (PM)—such as retrofits, engine repair, engine replacement, engine repower, idle reduction, and cleaner fuels—can also reduce CO_2 emissions through diesel fuel savings and help mitigate black carbon emissions. Black carbon, a component of PM, has been found to both increase atmospheric warming and speed Arctic melting. Removing PM may have a significant effect on slowing global warming due to the short-lived nature of black carbon.

The Diesel Emissions Reduction Act (DERA) provisions in EPAct are a significant funding source for NCDC. EPA DERA grants fund projects that provide immediate health and environmental benefits. During their lifetime, projects funded in fiscal years (FYs) 2008–2010 are estimated to reduce CO_2 by 2.3 MMt, as well as provide fuel savings of more than 205 million gallons.

Advanced Technology Vehicles Manufacturing Loan Program[11]

DOE's Loan Program Office (LPO) manages the Advanced Technology Vehicles Manufacturing (ATVM) Loan Program, authorized under Section 136 of EISA. ATVM provides direct loans to support re-equipping, expanding, or establishing manufacturing facilities

---

[8] *See* www.epa.gov/smartway.

[9] *See* www.epa.gov/carlabel.

[10] *See* www.epa.gov/cleandiesel.

[11] *See* https://lpo.energy.gov/?page_id=43.

in the United States to produce advanced-technology vehicles or qualifying components and to support U.S. engineering integration projects. Auto manufacturers and qualifying component manufacturers are eligible to apply for loans. For example, ATVM loans have been used to upgrade and retool several factories across the United States to produce advanced batteries and raise fuel efficiency in more than a dozen popular vehicle models.

Aviation Low Emissions, Fuel Efficiency, and Renewable Fuels Measures[12]

The Federal Aviation Administration (FAA) is pursuing a comprehensive approach to reduce GHG emissions from commercial aviation through aircraft and engine technology development, operational improvements, development and deployment of sustainable alternative jet fuels, and additional policies and measures. FAA's Next Generation Air Transportation System Plan, or NextGen, focuses its efforts on increasing efficient aircraft operations and reducing GHG emissions through airspace, operational, and infrastructure improvements.

FAA funds diverse programs to improve aviation energy and emissions performance, and coordinates with other agencies as appropriate, including the National Aeronautics and Space Administration. Following are some examples of FAA programs:

- The Continuous Lower Energy, Emissions, and Noise program is a collaborative partnership between FAA and five aviation manufacturers to develop technologies that will reduce emissions and fuel burn, and expedite the integration of these technologies into current aircraft.

- The Aviation Climate Change Research Initiative (ACCRI) is an FAA program that provides guidance to develop mitigation solutions based on state-of-the-art science results. ACCRI results are key to quantifying cost–benefit analyses of various policy options. ACCRI has reduced uncertainties, leading to overall improvement in understanding of the climate impacts of aviation. While ACCRI does not provide mitigation solutions on its own, recently completed ACCRI Phase II results can be used to help identify effective mitigation options.

- The Voluntary Airport Low Emissions Program (VALE) is a grant program that encourages airport sponsors to use Airport Improvement Program funds and Passenger Facility Charges to finance low-emission vehicles, refueling and recharging stations, gate electrification, and other airport air quality improvements. Under FAA's most recent reauthorization, VALE's work is supplemented by new programs that reduce airport emissions. FAA is creating a program where, following an assessment of airport energy requirements, FAA may make capital grants for airports to increase energy efficiency. FAA has also established a pilot program under which certain airports may acquire and operate zero-emission vehicles.

In addition, FAA is a founding member of the Commercial Aviation Alternative Fuels Initiative (CAAFI). CAAFI is a public–private partnership established in 2006 with the objective of advancing alternative jet fuels with equivalent safety/performance (drop-in) and comparable cost, environmental improvement, and security of energy supply for aviation. Work through CAAFI has also expanded internationally. Fuel production capability is beginning to emerge, including a recently announced airline and fuel producer agreement.

State and Alternative Fuel Provider Fleet Program[13]

Through its Vehicle Technologies Office (VTO), the U.S. Department of Energy (DOE) manages the State and Alternative Fuel Provider Fleet Program, which aims to reduce U.S. petroleum consumption by building a core market for alternative fuel vehicles (AFVs). The program requires covered fleets owned or controlled by states or by alternative fuel providers either to acquire AFVs as a percentage of their annual LDV acquisitions or to employ other petroleum-reduction methods in lieu of acquiring AFVs.

VTO also supports several key initiatives that accelerate the deployment of clean, cutting-edge advanced highway transportation technologies that reduce petroleum consumption and GHG emissions. The Electric Vehicle Everywhere Grand Challenge, a bold DOE-wide initiative, seeks by 2022 to make the United States the first country to produce a wide array of plug-in electric vehicle models that are as affordable and convenient as today's gasoline-powered vehicles. A

[12] *See* http://www.faa.gov/news/fact_sheets/news_story.cfm?newsId=10112; http://www.faa.gov/about/office_org/headquarters_offices/apl/environ_policy_guidance/policy/media/Aviation_Greenhouse_Gas_Emissions_Reduction_Plan.pdf; and http://www.faa.gov/airports/environmental/vale/.

[13] *See* http://www1.eere.energy.gov/vehiclesandfuels/epact/about.html and http://www1.eere.energy.gov/vehiclesandfuels/electric_vehicles/index.html.

companion Workplace Charging Challenge will encourage private-sector leadership in the deployment of convenient plug-in vehicle charging for consumers.

Federal Transit, Highway, and Railway Programs[14]

DOT's Federal Transit Administration (FTA) provides more than $10 billion a year in grants for the construction and operation of a range of transit services. While specific statutory authority for clean fuel buses and transit investments for GHG reductions was not continued as distinct programs in the surface transportation reauthorization (Moving Ahead for Progress in the 21st Century Act [MAP-21]), FTA continues to support the deployment of a range of advanced mitigation technologies for vehicles and stations, including hybrid and clean fuel transit buses, under its new authorities (Box 4-2).

Through its technical assistance efforts focused on transportation planning and transit-oriented development, FTA provides communities with the tools to effectively coordinate land-use and public transportation investment decisions. FTA also provides environmental management systems training to help transit agencies reduce the environmental impact of their operations. FTA's extensive research, development, and deployment program works to improve the efficiency and sustainability of public transportation, including supporting the demonstration and deployment of low-emission and no-emission vehicles to promote clean energy and improve air quality.

Administered by DOT's Federal Highway Administration and FTA, in consultation with EPA, the Congestion Mitigation and Air Quality Improvement Program apportions funds to states to reduce congestion and to improve air quality through transportation control measures and other transportation strategies that will contribute to attainment or maintenance of the national ambient air quality standards for ozone, carbon monoxide, and PM. Many of these projects also provide GHG emission reduction co-benefits. The projects vary by region, but typically include transit improvements, alternative fuel programs, shared-ride services, traffic flow improvements, demand management strategies, freight and intermodal facilities, diesel engine retrofits, pedestrian and bicycle programs, and inspection and maintenance programs.

DOT also uses available funding sources and opportunities to promote the development of improved passenger rail and the efficiency of freight rail transportation in the United States, notably through the Transportation Investment Generating Economic Recovery (TIGER) discretionary grant program. In addition, the Federal Rail Administration's Railroad Rehabilitation and Improvement Financing (RRIF) program authorizes up to $35 billion in direct loans and loan guarantees to improve or rehabilitate railroads. A recent RRIF loan of $54.6 million to Kansas City Southern Railway Company is enabling purchase of low-emission locomotives.

Energy Sector: Supply

Within the energy sector, numerous federal policies and measures are being implemented to reduce CO_2 emissions from energy supply sources, while also encouraging greater renewable energy resources. A mix of regulatory and economic instruments is being leveraged across multiple federal agencies. See Table 4-2 for estimates of GHG emission reductions in the energy sector.

Carbon Pollution Standard for Future Power Plants[16]

The President has directed EPA to work closely with states and other stakeholders to establish carbon pollution standards for both new and existing power plants. EPA is moving forward on the President's plan. For newly built power plants, EPA issued a new proposal on September 20, 2013. The new proposal, together with the ensuing rulemaking process, will ensure that carbon pollution standards for new power plants reflect recent developments and trends in the power sector. Also, the new proposal, comment period, and public hearings will allow an open and transparent review and robust input on the broad range of technical and legal issues contained among the more than 2.5 million comments generated by the first proposal submitted by EPA in April 2012. For existing power plants, the plan directs EPA to issue a draft rule by June 2014 and a final rule by June 2015.

Box 4-2 **Moving Ahead for Progress in the 21st Century Act**

Significant legislative activity has occurred in the transportation sector, affecting many of the federal climate mitigation measures implemented by DOT. On July 6, 2012, President Obama signed into law the Moving Ahead for Progress in the 21st Century Act (MAP-21), the first long-term highway authorization enacted since 2005.[15] MAP-21 is a milestone for the U.S. economy and the nation's surface transportation program. By transforming the policy and programmatic framework for investments to guide the transportation system's growth and development, MAP-21 creates a streamlined and performance-based surface transportation program and builds on many of the highway, transit, bike, and pedestrian programs and policies established in 1991.

[14] *See* http://www.fta.dot.gov/; http://www.fta.dot.gov/documents/PublicTransportationsRoleInRespondingToClimateChange.pdf; http://www.fra.dot.gov/Page/P0128; http://www.fta.dot.gov/grants_14835.html; http://www.fhwa.dot.gov/map21/cmaq.cfm; http://www.fta.dot.gov/13835_12125.html; and http://www.fta.dot.gov/12351_11424.html; and http://www.dot.gov/tiger.

[15] *See* Public Law (P.L.) 112-141.

[16] *See* http://epa.gov/carbonpollutionstandard/basic.html.

Box 4-3 **Climate Mitigation Co-Benefits of Legal Actions**
As part of legal actions to address CAA compliance issues at electric power plants, EPA and the U.S. Department of Justice may also include requirements to remedy, reduce, or offset harm caused by pollution previously emitted by power plants. Many of these actions provide climate mitigation co-benefits if they require investing in measures, such as renewable energy and end-use energy efficiency.

For example, the April 2011 settlement with the Tennessee Valley Authority (TVA) requires TVA to spend $350 million on environmental mitigation projects to address the impacts of past nitrogen oxide and sulfur dioxide emissions. Of this amount, TVA is allocating $280 million to energy efficiency and renewable energy programs, which TVA estimates will provide 30 million metric tons of CO_2 emission reduction co-benefits.[17]

Clean Energy Supply Programs[18]

Through the Clean Energy Supply Programs, EPA offers technical resources, develops nationally accepted standards, provides access to expertise, and recognizes environmental leadership. In turn, partner investments in clean energy yield significant environmental benefits by reducing GHG emissions and other air pollutants.

EPA's Green Power Partnership (GPP) encourages U.S. organizations to voluntarily purchase green power, offers recommended minimum levels of purchasing, and provides partners with information and recognition for their purchases and on-site renewable power systems. The program includes nearly 1,400 partners who have committed to purchasing about 25 billion kilowatt-hours of green power. In addition, the program recognizes towns, villages, cities, counties, and tribal governments that collectively buy green power in amounts that meet or exceed EPA's GPP community purchase requirements. These Green Power Communities also compete through an annual Green Power Community Challenge, which aims to increase the amount of green power used by communities nationwide.

The Combined Heat and Power Partnership (CHPP) reduces the environmental impact of power generation by encouraging the use of combined heat and power (CHP), an efficient, clean, and reliable approach to generating power and thermal energy from a single fuel source. Through the CHPP, EPA works closely with stakeholders to support the development of new projects, by providing tools and information resources, and to promote their energy, environmental, and economic benefits. The program now includes more than 450 partners and has assisted in the deployment of more than 5,500 megawatts (MW) of operational CHP. CHPP works to support balanced treatment of CHP in new or modified environmental regulations and documents, such as state and tribal air quality planning resources and output-based regulations.

Onshore Renewable Energy Development Programs[19]

The U.S. Department of the Interior (DOI) and its Bureau of Land Management (BLM) are working with communities, state regulators, industry, and other federal agencies in building a clean energy future by providing sites for environmentally sound development of renewable energy on public lands. Renewable energy projects on BLM-managed lands include wind, solar, geothermal, and biomass projects and the siting of transmission facilities needed to deliver this power to the consumer. As of May 2013, the BLM has approved 61 solar and wind energy projects with a total installed capacity of 1,421 MW, and another 11,000 MW under construction.

The BLM also manages 816 geothermal leases through its Geothermal Energy Development Program, of which 72 leases are in producing status and generating approximately 1,300 MW of capacity. In 2013, the BLM issued a regulation that allows for the segregation of lands from mining claim entry that will facilitate right-of-way applications for lands with wind and solar energy development potential. The BLM also released additional guidance documents for developing renewable energy projects on public lands, such as Best Management Practices for Reducing Visual Impacts of Renewable Energy Facilities. The BLM is also working on proposed regulations that will establish a competitive leasing process for offering lands within designated leasing areas (e.g., solar energy zones) for future solar or wind energy development.

[17] See http://www2.epa.gov/enforcement/tennessee-valley-authority-clean-air-act-settlement.

[18] See http://www.epa.gov/greenpower; and http://www.epa.gov/CHP.

[19] For solar, see http://www.blm.gov/wo/st/en/prog/energy/solar_energy.html); for geothermal, see http://www.blm.gov/wo/st/en/prog/energy/geothermal.html; for wind, see http://www.blm.gov/wo/st/en/prog/energy/wind_energy.html.

The Rural Energy for America Program[20]

USDA's Rural Energy for America Program (REAP) provides assistance to agricultural producers and rural small businesses to complete a variety of projects. By offering both loan guarantees and grants, REAP helps eligible applicants install energy systems, such as solar panels or anaerobic digesters; make energy efficiency improvements, such as installing irrigation pumps or replacing ventilation systems; and conduct energy audits and feasibility studies. The nearly 10,000 projects that have been awarded are reducing the demand for fossil fuels from conventional GHG-emitting sources. In addition, REAP has reduced GHG emissions by helping to install wind, geothermal, solar, small hydro, and anaerobic digester projects.

Carbon Capture and Storage Demonstration Plants and Large-Scale Geologic Storage Cooperative Agreements

By supporting research and development activities, DOE's Office of Fossil Energy seeks to reduce the cost of commercial deployment of carbon capture and storage (CCS) technology. DOE currently supports eight large-scale power plant and industrial CCS demonstration plants, three of which are under construction, and eight large-scale geologic storage cooperative agreements, four of which have reached the CO_2 injection stage. Cooperative agreements are a cost-shared collaboration between the federal government and private industry, aimed at stimulating investment in low-emission, coal-based power generation technology through successful commercial demonstrations.

Rural Development Biofuels Programs[21]

Several U.S. Department of Agriculture (USDA) programs support the development of new and emerging technologies for refining advanced biofuels and utilizing renewable biomass as an energy feedstock. For example, the Advanced Biofuel Payment Program provides payments to biorefineries to maintain and expand production of advanced biofuels (i.e., biofuels refined from renewable feedstocks other than corn kernel starch, such as cellulose, sugar, hemicelluloses, lignin, waste materials, and biogas). Similarly, the Biorefinery Assistance Program (BAP) supports the emerging commercialization of next-generation advanced biofuel facilities, plants capable of producing fuel and bio-products using nonedible feedstocks and organic wastes. BAP also emphasizes production of advanced biofuels, but focuses on facilities that produce at commercial scale. Finally, the Repowering Assistance Program (RAP) provides payments to eligible biorefineries to help offset the costs of converting existing fossil fuel refineries to renewable biomass fuel-powered systems.

While USDA's biofuels programs are primarily designed to promote energy independence and rural development, they reduce GHG emissions associated with energy production and/or fossil fuel use. For example, a large-scale anaerobic digester funded though RAP supplies enough biogas to a nearby ethanol plant to replace virtually all the fossil fuels previously used to power the refinery process.

Biofuel Regional Feedstock Partnerships[22]

Through such efforts as the Regional Feedstock Partnerships, DOE's Bioenergy Technologies Office is working to identify and analyze feedstock supply and regional logistics and conduct crop field trials in order to address barriers associated with the development of a sustainable and predictable supply of biomass feedstocks. In addition, DOE's Bioenergy Technologies Office is working through public–private, cost-sharing partnerships to address critical challenges in the deployment of technologies for integrated biorefineries. These partnerships undertake biorefinery projects to prove the viability of various feedstock and conversion pathways and reduce the associated technical and financial risks. Currently, DOE has awarded funds to 22 biorefinery projects.

Smart Grid Investment Grants[23]

As a result of the American Recovery and Reinvestment Act of 2009 (ARRA), DOE's Office of Electricity Delivery and Energy Reliability is applying approximately $4 billion, leveraging an additional $5 billion in cost-shared funds toward the modernization of the electric grid in 99 Smart Grid Investment Grant (SGIG) projects around the country through public–private

[20] See http://www.rurdev.usda.gov/energy.html.

[21] See http://www.rurdev.usda.gov/MN-RBS-AdvancedBiofuelPaymentProgram.html; http://www.rurdev.usda.gov/BCP_biorefinery.html; and http://www.rurdev.usda.gov/bcp_repoweringassistance.html.

[22] See http://www.sungrant.org/Feedstock+Partnerships/; http://www1.eere.energy.gov/biomass/integrated_biorefineries.html; and http://www1.eere.energy.gov/bioenergy/pdfs/ibr_portfolio_overview.pdf.

[23] See http://energy.gov/oe/technology-development/smart-grid/recovery-act-smart-grid-investment-grants.

partnerships. The projects are deploying smart grid technologies (e.g., automated controls on field devices, meters, sensors, communications infrastructure, consumer monitoring technology) within the transmission and distribution (T&D) systems and on customers' premises. Significant energy efficiency and stability improvements are expected primarily by demand reduction by customers, more efficient field operations, and optimized control of voltage and power.

In addition to the SGIG projects, significant resources are focused on coordinating transmission system planning and advancing energy storage technologies, as well as computational methods for grid modeling, to more effectively integrate renewable energy technologies into the electric grid, to improve operations, and to reduce the environmental footprint of energy generation and delivery.

Offshore Renewable Energy Program—Bureau of Ocean Energy Management[24]

The Outer Continental Shelf (OCS) has significant potential as a source of new domestic energy generation from renewable energy resources. In the foreseeable future, DOI's Bureau of Ocean Energy Management (BOEM) anticipates development of renewable energy on the OCS from three general energy sources: offshore wind, ocean waves, and ocean currents.

BOEM has achieved significant progress with respect to offshore wind development in recent years. In 2009, President Obama and Secretary of the Interior Salazar announced the final regulations for the OCS Renewable Energy Program, providing a framework for the issuance of renewable energy leases, easements, and rights-of-way. In November 2010, Secretary Salazar signed the nation's first commercial lease for wind energy development on the OCS for the Cape Wind Energy Project offshore Massachusetts. Then in late 2012, BOEM issued a commercial lease for a wind facility offshore Delaware. In 2013, BOEM announced the first competitive offshore lease sales for areas offshore Virginia, Rhode Island, and Massachusetts, and plans to announce additional competitive sales for a number of areas, including areas offshore New Jersey, Maryland, and Massachusetts.

Planning and environmental work continues on a number of unsolicited proposals for wind facilities and renewable energy transmission lines along the East Coast. BOEM is also working toward authorizing wind development off the Pacific Coast (e.g., offshore Oregon and Hawaii) and marine hydrokinetic testing activities offshore Florida and Oregon.

Nuclear Waste Management[25]

DOE has legal responsibility to manage nuclear waste under the Nuclear Waste Policy Act. DOE's Office of Nuclear Energy also provides funding for the SMR (Small Modular Reactor) Licensing Technical Support program, which is a cost-share program with industry to help make progress on design efforts that will enable SMRs to be evaluated by the Nuclear Regulatory Commission.

Energy Sector: Residential, Commercial, and Industrial End Use

Reducing the amount of energy used in homes, buildings, and industrial facilities is also critical to supporting efforts to reduce GHG emissions from the energy sector. At the federal level, DOE and EPA continue to make great progress implementing programs to increase energy efficiency through regulatory, voluntary, and economic instruments. See Table 4-2 for estimates of GHG emission reductions in the residential, commercial, and industrial end-use sectors.

Appliance and Equipment Energy Efficiency Standards[26]

DOE's Building Technologies Office (BTO) also implements minimum energy conservation standards for more than 50 categories of appliances and equipment in the residential, commercial, and industrial sectors. As a result of these standards, energy users saved about $40 billion on their utility bills in 2010. Since the 2010 CAR, 17 new or updated federal standards have been issued, which will help increase annual savings by more than 50 percent over the next decade (Box 4-4).

Products covered by standards represent about 90 percent of home energy use, 60 percent of commercial building use, and 29 percent of industrial energy use. Commercial and

[24] See http://www.boem.gov/Renewable-Energy-Program/index.aspx.

[25] See http://energy.gov/downloads/strategy-management-and-disposal-used-nuclear-fuel-and-high-level-radioactive-waste; and http://www.nrc.gov/reading-rm/doc-collections/fact-sheets/funds-fs.html.

[26] See https://www1.eere.energy.gov/buildings/appliance_standards/.

industrial standards were issued for air conditioners, heat pumps, ice makers, refrigerators, freezers, clothes washers, electric motors, boilers, and transformers. Residential standards were issued for boilers, dehumidifiers, cooking products, direct heating equipment, dishwashers, air conditioners, refrigerators, freezers, and clothes washers and dryers, among others.

Lighting Energy Efficiency Standards[28]

DOE's BTO implements lighting energy efficiency standards mandated by EISA. The standards will result in phasing out the 130-year-old incandescent light bulb by the middle of the next decade and phases out less efficient fluorescent tubes. New standards will also apply to reflector lamps—the cone-shaped bulbs used in recessed and track lighting.

ENERGY STAR Labeled Products[29]

As a national symbol for energy efficiency, ENERGY STAR makes it easy for consumers and businesses to purchase products that save them money and reduce GHGs. The program celebrated its 20th anniversary in 2012, with Americans purchasing more than 4.5 billion products across over 65 product categories. The level of public awareness of ENERGY STAR has increased to more than 85 percent of American households due to a combination of strategic efforts, including maintaining brand integrity, consumer education and outreach, and third-party verification and testing of products.

With support from DOE, EPA continues to identify new product categories for ENERGY STAR, as well as revise existing product specifications to more stringent levels. The ENERGY STAR qualification process requires that products be tested in EPA-recognized laboratories, with the results reviewed by an independent, accredited certification organization. EPA also continues to expand its new ENERGY STAR® Most Efficient recognition program, to increase demand for products that demonstrate cutting-edge efficiency.

ENERGY STAR Commercial Buildings[30]

EPA has continued to expand the ENERGY STAR program in the commercial market, offering thousands of businesses and other organizations a strategy for superior energy management, standardized measurement tools, and recognition for their efforts. More than 20,000 buildings have earned the ENERGY STAR label for top performance, and are using 35–40 percent less energy than average buildings. Since the 2010 CAR, EPA has expanded ENERGY STAR to include 16 different space types eligible to earn the certification, including senior care facilities and data centers.

In addition, approximately 40 percent of U.S. floor space has been rated using EPA's ENERGY STAR Portfolio Manager™ building tracking tool. Introduced in 1999, Portfolio Manager™ benchmarks the energy use of commercial buildings to help owners, facility managers, and tenants evaluate building energy efficiency and identify cost-effective opportunities for improvements. Unveiling the largest U.S. building energy benchmarking data analysis to date, EPA examined more than 35,000 buildings that consistently used the ENERGY STAR Portfolio Manager™ measurement tool from 2008 to 2011. The buildings showed an average of 7 percent energy savings and 6 percent GHG emission reductions over three years—with the buildings that were initially the lowest performers making the greatest improvements. In addition to this analysis, EPA released a series of ENERGY STAR Portfolio Manager Data Trends in 2012.[31]

ENERGY STAR for Industry[32]

EPA's ENERGY STAR for Industry program has continued to grow since the 2010 CAR. EPA's ENERGY STAR Industries in Focus, which directly addresses barriers to energy efficiency by providing industry-specific energy management tools and resources, had grown to include 24 industrial sectors and subsectors with the launch of the integrated pulp and paper mills in 2012. Energy-efficient industrial plants can earn the ENERGY STAR label by achieving energy performance in the top quartile nationally for their industry.

By 2012, EPA had awarded the ENERGY STAR label to more than 120 plants. EPA continues to expand the use of ENERGY STAR tools and reassess energy performance across sectors. Further, the 2012 ENERGY STAR Challenge engaged a record number of industrial sites that

---

Box 4-4 **Executive Order 13624: Accelerating Investment in Industrial Energy Efficiency**[27]
On August 30, 2012, President Obama issued an executive order that directs DOE, EPA, USDA, and other federal agencies to use their existing programs and authorities to advance industrial energy efficiency, in order to reduce costs for industrial users, improve U.S. competitiveness, create jobs, and reduce harmful air pollution. These efforts include (1) fostering a national dialogue through regional workshops to identify, develop, and encourage the adoption of best practice policies and investment models; and (2) providing technical assistance to states and manufacturers to encourage investment in industrial energy efficiency and combined heat and power. E.O. 13624 also sets a national goal to deploy 40 gigawatts of new combined heat and power capacity by 2020.

---

[27] *See* http://www.gpo.gov/fdsys/pkg/DCPD-201200674/pdf/DCPD-201200674.pdf.

[28] *See* https://www1.eere.energy.gov/buildings/appliance_standards/.

[29] *See* http://www.energystar.gov.

[30] *See* http://www.energystar.gov.

[31] *See* http://www.energystar.gov/datatrends.

[32] *See* http://www.energystar.gov.

committed to plant-specific energy savings goals, with 75 sites meeting or exceeding their targets of achieving a 10 percent reduction in energy intensity.

ENERGY STAR Certified New Homes[33]

Through ENERGY STAR, EPA works to increase the energy efficiency of new homes to cost-effectively reduce GHG emissions, while lowering Americans' utility bills and improving the comfort of their homes. More than 1.4 million ENERGY STAR-certified new homes have been built to date, with more than 100,000 ENERGY STAR new homes in 2012. More rigorous requirements for new homes to earn the ENERGY STAR label became fully effective in 2012, requiring homes that earn the ENERGY STAR label to be at least 15 percent more energy efficient than homes built to the 2009 International Energy Conservation Code (IECC). The new specifications also feature additional measures that deliver a total energy efficiency improvement of up to 30 percent compared with typical new homes.

In 2011, new and substantially rehabilitated multifamily high-rise buildings became eligible to earn the ENERGY STAR label. These buildings must meet EPA's energy efficiency guidelines and must be designated to be at least 15 percent more efficient than the buildings energy code. As of 2012, 40 multifamily high-rise buildings containing more than 3,800 individual units had been certified.

Home Performance with ENERGY STAR[34]

DOE's Home Performance with ENERGY STAR (HPwES) program provides homeowners with resources to identify trusted contractors who can help them understand their home's energy use, as well as identify home improvements that increase energy performance. Contractors who participate in HPwES are qualified by local sponsors, such as utilities, state energy offices, and other organizations, to ensure that they can offer high-quality, comprehensive energy audits. More than 300,000 residential retrofits have been completed to date.

Building Energy Codes[35]

DOE's Building Energy Codes Program (BECP) participates in the development of cost-effective building energy codes and provides technical support for adoption and compliance strategies. Through advancing building codes, DOE's BTO aims to improve building energy efficiency by 50 percent, and to help states achieve 90 percent compliance with their energy codes. Building energy code tools and resources include the current status of state energy codes, procedures and tools, technical assistance, commercial compliance software (COMcheck), residential compliance software (REScheck), and reference guides. BECP also provides technical assistance to states and localities as they adopt and enforce energy codes and establish regulations for energy efficiency in federal buildings and manufactured housing.

Additional DOE programs that promote building energy efficiency include the Better Buildings Alliance, which allows members in different market sectors to join DOE's exceptional network of commercial buildings research and technical experts. The Better Buildings Neighborhood Program is helping more than 40 competitively selected state and local governments develop sustainable programs to upgrade the energy efficiency of more than 100,000 buildings. Finally, DOE's Challenge Home Program is a new home construction program that recognizes builders who construct homes to the highest level of energy performance. The program provides voluntary guidelines, which achieve a minimum of 40 percent energy savings above the 2009 IECC. By 2017, the program aims to achieve a 10 percent incorporation rate of these voluntary standards in newly constructed U.S. housing.

Regional Combined Heat and Power Technical Assistance Partnerships and Industrial Assessment Centers[36]

Through its Advanced Manufacturing Office, DOE funds Regional Combined Heat & Power Technical Assistance Partnerships (CHP TAPs) and Industrial Assessment Centers (IACs) to provide technical assistance to end users. The CHP TAPs provide CHP project screenings and feasibility analyses to end users, as well as broader education and assistance to state policymakers and others.

[33] See http://www.energystar.gov.

[34] See http://www1.eere.energy.gov/buildings/residential/energystar.html.

[35] See http://www1.eere.energy.gov/buildings/codes.html; and http://energy.gov/better-buildings.

[36] See http://www1.eere.energy.gov/manufacturing/distributedenergy/chptaps.html; http://www1.eere.energy.gov/manufacturing/tech_deployment/iacs.html; http://www1.eere.energy.gov/manufacturing/tech_assistance/sep.html; and http://www1.eere.energy.gov/manufacturing/tech_deployment/betterplants/.

During FYs 2009–2012, CHP TAP assistance to end users resulted in more than 1.4 gigawatts of CHP under development or online. The IACs provide energy audits to end users that identify cost-saving opportunities. An average IAC assessment identifies about $55,000 in potential annual savings per manufacturer. More than 15,000 IAC assessments have been conducted. CHP TAPs also offer technical assistance to ensure that major sources burning coal or oil have information on cost-effective clean energy strategies, such as natural gas combined CHP, and to promote cleaner, more efficient boilers to cut harmful pollution and reduce operational costs to the more than 550 major source facilities affected by National Emission Standards for Hazardous Air Pollutants.

Industrial energy efficiency is also promoted through the following DOE programs:

- Superior Energy Performance provides a transparent system for verifying improvements in energy performance and management practices through application of the internationally accepted ISO (International Organization for Standardization) 50001 energy management standard.

- The Better Buildings Initiative seeks to make commercial and industrial buildings 20 percent more energy efficient by 2020 and accelerate private-sector investment in energy efficiency.

- The Better Plants Program is designed to encourage and recognize U.S. companies that are raising the bar for all manufacturing facilities by establishing and achieving ambitious energy efficiency goals. Companies joining the program sign a voluntary pledge to reduce energy intensity by 25 percent over 10 years and receive national recognition for their commitment and progress.

National Energy Information Surveys and Analysis[37]

DOE's Energy Information Administration (EIA) collects and publishes definitive, national end-use consumption data for commercial buildings, residential buildings, and manufacturing establishments. The end-use consumption surveys provide baseline information critical to understanding energy use, and serve as the basis for benchmarking and performance measurement for energy efficiency programs that provide policymakers with the tools to develop mitigation policies.

The Residential Energy Consumption Survey (RECS) collects information from a nationally representative sample of housing units, including data on energy characteristics of homes, energy use patterns, and household demographics. This information is combined with data from energy suppliers to estimate energy costs and use for heating, cooling, appliances, and other end uses, and is critical to meeting future energy demand and improving building efficiency and design.

The Commercial Buildings Energy Consumption Survey (CBECS) provides the only statistically reliable source of energy consumption, expenditures, and end uses in U.S. commercial buildings. CBECS is the only survey conducted by the U.S. government that collects data specifically about commercial buildings.

The Manufacturing Energy Consumption Survey (MECS) is a national sample survey that collects information on U.S. manufacturing establishments' energy consumption and expenditures, "nonfuel" use of energy sources, end uses, and other characteristics related to their use of energy.

The three surveys are conducted on a quadrennial basis. The most recent RECS data, for reference year 2009, have been posted on EIA's Web site over the last several years. Processing for CBECS 2012 is well underway, with preliminary data scheduled for release in FY 2014. MECS has been updated since the 2010 CAR with 2010 data. EIA is currently exploring ways to improve its energy consumption survey program by testing and implementing recommendations from a 2012 National Academies of Sciences study aimed at streamlining survey operations and improving data timeliness (Eddy and Marton 2012).

---

[37] See www.eia.gov.

EIA also provides regional and state data, including energy-related CO_2 emissions by state. These data provide input for an analysis of key emission factors by state, including energy intensity, the carbon intensity of the energy supply, and per capita emissions. The analysis has been performed on 2009 and 2010 data. In July 2013, EIA released the *State Energy Efficiency Program Evaluation Inventory,* which provides cost information for state-mandated energy efficiency program evaluations—e.g., for use in updating analytic and modeling assumptions used by EIA (U.S. DOE/EIA 2013m). The National Energy Modeling System is a key source of the projections presented in Chapter 5 of this report.

Industrial Processes (Non-CO$_2$) Sector

In addition to CO_2 emissions from energy use, the industrial sector contributes to CH_4 and fluorinated GHG emissions. Federal policies and measures are being implemented by EPA to reduce non-CO_2 emissions from various industries, utilizing a mix of regulatory, voluntary, and informational instruments. See Table 4-2 for estimates of non-CO_2 GHG emission reductions from industrial processes.

Federal Air Standards for the Oil and Natural Gas Industry[38]

On April 17, 2012, EPA issued regulations to reduce harmful air pollution from the oil and natural gas industry, while allowing continued, responsible growth in U.S. oil and natural gas production. The final rules include the first federal air standards for natural gas wells that are hydraulically fractured, along with requirements for several other sources of pollution in the oil and gas industry that currently are not regulated at the federal level. These other emission sources include storage vessels, pneumatic controllers, centrifugal compressors, reciprocating compressors, and equipment leaks at natural gas processing plants.

The final rules are expected to yield a nearly 95 percent reduction in volatile organic compound (VOC) emissions from more than 11,000 new hydraulically fractured gas wells each year. This significant reduction would be accomplished primarily through capturing natural gas that currently escapes into the air, and making that gas available for sale. Emissions of VOCs react with NO_x in the presence of sunlight to form ground-level ozone, commonly known as "smog." The rules also will reduce air toxics, which are known to cause or suspected of causing cancer and other serious health effects. Although these rules specifically regulate VOCs and air toxics, they significantly reduce CH_4 emissions, estimated at 32.6 Tg CO_2e in 2015 and 39.9 Tg CO_2e in 2020, as a co-benefit of VOC control.

Significant New Alternatives Policy Program[39]

Through its Significant New Alternatives Policy (SNAP) Program, EPA evaluates and regulates substitutes for the ozone-depleting chemicals that are being phased out nationally under the CAA and globally under the Montreal Protocol on Substances That Deplete the Ozone Layer. EPA evaluates a number of criteria for listing as acceptable those alternatives that reduce overall risk to human health and the environment, while placing restrictions or bans on others, thereby allowing for a safe and smooth transition. The SNAP Program lists are continually being revised, and consider the comparative risk of available and potentially available alternatives for a given use.

Since the 2010 CAR, SNAP has continued to identify substitutes for chlorofluorocarbons (CFCs), hydrochlorofluorocarbons (HCFCs), and other ozone-depleting substances (ODS). EPA has worked closely with industry to research, identify, and implement climate- and ozone-friendly alternatives, supporting a smooth transition to these new technologies. Many compounds with low global warming potentials (GWPs) have been found acceptable under SNAP, allowing for the uptake of such chemicals in place of both ODS and fluorinated GHGs, such as HFCs.

Natural Gas STAR Program[40]

Through its Natural Gas STAR Program, EPA works with oil and natural gas companies to promote proven, cost-effective technologies and practices that improve operational efficiency and reduce CH_4 (i.e., natural gas) emissions. CH_4 is emitted by oil production and by all sectors of the natural gas industry, from drilling and production, through processing and storage,

[38] *See* http://www.epa.gov/airquality/oilandgas/actions.html.

[39] *See* http://www.epa.gov/ozone/snap.

[40] *See* http://www.epa.gov/gasstar.

to T&D. Since its launch in 1993, Natural Gas STAR has been successful in working with U.S. oil and natural gas companies to reduce more than one trillion cubic feet of CH_4 emissions and bring more energy to markets.

Coalbed Methane Outreach Program[41]

EPA's voluntary Coalbed Methane Outreach Program (CMOP) has the goal of reducing CH_4 emissions from coal mining activities. CMOP's mission is to promote the profitable recovery and utilization of coal mine methane (CMM), a valuable fuel source. Since 1994, CMOP has worked cooperatively with the coal mining industry to reduce CMM emissions from underground, surface, and abandoned mines. The benefits of capturing and using CMM include improved worker safety, lower GHG emissions, an additional revenue stream for the mine, and a source of local clean energy. In recent years, new projects, such as ventilation air methane oxidation and electricity generation from drained gas, have come online.

Fluorinated Greenhouse Gas Programs[42]

EPA's voluntary fluorinated greenhouse gas (FGHG) partnership programs continue to make significant reductions in potent GHG emissions by working with participating industries. Through these programs, EPA identifies cost-effective emission reduction opportunities, recognizes industry accomplishments, and facilitates the transition toward environmentally friendlier technologies and best environmental practices. Partners include aluminum producers, electrical T&D system operators, supermarkets, utilities, and appliance retailers and manufacturers.

Although FGHGs account for a small portion of total U.S. GHG emissions, they have very high GWPs, and emissions on a per-facility basis tend to be high. PFCs and sulfur hexafluoride (SF_6) also have extremely long atmospheric lifetimes, making climate impacts essentially irreversible.

SF_6 Emission Reduction Partnership for Electric Power Systems—Through its SF_6 Emission Reduction Partnership for Electric Power Systems, EPA works with electric power T&D companies to reduce emissions of SF_6, which is used as a gaseous dielectric in high-voltage circuit breakers and gas-insulated substations. The program promotes best management practices and cost-effective operational improvements, such as leak detection and repair, use of recycling equipment, and employee education and training. The program also engages stakeholders, such as equipment manufacturers, gas distributors, and recyclers, to improve SF_6 handling during installation, servicing, and decommissioning of equipment.

GreenChill Advanced Refrigeration Partnership—Through the GreenChill Advanced Refrigeration Partnership, EPA works with supermarkets to reduce the amount of refrigerants they use in their stores and emit to the atmosphere. Refrigerants have high-GWPs and are often ozone-depleting gases, so their minimization is especially beneficial for the environment. GreenChill now has 50 partners with more than 8,000 supermarkets (more than 21 percent of all U.S. supermarkets) in all 50 states. On average, more than 20 percent of the refrigerant used each year in the supermarket industry is released into the atmosphere in the form of harmful GHGs. Since the start of the program in 2007, GreenChill partners have reduced their aggregate total corporate emission rate to below 12 percent per year—about half the national average.

Responsible Appliance Disposal Program—Through EPA's voluntary Responsible Appliance Disposal Program, partners—utilities, retailers, manufacturers, and state affiliates—ensure responsible disposal of refrigerant-containing appliances in order to recover and recycle refrigerants and recycle or properly destroy GHGs from foam, thereby reducing emissions of high-GWP gases. The partners also prevent the release of hazardous materials (e.g., used oil, polychlorinated biphenyls, and mercury), and save landfill space and energy by recycling durable materials.

Voluntary Aluminum Industry Partnership—EPA's Voluntary Aluminum Industry Partnership works with industry to reduce PFCs, tetrafluoromethane, and hexafluoroethane where cost-effective technologies and operational practices are technically feasible. The partnership works to reduce PFC emissions through training and implementing best practices in

[41] *See* http://www.epa.gov/coalbed/.

[42] *See* http://www.epa.gov/climatechange/EPAactivities/voluntaryprograms.html; http://www.epa.gov/climatechange/EPAactivities/voluntaryprograms.html; http://www.epa.gov/greenchill; http://www.epa.gov/rad; and http://www.epa.gov/ozone/snap/fire/vcopdocument.pdf.

aluminum smelter pot rooms. In addition, the partnership has advanced the scientific understanding of PFC emissions from primary aluminum production. Work has included evaluating smelter cell conditions at the initiation of PFC-emitting anode effects and documenting low-voltage PFC emissions in different technology types.

Voluntary Code of Practice for the Reduction of Emissions of HFC & PFC Fire Protection Agents—EPA also works with manufacturers and distributors in the U.S. fire protection industry to advance the Voluntary Code of Practice for the Reduction of Emissions of HFC & PFC Fire Protection Agents (VCOP). VCOP minimizes nonfire emissions of HFCs and PFCs (predominantly HFCs), while effectively protecting people and property from the threat of fire. Approximately 14 manufacturers and distributors annually report to the HFC Emissions Estimating Program, tracking industry-wide emissions of HFCs and progress under VCOP.

Agricultural Sector

The federal government is utilizing voluntary, economic, and informational instruments to reduce GHG emissions from the agricultural sector. USDA and EPA implement policies and measures to reduce CO_2, CH_4, and NO_x emissions from this sector. See Table 4-2 for estimates of GHG emission reductions in the agricultural sector.

Conservation Reserve Program[43]

USDA's Conservation Reserve Program (CRP) pays farmers to voluntarily convert environmentally sensitive land to native grasses, wildlife plantings, trees, restored wetlands, filter strips, or riparian buffers under 10–15-year contracts. Administered by USDA's Farm Service Agency (FSA), the CRP sequesters more carbon on private lands than any other federally administered program. FSA also facilities the potential for private sale of carbon credits for lands enrolled in the CRP, as USDA does not claim ownership to related credits. FSA includes carbon sequestration potential in its ranking process, by which offers are selected for enrollment. In addition to increasing carbon sequestration, CRP lands produce GHG benefits in the form of reduced CO_2 emissions from fewer field operations and reduced NO_x emissions from avoided fertilizer applications.

Natural Resources Conservation Service Programs[44]

USDA's Natural Resources Conservation Service (NRCS) administers several conservation programs designed to address specific natural resource concerns on working agricultural lands—i.e., lands in active crop, livestock, or forestry production. The concerns include reducing soil erosion, enhancing water supplies, improving water quality, increasing wildlife habitat, and reducing damages caused by floods and other natural disasters. In each program, participation by producers and other land owners is voluntary.

Typically, participants enter into fixed-term contracts with USDA, in which they receive financial and technical assistance in exchange for agreeing to implement specified conservation practices or measures within their operation. Contracts identify the natural resource concerns to be addressed and require producers to develop a plan of operations that identifies the conservation practices or measures needed to address identified concerns.

GHG mitigation is a resource concern identified under NRCS conservation programs, and many of the practices and measures encouraged in the programs reduce GHG emissions and/or increase carbon sequestration. In terms of addressing GHG emissions and encouraging carbon sequestration, the principal NRCS conservation programs are the Environmental Quality Incentives Program and Conservation Technical Assistance Program.

Environmental Quality Incentives Program (EQIP)—EQIP provides financial and technical assistance to eligible producers based on a portion of the average cost associated with practice implementation. Additional payments may be available to help producers develop conservation plans, which are required to obtain financial assistance. Program contracts can cover periods of up to 10 years, and total program payments to any participant are generally capped at $300,000 during any 6-year period. NRCS has identified 23 EQIP conservation practices that result in quantifiable carbon sequestration or emission reductions. Between 2010 and 2012, annual GHG mitigation benefits associated with these practices ranged between 3.2 and 4.0 Tg CO_2e.

[43] See http://fsa.usda.gov/FSA/webapp?area=home&subject=copr&topic=crp.

[44] See www.nrcs.usda.gov/.

Conservation Technical Assistance (CTA) Program—The CTA Program provides technical assistance to landowners and other individuals and groups responsible for managing nonfederal lands. The program addresses opportunities, concerns, and problems related to the use of natural resources and helps land users make sound natural resource management decisions on private, tribal, and other nonfederal lands. Many of the changes in land management that have been facilitated through the CTA Program reduce GHG emissions and/or increase carbon sequestration. Between 2010 and 2012, the annual GHG mitigation benefits of the program ranged between 7.9 and 8.4 Tg CO_2e.

Other NRCS Programs—Other NRCS conservation programs include the Conservation Stewardship Program, Wildlife Habitat Incentive Program, Wetlands Reserve Program, Farm and Ranchland Protection Program, and Grassland Reserve Program. These programs target more specific conservation objectives than EQIP and the CTA Program, but similarly contribute to addressing GHG resource concerns. Between 2010 and 2012, the estimated aggregate annual GHG mitigation for these programs ranged between 0.38 and 0.75 Tg CO_2e.

AgSTAR[45]

AgSTAR was launched as a voluntary effort between EPA and USDA in 1993. Run by EPA with support from USDA, AgSTAR encourages the use of methane (biogas) recovery technologies at confined animal feeding operations that manage manure as liquids or slurries. These technologies reduce methane emissions while achieving other environmental benefits. The practices recommended under AgSTAR have been incorporated into USDA's broader technical, conservation, and cost-share programs. AgSTAR also works at a national level to identify and address barriers to these biogas recovery projects, as well as to provide information and training to state and local government agencies that permit these projects and the private-sector organizations that implement them. Key benefits promoted by AgSTAR include sustainable management of manure, reduced GHG emissions, and the development of value-added by-products.

Forestry Sector

USDA's Forest Service (USFS) continues to implement federal programs for climate mitigation, utilizing voluntary, economic, and informational instruments.

Woody Biomass Utilization Grant Program[46]

The Woody Biomass Utilization Grant Program focuses on creating markets for small-diameter woody material and low-value trees removed during forest restoration activities. Grants range from $50,000 to $250,000, can be in place for up to three years, and require a nonfederal match of at least 20 percent. Grantees report on the amount of green tons of woody biomass that is removed and utilized each year. Since most of this biomass would have otherwise been piled and burned in the open, GHG mitigation benefits accrue in the form of reduced CO_2 emissions associated with open residue burning.

In 2011 and 2012, the program's focus shifted to assisting wood energy facilities to develop the engineering design and detailed cost estimates critical to obtaining and leveraging funding. These facilities are not yet operational, so biomass removals and GHG benefits are not reported for these years in Table 4-2.

Forest Ecosystem Restoration and Hazardous Fuels Reduction Programs[47]

Since 2000, several USFS policies and initiatives (e.g., the National Fire Plan, Healthy Forests Initiative, Healthy Forests Restoration Act, National Cohesive Wildland Fire Management Strategy, and the Collaborative Forest Landscape Restoration Program) have aimed to reduce wildfire risk near communities and elsewhere, and to restore or increase forest resilience to climate-related stressors, such as drought, wildfire, insects, and disease. These programs and initiatives have applied restoration treatments to 10.6 million hectares (27.6 million acres). The net CO_2 mitigation impacts of these programs and initiatives are difficult to quantify because they largely take the form of an enhanced ability of treated areas to sequester carbon over the long term.

---

[45] *See* www.epa.gov/agstar.

[46] *See* http://www.fpl.fs.fed.us/research/units/tmu/tmugrants_goals.shtml.

[47] For forest ecosystem restoration, see http://www.fs.fed.us/restoration/; for hazardous fuels, see http://fsweb.wo.fs.fed.us/fire/fam/fuels/hazardous.html; for restoring and maintaining landscapes, see http://www.forestsandrangelands.gov/strategy/goals.shtml.

Waste and Waste Management Sector

EPA implements federal policies and measures to reduce GHGs from the waste management/waste sector. Regulatory and voluntary efforts are reducing CH_4 emissions from landfills, while CO_2 emission reductions result from sustainable materials management programs. See Table 4-2 for estimates of GHG emission reductions.

Landfill Air Regulations[48]

Municipal solid waste landfills are the third-largest source of U.S. anthropogenic CH_4 emissions. Promulgated in 1996, the New Source Performance Standards and Emission Guidelines require large landfills to collect and control their gas emissions. Landfill gas is comprised of approximately 50 percent CH_4, 50 percent CO_2, and trace amounts of nonmethane organic compounds. Although the emission thresholds in both rules are based on nonmethane organic compounds, significant CH_4 co-benefits are also achieved. EPA estimates that the 1996 rules will reduce emissions by about 183 Tg CO_2e in 2020.

Landfill Methane Outreach Program[49]

EPA's Landfill Methane Outreach Program (LMOP) reduces GHG emissions at landfills by supporting the recovery and use of landfill gas for energy. Capturing and using landfill gas reduce CH_4 emissions directly and reduce CO_2 emissions indirectly by displacing the use of fossil fuels through the utilization of landfill gas as a source of energy. LMOP focuses its efforts on smaller landfills that are not required to collect and combust their landfill gas, as well as larger regulated operations that are combusting their gas but not utilizing it as a clean energy source. LMOP has developed a range of technical resources and tools to help the landfill gas industry overcome barriers to energy development, including feasibility analyses, project evaluation software, a database of approximately 500 candidate landfills across the country, a project development handbook, and industrial sector analyses.

Sustainable Materials Management Programs[50]

Historically, most of the nation's resource conservation efforts have focused on decisions to reuse or recycle materials that would otherwise be disposed of as waste. Although these remain important resource conservation practices, they only represent a fraction of all the opportunities available to conserve resources.

Sustainable Materials Management—Through a sustainable materials management (SMM) approach, EPA is helping change the way Americans protect the environment and conserve resources for future generations. SMM is a systemic approach that seeks to reduce materials use and their associated environmental impacts over their entire life cycle, starting with extraction of natural resources and product design and ending with decisions on recycling or final disposal. EPA is playing a leadership role in advancing SMM by convening dialogues with key SMM stakeholders, providing sound science and information to the public, and establishing challenges to specific sectors to achieve shared goals. EPA is collaborating with other federal agencies, businesses, and schools in key SMM challenges, including Federal Green Challenge, Food Recovery Challenge, and Electronics Challenge.

WasteWise—EPA is also working with organizations and businesses to reduce municipal and select industrial wastes via the WasteWise program. Launched in 1994, WasteWise has become a mainstay in environmental stewardship and continues to evolve to address tomorrow's environmental needs.

Federal Government Leading by Example

Since the federal government is the largest single user of energy in the United States, a great potential for GHG emission reductions exists from federal facilitates themselves. Implementation of efforts to reduce CO_2 emissions from federal facilities continues since the 2010 CAR, with great progress being made. See Table 4-2 for estimates of GHG emission reductions.

E.O. 13514: Federal Leadership in Environmental, Energy, and Economic Performance[51]

In October 2009, President Obama signed Executive Order (E.O.) 13514, setting sustainability goals for federal agencies and focusing on improving each agency's environmental, energy,

[48] See http://www.epa.gov/ttn/atw/landfill/landflpg.html.

[49] See www.epa.gov/lmop.

[50] See http://www.epa.gov/smm; and http://www.epa.gov/wastewise.

[51] See http://sustainability.performance.gov/.

and economic performance. E.O. 13514 required federal agencies to establish a 2020 GHG emission reduction target, increase energy efficiency and renewable energy use, reduce fleet petroleum consumption, conserve water, reduce waste, support sustainable communities, and leverage federal purchasing power to promote sustainable products and technologies.

E.O. 13514 requires federal agencies to meet a number of energy, water, and waste reduction targets, relative to 2005, including:

- 30 percent reduction in vehicle fleet petroleum use by 2020;

- 26 percent improvement in water efficiency by 2020;

- 50 percent recycling and waste diversion by 2015;

- 95 percent of all applicable contracts in compliance with sustainability requirements;

- Implementation of the 2030 net-zero-energy building requirement;

- Implementation of the stormwater provisions of Section 438 of EISA; and

- Development of guidance for sustainable federal building locations in alignment with the Livability Principles put forward by the U.S. Department of Housing and Urban Development (HUD), DOT, and EPA.

In 2010, President Obama announced a federal government-wide target of a 28 percent reduction by 2020 in direct GHG emissions, such as those from fuels and building energy use, and a target 13 percent reduction by 2020 in indirect GHG emissions, such as those from employee commuting and landfill waste. Implementation of E.O. 13514 has focused on integrating the pursuit of sustainability goals with agency missions and strategic planning, to optimize performance and minimize implementation costs. Under E.O. 13514, federal agencies are required to develop, implement, and annually update a plan that prioritizes actions based on a positive return on investment for the American taxpayer and to meet GHG emission, energy, water, and waste reduction targets (Box 4-5).

On February 7, 2013, federal agencies released their third annual Sustainability Plans.[52] In these updated plans, agencies discuss highlights and challenges from the previous year and explain how they will refine their strategies, expand on successes, and plan new initiatives to meet the goals of E.O. 13514. Implementation by agencies is managed through the previously established Office of the Federal Environmental Executive, working in close partnership with the Office of Management and Budget, the White House Council on Environmental Quality, and other federal agencies.

Federal Energy Management Program[53]

DOE's Federal Energy Management Program (FEMP) works with federal leaders to accomplish energy change within organizations by bringing expertise from all levels of project and policy implementation to enable federal agencies to meet energy-related goals and to provide energy leadership to the nation. FEMP assists agencies in identifying, obtaining, and implementing project-funding mechanisms, guiding them to use funding more effectively to meet

---

[52] Ibid.

[53] See http://www1.eere.energy.gov/femp/index.html.

[54] Scope 1 includes all direct GHG emissions; Scope 2 includes indirect GHG emissions from consumption of purchased electricity, heat, or steam; and Scope 3 includes all other indirect emissions.

Box 4-5 U.S. Department of Defense GHG Emissions Reductions under E.O. 13514

The U.S. Department of Defense (DoD) intends to achieve its goal for GHG emission reductions under E.O. 13514 primarily by reducing consumption of fossil fuels by facilities and vehicles, and increasing the use of renewable energy. The DoD target of 34 percent reduction in GHG emissions from FY 2008 levels by 2020 includes cumulative Scope 1 and 2 GHG emissions.[54] In FY 2012, DoD reduced annual GHG emissions by 1.29 million metric tons of CO_2e, a 9.2 percent reduction from 2008 levels.

DoD continues to pursue an investment strategy designed to reduce energy demand in fixed installations managed by its military departments, while increasing the supply of renewable energy sources. Efforts to curb demand for energy—through conservation and improved energy efficiency—are by far the most cost-effective ways to improve an installation's energy profile. A large fraction of DoD energy efficiency investments goes to retrofit existing buildings. Typical retrofit projects install high-efficiency heating, ventilation, and cooling systems; energy management control systems; more efficient lighting; and green roofs.

federal and agency-specific energy management objectives. FEMP provides technical support in sustainable design, energy efficiency, renewable energy, water conservation, fleet management, product procurement, technology deployment, and laboratory and data center best practices. FEMP also helps federal agencies comply with applicable energy, water, and fleet requirements by advising on energy management authorities, developing rules and guidance, evaluating reported data, tracking agency progress, providing training, developing interagency collaboration, and motivating federal staff through awards and incentives.

National Park Service Programs[55]

The National Park Service (NPS) is committed to reducing its impact on the environment, mitigating the effects of climate change, and integrating sustainable practices within and across its borders. The NPS Director's *Call to Action* lays out how the NPS will prepare for America's second century of stewardship and engagement and calls on NPS staff to "Go Green" by reducing GHG emissions (U.S. DOI/NPS 2011).

In 2012, the NPS released the *Green Parks Plan* (GPP) to define a collective vision and long-term strategic plan for sustainable management of NPS operations (U.S. DOI/NPS 2012). Within the first year of the GPP's release, the NPS has made significant progress toward meeting many of the plan's goals, including reducing emissions, energy and water use and intensity, and waste production. Through the GPP's implementation, the NPS has succeeded in:

- Decreasing Scope 1 and 2 GHG emissions by 13 percent and Scope 3 GHG emissions by 7 percent.[56]

- Reducing NPS-wide building energy intensity by 18 percent.

- Decreasing potable water use by 22 percent.

- Increasing waste diversion by 28 percent.

Climate Friendly Parks—To support GPP goals, the Climate Friendly Parks (CFP) Program continues to engage NPS staff in the climate change and sustainability conversation. With more than 100 member parks, CFP assists parks in measuring GHG emissions; provides educational opportunities for staff and the public to learn about climate change and sustainability-related topics; and aids in the development of park-based strategies and specific actions to reduce GHG emissions, address sustainability challenges, and anticipate the effects of climate change on park resources.

Clean Cities National Parks Initiative—Also in support of the GPP and Go Green challenge, the NPS and DOE partnership, Clean Cities National Parks Initiative, takes the NPS yet another step further in reducing GHG emissions associated with transportation in and around national parks. This unique partnership supports transportation projects that help to educate park visitors on the benefits of reducing dependence on petroleum, cutting GHG emissions, and easing traffic congestion. Participant parks and projects include Mammoth Cave National Park's propane-powered school buses and pickup trucks, and electric utility vehicles; San Antonio Mission National Historical Park's propane-powered mowers and pickup trucks, and installation of two 220-volt electric vehicle chargers with data collection capabilities; and Yellowstone National Park's electric utility and hybrid vehicles, and implementation of a no-idling campaign for visitors and employees.

Cross-Cutting Policies and Measures

Several federal policies and measures seek to mitigate climate change across multiple sectors. Multiple federal agencies implement cross-cutting programs, utilizing regulatory, economic, and informational instruments. See Table 4-2 for estimates of GHG cross-cutting emission reductions.

Best Available Control Technology for GHG Emissions[57]

The CAA requires large stationary sources of air pollution to apply for and receive permits before building a new facility or modifying an existing facility. These permits include information on the amount of GHGs a facility can emit, how often a facility can be run, and any other

---

[55] *See* http://www.nps.gov/calltoaction/; http://www.nps.gov/greenparksplan/; http://www.nps.gov/climatefriendlyparks/; and http://www1.eere.energy.gov/cleancities/national_parks.html.

[56] Scope 1 includes all direct GHG emissions; Scope 2 includes indirect GHG emissions from consumption of purchased electricity, heat, or steam; and Scope 3 includes all other indirect emissions.

[57] *See* http://www.epa.gov/nsr/ghgpermitting.html.

requirements that would ensure public health and the environment continues to be protected after the facility begins to operate. A key component of these permits is the requirement for large sources of emissions to use the best available technology for controlling GHG emissions. EPA anticipates that, in most cases, this requirement will be met through energy efficiency improvements.

In May 2010, EPA issued a regulation establishing a common-sense approach to permitting GHG emissions. EPA continues to focus GHG permitting on the largest emitters and has worked with states and industry to make a number of important updates that streamline the permitting process. As of September 2013, EPA and states had issued more than 130 permits to large industrial sources that cover GHG emissions. In addition, EPA is processing or tracking permit applications from across the United States that have not yet been issued.

Mandatory Greenhouse Gas Reporting Rule[58]

In 2009, EPA issued the Greenhouse Gas Reporting Rule. The rule requires reporting of GHG emissions from 41 U.S. industry groups that, in general, emit 25,000 metric tons (t) or more of CO_2e per year. The 25,000-t reporting threshold is roughly equivalent to the annual GHG emissions from just over 5,200 passenger vehicles or the carbon equivalent of burning 107 rails cars of coal.[59]

The GHG Reporting Rule is intended to collect accurate and timely emissions data to inform future policy decisions. Under the rule, direct emitters and suppliers of certain products that would result in GHG emissions if released, combusted, or oxidized or facilities that inject CO_2 underground (e.g., for geologic sequestration) are required to submit electronic annual reports to EPA. The gases covered by the GHG Reporting Rule are CO_2, CH_4, nitrous oxide (N_2O), HFCs, PFCs, SF_6, and other fluorinated gases, including nitrogen trifluoride and hydro-fluorinated ethers.

The reporting program covers about 85–90 percent of total U.S. emissions from approximately 8,000 facilities. Annual reporting began in 2011 for calendar year 2010 emissions. EPA now has three years of data for 29 industry groups and two years of data for an additional 12 industry categories. Publicly available GHG data are published in EPA's user-friendly FLIGHT (Facility Level Information on Greenhouse gases Tool) and in Envirofacts.[60]

State Energy Program[61, 62]

Through its State Energy Program, DOE provides financial and technical assistance to states through formula and competitive grants. States use their formula grants to develop state strategies and goals to address their energy priorities. Competitive grant solicitations for the adoption of energy efficiency/renewable energy products and technologies are issued annually based on available funding.

Indian Energy Programs provide financial and technical assistance that enables American Indian and Alaska Native tribes to deploy renewable energy resources, reduce their energy costs through efficiency and weatherization, and increase energy security for tribes and villages.

Energy Efficiency and Conservation Block Grants[63]

DOE's Energy Efficiency and Conservation Block Grant Program has provided more than $2.7 billion in funding to local and state governments, tribal governments, and territories. The program assists eligible entities in implementing strategies that will improve energy efficiency in the transportation, building, and other appropriate sectors, and reduce fossil fuel emissions and total energy use in an environmentally sustainable manner. Activities that may use grant funds range from strategic planning, information sharing, and developing building codes, to installing renewable energy technologies, to implementing technologies to reduce, capture, and use GHGs from landfills or similar sources.

In addition, the Community Renewable Energy Deployment grant program leveraged $20.5 million in ARRA funding, with approximately $167 million in local government and private industry funding to complete five projects nationwide. The projects receive technical assistance

[58] See http://www.whitehouse.gov/sites/default/files/omb/inforeg/social_cost_of_carbon_for_ria_2013_update.pdf.

[59] See www.epa.gov/ghgreporting/index.html.

[60] See http://www.epa.gov/cleanenergy/energy-resources/calculator.html.

[61] See http://ghgdata.epa.gov/ghgp/main.do and http://www.epa.gov/enviro/.

[62] See http://www1.eere.energy.gov/wip/sep.html; and http://www.eere.energy.gov/tribalenergy.

[63] See http://www1.eere.energy.gov/wip/eecbg.html.

from DOE's National Renewable Energy Laboratory in the areas of concepts, best practices, planning, financial approaches, and policy guidance.

Section 1703/1705 Loan Guarantee Programs[64]

DOE's Loan Guarantee Programs enable DOE to work with private companies and lenders to mitigate the financing risks associated with innovative and advanced energy technologies, thereby fostering their deployment on a broader, commercial scale. DOE's LPO provides loan guarantees to qualifying projects that employ new or significantly improved energy technologies that avoid, reduce, or sequester air pollutants or GHGs. There are 24 active loan guarantees.

Weatherization Assistance Program[65]

DOE's Weatherization Assistance Program (WAP) enables low-income families to permanently reduce their energy bills by making their homes more energy efficient. Funds are used to improve the energy performance of dwellings of needy families, using the most advanced technologies and testing protocols available in the housing industry. WAP provides funding, primarily through formula grants, to states, U.S. overseas territories, and Indian tribal governments, which manage the day-to-day details of the program. These governments, in turn, fund a network of community action agencies, nonprofit organizations, and local governments that provide these weatherization services in every state, the District of Columbia, U.S. territories, and among Native American tribes.

The energy conservation resulting from these efforts of state and local agencies helps reduce U.S. dependence on foreign oil and decrease the cost of energy for families in need, while improving the health and safety of their homes. Because the energy improvements that make up weatherization services are long lived, the savings add up over time to substantial benefits for weatherization clients and their communities, and the nation as a whole.

Tax Provisions[66]

Several existing federal energy tax provisions and energy grants may reduce GHGs. Combined, these provisions had estimated federal tax expenditures for FY 2012 of more than $10 billion. This includes estimated payments from the U.S. Department of the Treasury authorized by ARRA Section 1603. Tax expenditures are exceptions to baseline provisions of the tax structure that usually result in a reduction in the amount of tax owed.

Federal energy tax provisions capture various objectives that help reduce GHG emissions across transportation, energy, and industrial sectors:

- Providing an incentive for alternative fuel vehicles through the credit and deduction for clean fuel-burning vehicles.

- Providing an incentive for renewable and alternative energy production—as an incentive either directly for production, or indirectly through property and manufacturing projects that help support production. Incentives include the Residential Energy Efficient Property Credit; the Energy Production Tax Credit (for renewable and alternative energy only); the Business Energy Investment Tax Credit; the Energy Grant (in lieu of the Business Energy Investment Tax Credit and the Energy Production Tax Credit); the credit for holding Clean Renewable Energy Bonds and Qualified Energy Conservation Bonds (which also encourages energy conservation); and the Qualifying Advanced Energy Property Credit.

- Encouraging energy conservation through the Deduction for Energy Efficient Commercial Buildings, Credit for Construction of New Energy Efficient Homes, Credit for Energy Efficient Improvements to Existing Homes, the Manufacturers' Energy Efficient Appliance Credit, and Exclusion of Utility Conservation Subsidies.

- Encouraging carbon sequestration through the Industrial CO_2 Capture and Sequestration Tax Credit.

Interagency Partnership for Sustainable Communities[67]

Through the interagency Partnership for Sustainable Communities, DOT, HUD, and EPA are aligning federal policies and investments for transportation, environmental protection, and

[64] See http://www.lgprogram.energy.gov/.

[65] See http://www.eere.energy.gov/weatherization/.

[66] See http://www.whitehouse.gov/sites/default/files/omb/budget/fy2014/assets/receipts.pdf.

[67] See http://www.sustainablecommunities.gov/index.html.

housing. Partnership agencies support communities that want to give Americans more choices in housing and transportation, and build healthy and economically vibrant neighborhoods. Through these efforts, the partnership is helping communities make it convenient for residents to walk, bike, take transit, or drive short distances to daily destinations.

Between 2009 and 2012, the partnership provided more than $3.5 billion in assistance to more than 700 communities, and funded 744 projects with approximately $3.51 billion. Partnership grant and technical assistance recipients are located in all 50 states, the District of Columbia, and Puerto Rico.

Partnership agency efforts include the following:

• Between 2009 and 2012, HUD awarded 152 grants in 48 states as part of its Sustainable Communities Initiative. The $240 million in federal investment leveraged almost $253 million in private investment and commitments from local partners.

• In 2012, partnership agencies announced support for the Governors' Institute on Community Design to provide enhanced technical guidance to governors seeking to tackle housing, transportation, environmental, and health challenges. Facilitated by EPA, the Institute brings together leading practitioners and academics in government, design, development, and regional economics to help governors make informed choices about growth and development.

• Since 2009, DOT has awarded $3.1 billion in TIGER Discretionary Grants to 218 projects in all 50 states, the District of Columbia, and Puerto Rico. The program's competitive review process allows DOT to choose projects that will improve energy efficiency and make significant investments in expanding transportation choices for communities across the nation.

Key benefits of this partnership include reduced vehicle miles traveled (VMT), lower per-capita GHG emissions, and reduced dependence on fossil fuels.

Center for Corporate Climate Leadership[68]

EPA's Center for Corporate Climate Leadership was launched in 2012 to establish norms of climate leadership by encouraging organizations with emerging climate objectives to identify and achieve cost-effective GHG emission reductions, while helping more advanced organizations reduce their GHG impacts outside of their operations (e.g., in their supply chains). The Center serves as a comprehensive resource to help organizations measure and manage GHG emissions, providing technical tools, ground-tested guidance, educational resources, and opportunities for information sharing and peer exchange among organizations. The Center also recognizes exemplary corporate, organizational, and individual leadership in addressing climate change by co-sponsoring the Climate Leadership Awards.

Measuring Progress

The U.S. government is continuing to make important progress toward reducing GHG emissions through policies and measures that promote increased investment in technologies and practices that reduce CO_2, methane, and other GHG emissions across all sectors. Table 4-2 summarizes the U.S. policies and measures discussed above, and provides their estimated annual GHG mitigation impacts in 2011 and expected annual reductions in 2015 and 2020. The estimates are not cumulative reductions; rather, they are a snapshot of estimated annual reductions.

Mitigation levels and projections are estimated using a range of methodologies and assumptions, as appropriate, given sector affected, type of effort, and statutory requirements. Levels and projections are subject to change in the future and may have changed relative to those presented in past reports due to improvements in calculation methodologies. GHG mitigation estimates are offered to demonstrate progress made by individual policies and measures, should not be aggregated to the sectoral level, and may not be directly comparable, due to differences in calculation methodology and possible synergies and interactions among policies and measures that may result in double counting.

---

[68] *See* www.epa.gov/climateleadership.

The policies and measures in this chapter highlight the successful U.S. government initiatives focused on reducing GHG emissions. Although many of them include projections for reducing GHGs, several do not for a variety of reasons, such as potential for double counting, lack of quality data, lack of data specific to program actions, and varying stages of implementation and types of measures. For example, policies to encourage greater transparency and improved measurement of GHG emissions may not reduce emissions directly, but the existence of such policies is key to enabling additional actions to reduce GHG emissions. Further, the projections presented in this chapter should not be compared with the information presented in Chapter 5, which is inclusive of actions from the full suite of U.S. policies and measures, and avoids double counting.

Table 4-2 **Summary of Federal Policies and Measures by Sector**

The U.S. government deploys a robust set of policies and measures to reduce GHG emissions across sectors.

Name of Policy or Measure	Objective and/or Activity Affected	GHGs Affected	Types of Instrument	Status	Implementing Entities	Estimated Mitigation Impacts (Tg CO$_2$e)		
						2011	2015	2020
Transportation								
National Program for Light-Duty Vehicle GHG Emissions and CAFE Standards	Establishes corporate average fuel economy and GHG emission standards for new light-duty vehicles (LDVs) produced for sale in the U.S.	CO$_2$, HFCs	Regulatory	Implemented	DOT/EPA	35.0	92.0	236.0
Renewable Fuel Standard	Increases the share of renewable fuels used in transportation via implementation of the Renewable Fuel Standard program.	CO$_2$	Regulatory	Implemented	EPA	n/a	n/a	138.4
National Program for Heavy-Duty Vehicle GHG Emissions and Fuel Efficiency Standards	Establishes fuel efficiency and GHG emission standards for work trucks, buses, and other heavy-duty vehicles.	CO$_2$, N$_2$O, CH$_4$, HFCs	Regulatory	Implemented	DOT/EPA	n/a	n/a	37.7
SmartWay Transport Partnership	Promotes collaboration with businesses and other stakeholders to decrease climate-related and other emissions from movement of goods.	CO$_2$	Voluntary	Implemented	EPA	23.6	37.0	43.0
Light-Duty Vehicle Fuel Economy and Environment Label	Provides comparable information on new LDVs' fuel economy, energy use, fuel costs, and environmental impacts.	CO$_2$	Regulatory, Information	Implemented	EPA/DOT/ DOE	n/a	n/a	n/a
National Clean Diesel Campaign	Reduces diesel emissions through the implementation of proven emission control technologies and innovative strategies.	CO$_2$	Voluntary/ Negotiated Agreements	Implemented	EPA	n/a	n/a	n/a
Advanced Technology Vehicle Manufacturing Loan Program	Provides direct loans to qualifying U.S. advanced technology vehicles or component and engineering integration projects.	CO$_2$	Economic	Implemented	DOE	1.5	2.5	2.5

Table 4-2 (Continued)   **Summary of Federal Policies and Measures by Sector**

Name of Policy or Measure	Objective and/or Activity Affected	GHGs Affected	Types of Instrument	Status	Implementing Entities	Estimated Mitigation Impacts (Tg CO$_2$e)		
						2011	2015	2020
Next Generation Air Transportation Systems	Achieves more efficient aircraft operations and reduced GHG emissions through airspace, operational, and infrastructure improvements. The Continuous Lower Energy, Emissions, and Noise Program is an element of NextGen.	CO$_2$	Economic, Research	Implemented	DOT	n/a	1.0	3.8
Other Aviation Low-Emission, Fuel Efficiency, and Renewable Fuels Measures	Implement strategies that reduce GHG emissions from the aviation sector.	CO$_2$	Economic, Voluntary, Research	Implemented	DOT	n/a	n/a	n/a
State and Alternative Fuel Provider Fleet Program	Requires covered fleets either to acquire alternative fuel vehicles as a percentage of their annual LDV acquisitions or to employ other petroleum-reduction methods.	CO$_2$	Regulatory	Implemented	DOE	n/a	n/a	n/a
Federal Transit, Highway, and Railway Programs	Help public transportation providers, railways, and other key stakeholders to implement strategies that reduce GHGs.	All	Fiscal, Voluntary, Research	Implemented	DOT	n/a	n/a	n/a
On-road GHG Assessment Tools	Supports and encourages state and local governments to estimate future GHG emissions from the on-road portion of the transportation sector and find strategies to mitigate these effects.	CO$_2$	Information	Implemented	DOT	n/a	n/a	n/a
Energy: Supply								
Clean Energy Supply Programs	Green Power Partnership encourages U.S. organizations to voluntarily purchase green power, and Combined Heat and Power Partnership reduces the environmental impact of power generation by encouraging the use of CHP.	CO$_2$	Voluntary/ Negotiated Agreements	Implemented	EPA	29.6	44.0	73.3
Onshore Renewable Energy Development Programs	Provide opportunities for and encourage use of federal public lands for the development of wind, solar, and geothermal energy.	CO$_2$	Economic, Voluntary	Implemented	DOI/BLM	6.7	25.6	41.5
Rural Energy for America Program	Provides grants and loan guarantees to agricultural producers and rural businesses for energy efficiency and renewable energy systems.	CO$_2$	Voluntary, Economic	Implemented	USDA	1.9	10.2	17.5

Table 4-2 (Continued)   **Summary of Federal Policies and Measures by Sector**

Name of Policy or Measure	Objective and/or Activity Affected	GHGs Affected	Types of Instrument	Status	Implementing Entities	Estimated Mitigation Impacts (Tg CO_2e)		
						2011	2015	2020
CCS Demonstration and Large-Scale Geologic Storage Cooperative Agreements	The power plant, industrial, and geologic storage large-scale carbon capture and storage (CCS) demonstrations are cost-shared cooperative agreements between the government and industry to increase investment in CCS.	CO_2	Economic	Implemented	DOE	1.0	7.0	16.2
Rural Development Biofuels Programs	Supports expansion of biofuels by providing payments to biorefineries and biofuel producers, and providing loan guarantees for biorefineries. Programs include the Bioenergy Program for Advanced Biofuels, Biorefinery Assistance Program, and Repowering Assistance Program.	CO_2	Voluntary, Economic	Implemented	USDA	0.0	0.1	0.1
Biofuel Regional Feedstock Partnerships	Identify and analyze feedstock supply and regional logistics, and conduct crop field trials to address barriers to the development of a sustainable and predictable supply of biomass feedstocks.	CO_2	Economic	Implemented	DOE	n/a	n/a	n/a
Smart Grid Investment Grants	Provide approximately $9 billion toward the modernization of the electric grid in 131 Smart Grid Investment Grant projects around the country through public–private partnerships.	CO_2	Economic	Implemented	DOE	n/a	n/a	n/a
Offshore Renewable Energy Program— Bureau of Ocean Energy Management	Advances a sustainable Outer Continental Shelf renewable energy future through site planning and environmentally responsible operations and energy generation.	CO_2	Regulatory	Implemented	DOI	n/a	n/a	n/a
Price-Anderson and Nuclear Waste Policy Acts	Establish legal responsibility to manage nuclear waste and support the deployment of nuclear power by limiting nuclear plant operators' liability in the event of an accident.	CO_2	Economic	Implemented	DOE	n/a	n/a	n/a
Energy: Residential, Commercial, and Industrial End Use								
Appliance and Equipment Energy Efficiency Standards	Establish minimum energy conservation standards for more than 50 categories of appliances and equipment.	CO_2	Regulatory	Implemented	DOE	156.0	195.0	216.0

Table 4-2 (Continued)   **Summary of Federal Policies and Measures by Sector**

Name of Policy or Measure	Objective and/or Activity Affected	GHGs Affected	Types of Instrument	Status	Implementing Entities	Estimated Mitigation Impacts (Tg CO_2e)		
						2011	2015	2020
ENERGY STAR Labeled Products	Labels distinguish energy-efficient products in the marketplace.	CO_2	Voluntary	Implemented	EPA/DOE	99.7	113.6	141.2
ENERGY STAR Commercial Buildings	Promotes improvement in energy performance in commercial buildings.	CO_2	Voluntary	Implemented	EPA	86.6	75.0	93.5
Lighting Energy Efficiency Standards	Lighting component of DOE's comprehensive Appliance and Equipment Energy Efficiency Standards program.	CO_2	Regulatory	Implemented	DOE	19.0	38.0	41.0
ENERGY STAR for Industry	Promotes improvement in energy performance across industry.	CO_2	Voluntary	Implemented	EPA	32.2	25.6	36.6
ENERGY STAR Certified New Homes	Promotes improvement in energy performance in residential buildings beyond the labeling of products.	CO_2	Voluntary	Implemented	EPA	2.7	3.2	3.8
Home Performance with ENERGY STAR	Provides homeowners with resources to identify trusted contractors for high-quality, comprehensive energy audits and residential retrofits.	CO_2	Economic	Implemented	DOE	0.2	0.8	2.8
Building Energy Codes	Develops cost-effective building energy codes with adoption and compliance strategies.	CO_2	Regulatory	Implemented	DOE	n/a	n/a	n/a
Combined Heat & Power Technical Assistance Partnerships and Industrial Assessment Centers	Provide technical assistance, including energy audits, to increase energy efficiency and reduce costs for CHP plants and industrial processes.	CO_2	Economic	Implemented	DOE	n/a	n/a	n/a
Industrial Processes (Non-CO_2)								
Significant New Alternatives Policy Program	Facilitates smooth transition away from ozone-depleting chemicals in industrial and consumer sectors.	HFCs, PFC, SF_6	Regulatory, Information	Implemented	EPA	206.9	252.0	311.1
Federal Air Standards for Oil and Natural Gas Sector	The new source performance standards control volatile organic compound emissions from various sources, substantially reducing methane emissions as a co-benefit.	CH_4	Regulatory	Adopted	EPA	n/a	32.6	39.9

Table 4-2 (Continued)   **Summary of Federal Policies and Measures by Sector**

Name of Policy or Measure	Objective and/or Activity Affected	GHGs Affected	Types of Instrument	Status	Implementing Entities	Estimated Mitigation Impacts (Tg CO$_2$e)		
						2011	2015	2020
Natural Gas STAR Program	Works with oil and natural gas companies to promote proven, cost-effective technologies and practices that improve operational efficiency and reduce methane (i.e., natural gas) emissions.	CH$_4$	Voluntary, Information	Implemented	EPA	35.3	20.6	22.1
Coalbed Methane Outreach Program	Voluntary program with the goal of reducing methane emissions from coal mining activities.	CH$_4$	Voluntary, Information	Implemented	EPA	8.5	9.3	9.4
SF$_6$ Emission Reduction Partnership for Electric Power Systems	Partners with electric power transmission and distribution companies to reduce emissions of SF$_6$, which is used as a gaseous dielectric in high-voltage circuit breakers and gas-insulated substations.	SF$_6$	Voluntary, Information	Implemented	EPA	6.4	9.0	9.3
GreenChill Advanced Refrigeration Partnership	Reduces ozone-depleting and GHG refrigerant emissions from supermarkets.	HFCs	Voluntary/ Negotiated Agreements, Information, Education	Implemented	EPA	3.8	5.4	8.8
Responsible Appliance Disposal Program	Reduces emissions of refrigerant and foam-blowing agents from end-of-life appliances.	HFCs	Voluntary/ Negotiated Agreements	Implemented	EPA	0.3	0.4	0.6
Voluntary Aluminum Industry Partnership	Partners with industry to reduce PFCs, tetrafluoro-methane, and hexafluoro-ethane where cost-effective technologies and operational practices are technically feasible.	PFCs	Voluntary, Information	Implemented	EPA	6.3	0.4	0.4
Voluntary Code of Practice for the Reduction of Emissions of HFC and PFC Fire Protection Agents	Minimizes nonfire emissions of HFCs and PFCs used as fire-suppression alternatives and protects people and property from the threat of fire through the use of proven, effective products and systems.	HFCs, PFCs	Voluntary/ Negotiated Agreements	Implemented	EPA	n/a	n/a	n/a
Agricultural								
Conservation Reserve Program	Encourages farmers to convert highly erodible cropland or other environ-mentally sensitive acreage.	CO$_2$	Economic, Information	Implemented	USDA	51.6	41.5–61.2	41.5–61.2
Natural Resources Conservation Service	Helps landowners to implement practices or measures that address natural resource concerns.	CO$_2$, CH$_4$, N$_2$O	Voluntary, Economic, Information	Implemented	USDA	11.9	20.1	27.6

Table 4-2 (Continued)   **Summary of Federal Policies and Measures by Sector**

Name of Policy or Measure	Objective and/or Activity Affected	GHGs Affected	Types of Instrument	Status	Implementing Entities	Estimated Mitigation Impacts (Tg CO₂e)		
						2011	2015	2020
AgSTAR	Encourages the use of methane (biogas) recovery technologies at confined animal feeding operations that manage manure as liquids or slurries.	CH_4	Voluntary, Information	Implemented	EPA/USDA	1.2	0.9	0.9
Forestry								
Woody Biomass Utilization Grants Program	Creates markets for small-diameter woody material and low-valued trees removed from forest restoration activities.	CO_2	Voluntary, Economic, Information	Implemented	USDA	n/a	n/a	n/a
Forest Ecosystem Restoration and Hazardous Fuels Reduction Programs	Restore the health of the nation's forests, woodlands, and rangelands.	CO_2	Voluntary	Implemented	USDA/DOI	n/a	n/a	n/a
Waste Management/Waste								
Landfill Air Regulations	Limit GHG emissions by limiting landfill gas emissions from landfills that are at least 2.5 million megagrams in size. Landfill gas is approximately 50 percent methane.	CH_4	Regulatory	Implemented (under 8-year review)	EPA	n/a	162.7	183.1
Landfill Methane Outreach Program	Reduces GHG emissions at landfills by supporting the recovery and use of landfill gas for energy.	CH_4	Voluntary, Information	Implemented	EPA	15.8	14.3	15.7
Sustainable Materials Management	Provides a systemic approach to reduce the use of materials and their associated environmental impacts over their entire life cycle.	CO_2	Voluntary/ Negotiated Agreements, Information, Education	Implemented	EPA	n/a	<0.1	<0.1
Wastewise	Helps organizations and businesses apply sustainable material management practices to reduce municipal and select industrial wastes.	CO_2	Voluntary/ Negotiated Agreements, Information, Education	Implemented	EPA	23.2	n/a	n/a
Federal Government								
Federal Energy Management Program	Promotes energy efficiency and renewable energy use in federal buildings, facilities, and operations.	CO_2	Regulatory	Implemented	DOE	4.2	10.0	14.4
National Parks Service Programs	Support efforts to mitigate the effects of climate change and integrate sustainable practices.	CO_2	Economic, Voluntary, Educational	Implemented	DOI	<0.1	0.1	0.2

Table 4-2 (Continued)   **Summary of Federal Policies and Measures by Sector**

Name of Policy or Measure	Objective and/or Activity Affected	GHGs Affected	Types of Instrument	Status	Implementing Entities	Estimated Mitigation Impacts (Tg CO_2e)		
						2011	2015	2020
Cross-Cutting								
State Energy Program	Provides funding to state energy offices to reduce market barriers to the cost-effective adoption of renewable energy and energy efficiency technologies.	CO_2	Economic	Implemented	DOE	8.6	14.9	16.2
Energy Efficiency and Conservation Block Grants	Assist eligible entities in implementing strategies that will improve energy efficiency in the transportation, building, and other sectors, and reduce fossil fuel emissions and total energy use.	CO_2	Economic	Implemented	DOE	7.1	11.3	11.3
Section 1703/1705 Loan Guarantee Program	Mitigates the financing risks associated with innovative and advanced energy.	CO_2	Economic	Implemented	DOE	0.4	7.3	7.3
Weatherization Assistance Program	Provides funding and technical support to states, U.S. territories, and tribes, which in turn work with a network of about 900 local agencies to provide trained crews to perform residential weatherization services for income-eligible households.	CO_2	Economic	Implemented	DOE	1.9	2.9	3.3
Indian Energy Policy and Programs/Tribal Energy Program	Provides financial and technical assistance that enables American Indian and Alaska Native tribes to deploy renewable energy resources, reduce their energy costs through efficiency and weatherization, and increase energy security for tribes and villages.	CO_2	Economic	Implemented	DOE	0.1	0.2	0.4
Climate Showcase Communities Grant Program	In 2009 and 2010, EPA awarded $20 million in grants to help local and tribal governments take steps to reduce GHG emissions while achieving additional environmental, economic, and social benefits.	CH_4, CO_2	Economic, Information	Implemented	EPA	<0.1	0.4	0.4

Table 4-2 (Continued)   **Summary of Federal Policies and Measures by Sector**

Name of Policy or Measure	Objective and/or Activity Affected	GHGs Affected	Types of Instrument	Status	Implementing Entities	Estimated Mitigation Impacts (Tg CO_2e)		
						2011	2015	2020
Community Renewable Energy Deployment Grants	Create up to a 50% matching grant for the construction of small renewable energy projects that will have commercial electrical generation capacity of less than 15 megawatts. Types of renewable energy sources include solar, wind, geothermal, ocean, biomass, and landfill gas.	CO_2	Economic	Implemented	DOE	n/a	n/a	n/a
Tax Provisions	Provide incentives for alternative fuel vehicles and renewable/alternative energy production. Encourage energy conservation, production of renewable energy and energy efficiency manufacturing projects, and carbon sequestration.	CO_2	Economic	Adopted	Treasury	n/a	n/a	n/a
Interagency Partnership for Sustainable Communities	Encourages integrated regional planning by aligning federal policies for housing, transportation, and the environment.	All	Voluntary, Economic, Information	Implemented	EPA/DOT/ HUD	n/a	n/a	n/a
Center for Corporate Climate Leadership	Serves as a resource center for organizations interested in GHG measurement and management.	All	Information	Implemented	EPA	n/a	n/a	n/a

Notes:

• n/a (i.e., not applicable) indicates either the value does not apply for the given year or quantifying GHG emissions does not apply.

• The methodologies in this chapter are estimates and are not intended to be aggregated for the purpose of understanding the "with measures" trajectory reflected in Chapter 5.

• The estimated mitigation impacts are an annual estimate, but are calculated from the year the policy or measure was implemented. The start year can vary significantly from one policy or measure to the next.

• BLM = Bureau of Land Management; CAFE = corporate average fuel economy; CCS = carbon capture and storage; CH_4 = methane; CHP = combined heat and power; CO_2 = carbon dioxide; CO_2e = carbon dioxide equivalent; DOE = U.S. Department of Energy; DOI = U.S. Department of the Interior; DOT = U.S. Department of Transportation; EPA = U.S. Environmental Protection Agency; GHG = greenhouse gas; HFCs = hydrofluorocarbons; HUD = U.S. Department of Housing and Urban Development; LDV = light-duty vehicle; N_2O = nitrous oxide; PFCs = perfluorocarbons; SF_6 = sulfur hexafluoride; Tg = teragram; USDA = U.S. Department of Agriculture.

NONFEDERAL POLICIES AND MEASURES

In the United States, local, tribal, state, and federal governments share responsibility for the nation's economic development, energy, natural resource, and many other issues that affect climate mitigation. The federal government supports state and local government actions to reduce GHG emissions by sponsoring policy dialogues, issuing technical documents, facilitating consistent measurement approaches and model policies, and providing direct technical assistance. Table 4-3 summarizes key federal programs that provide support to state and local activities across four sectors. Such federal support helps state and local governments learn from each other to leverage best practices, helping reduce overall time and cost for both policy adoption and implementation. The federal government also helps state and local governments learn from each other to leverage policy and program best practices for climate mitigation.

State Policies and Measures

Numerous state policies and measures complement federal efforts to reduce GHG emissions. A wide range of key policies affects the electricity and transportation sectors, from actions that regulate GHG emissions to complementary policies that indirectly reduce emissions (Figure 4-1 and Table 4-4).

Carbon Markets Initiatives

Regional Greenhouse Gas Initiative[69]—Launched on January 1, 2009, the Regional Greenhouse Gas Initiative (RGGI) is the first mandatory, market-based U.S. cap-and-trade program to

Figure 4-1 and Table 4-4 **States implementing Renewable Energy, Energy Efficiency, and Greenhouse Gas Policies and Measures**

Key state policies and measures are complementing federal efforts to reduce GHG emissions.

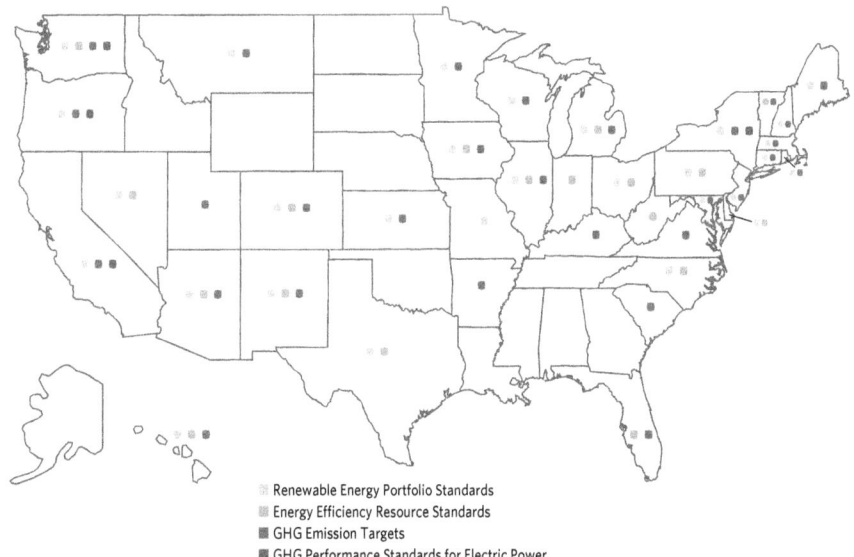

Renewable Energy Portfolio Standards
Energy Efficiency Resource Standards
GHG Emission Targets
GHG Performance Standards for Electric Power

Key Policies and Measures	Number of States
Renewable Energy Portfolio Standards	29
Energy Efficiency Resource Standards	18
GHG Emission Targets	29
GHG Performance Standards for Electric Power	4

Note: The count is inclusive of mandatory portfolio and resource standards only.
Source: U.S. EPA State and Local Climate and Energy Program. <http://www.epa.gov/statelocalclimate/>

[69] See www.rggi.org.

Table 4-3 **Federal Programs Supporting State and Local Policies and Measures**

Key federal programs are helping state and local governments to leverage best practices to reduce the overall time and cost for adopting and implementing policies and measures for reducing GHG emissions across four sectors.

Name of Policy or Measure	Overview	Sectors	GHGs Affected
New On-Road GHG Assessment Tools (EPA)	Supports and encourages state and local governments to estimate future GHG emissions from the on-road portion of the transportation sector and find strategies to mitigate these effects.	Transport	CO_2
Federal Transit Program (DOT)	Provides communities with the tools to effectively coordinate land use and transportation decisions, as well as provide training in environmental management systems to help transit agencies reduce the environmental impact of their operations.	Transport	CO_2
Climate Showcase Communities Grant Program (EPA)	In 2009 and 2010, EPA awarded $20 million in grants through this program to help local and tribal governments take steps to reduce GHG emissions, while achieving additional environmental, economic, and social benefits.	Transport; Energy: Supply, Residential, and Commercial; Waste Management	CO_2 and CH_4
State Energy Program (DOE)	Provides financial and technical assistance to states through formula and competitive grants.	Energy: Supply, Residential, Commercial, and Industrial	CO_2
Energy Efficiency and Conservation Block Grant Program (DOE)	Provided more than $2.7 billion in funding to local and state governments, Indian tribes, and territories to develop and implement projects to improve energy efficiency and reduce energy use and fossil fuel emissions in their communities.	Energy: Supply, Residential, Commercial, and Industrial	CO_2
Indian Energy Policy and Programs/Tribal Energy Program (DOE)	Provide financial and technical assistance that enables American Indian and Alaska Native tribes to deploy renewable energy resources, reduce their energy costs through efficiency and weatherization, and increase their energy security.	Energy: Supply, Residential, Commercial, and Industrial	CO_2
Better Buildings, Better Plants Program (DOE)	Shares implementation models among participants, including state and local governments, as a part of its broader efforts.	Energy: Residential, Commercial, and Industrial	CO_2
Regional Clean Energy Application Centers (DOE)	Promote and assist in transforming the market for CHP, waste heat to power, and district energy technologies and concepts. As part of this work, CEACs provide information on the benefits and applications of CHP to state and local policy makers.	Energy: Residential, Commercial, and Industrial	CO_2
Industrial Assessment Centers (DOE)	Provide in-depth assessments of a plant's site and its facilities, services, and manufacturing operations.	Energy: Industrial	CO_2
Weatherization Assistance Program (DOE)	Reduces energy costs for low-income households by increasing the efficiency of their homes, and provides technical assistance to local governments.	Energy: Residential and Commercial	CO_2
Building Energy Codes Program (DOE)	Provides technical assistance to states and localities as they adopt and enforce energy codes.	Energy: Residential and Commercial	CO_2
Home Performance with ENERGY STAR (DOE)	Provides a directed set of resources to more than 50 Program Sponsors who are represented by local organizations (utilities, state energy offices, etc.).	Energy: Residential	CO_2

Note: CEACs = Clean Energy Application Centers; CHP = combined heat and power; CH_4 = methane; CO_2 = carbon dioxide; DOE = U.S. Department of Energy; DOT = U.S. Department of Transportation; EPA = U.S. Environmental Protection Agency; GHG = greenhouse gas.

reduce GHG emissions. RGGI currently applies to 168 electricity generation facilities in nine Northeast and Mid-Atlantic states, which account for approximately 95 percent of CO_2 emissions from electricity generation in the region. In February 2013, the participating states agreed to significantly revise the program. Under these revisions, CO_2 emissions will be capped at 91 million short tons per year in 2014, a 45 percent reduction from the previous cap of 165 million short tons. The cap will then be reduced by 2.5 percent each year from 2015 through 2020. Under the program, nearly 90 percent of allowances are distributed through auction. As of March 2013, cumulative auction proceeds exceeded $1.2 billion. Participating states have invested approximately 80 percent of auction proceeds in consumer benefit programs, including investments in state and local government end-use energy efficiency and renewable energy deployment programs.

California Global Warming Solutions Act[70]—California's Global Warming Solutions Act (Assembly Bill [AB] 32) was signed into law in 2006, establishing a statewide GHG emissions limit of 1990 levels to be achieved by 2020. As part of a portfolio of measures implemented to achieve this statewide limit, the California Air Resources Board (ARB) adopted cap-and-trade regulations in 2011. The program established a declining cap limit on emission sources responsible for approximately 85 percent of statewide GHG emissions, including refineries, power plants, industrial facilities, and transportation fuels. In addition to the cap-and-trade program, the portfolio of programs implemented to achieve the statewide GHG emissions limit under AB 32 includes a mandatory GHG emissions reporting program for large emitters, a renewable energy portfolio standard (RPS), and various energy efficiency measures and incentives.

Other State GHG Policies and Measures

State Emission Targets—As of August 2013, 29 states had adopted some sort of state GHG reduction target or limit, although these vary in stringency, timing, and enforceability. Statewide GHG targets are nonregulatory commitments to reduce GHG emissions to a specified level in a certain timeframe (e.g., 1990 levels by 2020). Such targets can be included in legislation, but are more typically established by the governor in an executive order or a state advisory board in a climate change action plan.

Statewide GHG caps also commit to reduce emissions in a certain time frame, but are regulatory in nature and more comprehensive than emission targets. These policies can include regulations to require GHG emission reporting and verification, and may establish authority for monitoring and enforcing compliance. An emission cap can be combined with emission trading into a cap-and-trade program.

Performance Standards for Electric Power—As of February 2013, three states (New York, Oregon, and Washington) have GHG emission standards for electric-generating utilities, requiring power plants to have emissions equivalent to or lower than the established standard. For example, in New York, new or expanded baseload plants (25 MW and larger) must meet an emission rate of either 925 pounds (lb) of CO_2 per megawatt-hour (MWh) (output based) or 120 lb CO_2/per million British thermal units (MMBtu) (input based). Non-baseload plants (25 MW and larger) must meet an emission rate of either 1,450 lb CO_2/MWh (output based) or 160 lb CO_2/MMBtu (input based).

Three states (California, Oregon, and Washington) also have standards that apply to electric utilities that provide electricity to retail customers. These standards place conditions on the emission attributes of electricity procured by electric utilities. For example, in Oregon and Washington, electric utilities may only enter into long-term power purchase agreements for baseload power if the electric generator supplying the power has a CO_2 emission rate of 1,100 lb CO_2/MWh or less.

Integrated Utility Emission Reduction Plans—In addition to state GHG emission reduction policies, states are finding other ways to reduce emissions. One example is Integrated Utility Emission Reduction Plans, where utilities partner with state governments to develop plans to reduce emissions. The most notable example is Colorado's Clean Air–Clean Jobs Act, which requires utilities to consider current and reasonably foreseeable air pollution regulations, and create a plan that could include emission controls, generation plant refueling, or retirement of certain units.

[70] See http://www.arb.ca.gov/cc/ab32/ab32.htm.

Transportation Policies

Transportation Climate Initiative—The counterpart to RGGI for the transportation sector is the Transportation Climate Initiative (TCI), a regional collaboration of 12 Northeast and Mid-Atlantic jurisdictions that seeks to stimulate sustainable economic development and improve the environment by supporting innovative technologies and smart planning, and by finding greater efficiencies within the transportation sector. TCI's core work areas are expediting the deployment of electric vehicles and alternative fuels; creating sustainable communities; adopting innovative communications technologies, e.g., to promote public transit and expand the use of real-time information on traffic and alternative routes; and advancing more efficient freight movement. Already, TCI jurisdictions have taken action by forming the Northeast Electric Vehicle Network, and all TCI states have agreed to regional sustainability principles that make sustainable development a top regional transportation goal.

California's Senate Bill 375—To achieve the GHG reduction goals set out in California's AB 32, California Senate Bill 375 (SB 375) focuses on reducing VMT and urban sprawl. SB 375 became law on January 1, 2009, to more specifically address the transportation and land-use components of GHG emissions. SB 375 prompts California regions to work together to reduce GHG emissions from cars and light trucks, and requires integration of planning processes for transportation, land use, and housing. The goal is for integrated planning to lead to more efficient communities that provide residents with alternatives to using single-occupant vehicles.

Specifically, SB 375 requires the California ARB to develop regional reduction targets for automobile and light-truck GHG emissions for each region. California metropolitan planning organizations, which are traditionally responsible for transportation planning, are tasked with creating a Sustainable Communities Strategy that combines transportation and land-use elements to achieve the emission reduction target set by ARB, if feasible. SB 375 also offers local governments regulatory and other incentives to encourage more compact new development and transportation alternatives.

Renewable Energy and Energy Efficiency

Renewable Portfolio Standards[71]—A mandatory RPS requires utilities to supply a certain amount of electricity to customers from renewable energy sources or install a certain amount of electricity-generating capacity from renewable energy sources in a set time frame. As of January 2013, 29 states had an RPS.

Energy Efficiency Resource Standards[72]—Energy efficiency resource standards (EERSs) require utilities to reduce energy use by a certain percentage or amount each year. Standards can vary, with annual or cumulative targets. As of August 2013, 18 states had a mandatory EERS program in place. The Lawrence Berkeley National Laboratory considers EERS policies to be one of the most significant drivers for state spending on energy efficiency.[73]

Government Lead-by-Example Procurement Activity—Many state and local governments lead by example by establishing programs that achieve substantial energy cost savings within their own operations and buildings (owned or leased). These lead-by-example programs include energy standards for new buildings, binding usage reductions for existing buildings, and innovations in financing efficiency projects. In addition to reducing state energy bills and emissions, these efforts demonstrate the feasibility and benefits of clean energy and serve as a model to others.

Public Benefit Funds[74]—As of August 2013, 19 states, Washington, D.C., and Puerto Rico have some form of public benefit funds, in which utility consumers pay a small charge to a common fund, often as part of the monthly billing cycle. The utility uses these funds to invest in energy efficiency and renewable energy projects and programs, such as home weatherization and renewable technologies. Existing funds are anticipated to generate $7.7 billion by 2017.

Local Policies and Measures[75]

Local governments are also making a significant contribution to overall U.S. GHG reductions (Box 4-6). Actions taken by local governments are complementary to and supported by state and federal government policies and programs. While local governments often best understand and can directly control the local factors that influence GHG reductions, the creation and implementation of their reduction policies and programs can benefit from support at the state and federal levels. EPA provides such resource support in many forms, including peer exchange; training opportunities; and planning, policy, technical, and analytical support.

[71] See http://www.dsireusa.org/summarytables/rrpre.cfm.

[72] See http://emp.lbl.gov/publications/future-utility-customer-funded-energy-efficiency-programs-united-states-projected-spend.

[73] See http://emp.lbl.gov/publications/future-utility-customer-funded-energy-efficiency-programs-united-states-projected-spend.

[74] See http://dsireusa.org/documents/summarymaps/PBF_Map.pdf.

[75] See http://www.epa.gov/statelocalclimate/local/index.html.

Box 4-6 **U.S. Cities as Global Leaders**[76]

Eleven U.S. cities are members of C40 Cities, a climate leadership group comprised of 58 mega-cities from around the world. These cities are implementing innovative and effective policies and programs in buildings, renewable energy, lighting, ports, transport, and waste that can serve as a model to other communities worldwide.

New York, New York—New York Mayor Michael Bloomberg chairs the C40 and was the driving force behind PlaNYC—a plan to reduce GHGs by 30 percent by 2030. The city has achieved a 13 percent reduction in GHGs by enacting stringent building energy efficiency laws, increasing transit options, and improving infrastructure—all while continuing to grow its economy and addressing climate resiliency in the wake of Hurricane Sandy.

Houston, Texas—Houston has recently emerged as a new leader in sustainability. The city was one of five winners of the Bloomberg Philanthropies' Mayors Challenge for an innovative single-stream waste management approach. Houston is the largest municipal purchaser of green power in the United States, provides financial incentives for commercial building efficiency improvements, implemented the nation's first municipal electric vehicle fleet-sharing system, and recently expanded the city's pilot bike-share program.

San Francisco, California—San Francisco is on the path to achieving its long-term goal of reducing GHGs by 80 percent from 1990 levels by 2050. Since 2010, San Francisco has reduced its emissions by 14.5 percent below 1990 levels through a combination of initiatives, such as the San Francisco Carbon Fund and Community Climate Acton Advisory Panels. San Francisco has the largest municipally owned solar power system in the United States, generating 826 MWh annually; runs one of the largest clean air fleets; and is aggressively pursuing a goal of zero waste by 2020.

Seattle, Washington—Seattle initiated the United States Mayors Climate Protection Agreement in 2005. Now pursuing a goal of climate neutrality by 2050, the city has the nation's first carbon-neutral electric utility, strong green building efficiency mandates, and expansive light-rail and transit options, and is working with the private sector to address port-related emissions.

Sustainability and Energy Planning

Local governments are addressing GHG emissions through integrated energy and environmental planning. This approach considers both energy supply and demand to ensure long-range energy policies are both environmentally sensible and economically feasible. El Cerrito, California, is working with neighborhoods to monitor energy use and identify opportunities for energy savings. By creating a multi-jurisdictional GHG planning and management framework for small communities, El Cerrito is demonstrating how small governments can partner to share resources and best practices. By aggregating resources, these small local governments can overcome barriers to climate change mitigation and achieve economies of scale that make mitigation easier and more cost-effective.

Transportation and Land Use Planning

Local governments are addressing GHG emissions through a variety of transportation and land-use planning initiatives. For example, Salt Lake City, Utah, aims to reduce GHG emissions by reducing VMTs using a community-based social marketing campaign to promote public transit, walking, biking, carpooling, and teleworking. And Tompkins County, New York, is creating models in three pilot projects for new building codes, policies, and zoning ordinances to support sustainable development and decrease emissions (Box 4-7).

Sustainable Materials Management

Reducing solid waste through sustainable materials management is a method that communities are undertaking to reduce GHG emissions. The Alameda County Waste Management Authority in California has launched a project to reduce limited-use transport packing materials (such as wooden pallets and cardboard boxes) by helping businesses convert to sustainable and reusable alternatives. Switching to reusable alternatives not only reduces solid waste, but also reduces GHG emissions from raw materials extraction and the production, transport, and landfilling of packaging.

---

[76] For additional examples of C40 actions, see http://www.c40cities.org/.

Box 4-7 **EPA's New On-Road GHG Assessment Tools**

By providing models, tools, and guidance, EPA supports and encourages state and local governments to estimate future GHG emissions from the on-road portion of the transportation sector, and find strategies to mitigate these effects. To fulfill its mission of protecting air quality and public health, EPA develops on-road emissions models to project future levels of emissions of all types of air pollutants from all on-road vehicles, including cars, trucks, and buses. In 2012, EPA updated its state-of-the-art model on-road emissions model, MOVES (Motor Vehicle Emissions Simulator).[77]

In 2011, EPA documented an approach called the Travel Efficiency Assessment Method (TEAM) for assessing the potential of on-road travel efficiency strategies to reduce pollution and GHG emissions. Travel efficiency strategies affect travel activity, such as travel demand management (telecommuting, transit subsidies); public transit fare changes and service improvements; road and parking pricing; and land use/smart growth. TEAM uses regionally derived travel model data and other travel activity information with EPA's MOVES model to estimate emissions reduced.

EPA has developed a guide for planners to apply the method locally.[78] EPA has also released information on transportation control measures that have been implemented across the country for a variety of purposes, including reducing GHGs and the six common air pollutants for which EPA sets National Ambient Air Quality Standards.[79]

Residential Energy Efficiency

Considering the sizable contribution the residential sector makes to overall GHG emissions, residential energy efficiency measures represent an important strategy for reducing emissions. Many communities are making efforts to improve residential energy efficiency in a variety of ways. For example, Durham City-County, North Carolina, is instituting a neighborhood-based residential energy efficiency program targeting at least 344 residences. By leveraging existing neighborhood relationships, focusing on streamlining the residential upgrades, and targeting households ineligible for other retrofit funding, Durham has demonstrated an effective strategy to achieve cost-effective and timely reductions of GHGs.

Energy Efficiency in Government Operations

Improved energy efficiency in local government operations also represents a way communities can lead by example and reduce their GHG emissions. The Delaware Valley Regional Planning Commission is providing training assistance to the governments of small and medium-sized municipalities in four suburban counties in southeastern Pennsylvania to develop and implement strategies to reduce energy use and GHG emissions associated with their operations.

Commercial Energy Efficiency

Communities are also reducing emissions by improving energy efficiency in the commercial sector. The Tri-County Small Business Efficiency Program is working to educate small business owners in three counties in Montana about strategies to reduce their energy and water use. The program is also helping small business owners make energy efficiency improvements by partnering with local energy and water utilities to offer free energy audits and financial assistance to implement audit recommendations.

A growing number of communities (including Austin, New York City, Seattle, San Francisco, Philadelphia, and Minneapolis) have adopted policies that require benchmarking and disclosure of commercial building energy use using EPA's Portfolio Manager™. Mandatory benchmarking allows building owners to compare energy use and efficiency among comparable buildings, and mandatory disclosure provides information on energy use for potential building purchasers and renters. The availability of this information highlights the value of energy efficiency in the commercial building market.

Renewable Energy Programs

Renewable energy programs reduce GHG emissions by providing energy from nonemitting and lower-emitting sources of energy, such as solar, wind, geothermal, and low-impact hydropower. Many communities are realizing the benefits of utilizing renewable energy. For

[77] See http://www.epa.gov/otaq/ models/moves/index.htm.

[78] See http://www.epa.gov/otaq/ stateresources/ghgtravel.htm.

[79] See http://www.epa.gov/otaq/ stateresources/policy/430r09040.pdf.

example, West Union, Iowa, is installing a geothermal heating and cooling system for six blocks of the downtown area. When finished, this system will provide heating and cooling to 80 percent of the building space in this area.

Some states have laws allowing community choice aggregation (CCA) through which local governments aggregate electricity demand within their jurisdictions to procure alternative energy supplies, while maintaining the existing electricity provider for transmission and distribution services. A small but growing number of municipalities (e.g., Cincinnati and Cleveland, Ohio; Normal, Evanston, Oak Park, Peoria, Urbana, and Chicago, Illinois; and Marin, California) have used CCA to purchase electricity from renewable energy sources.

5

Projected Greenhouse Gas Emissions

This chapter provides projections of U.S. greenhouse gas (GHG) emissions through 2030, including the effects of policies and measures in effect as of September 2012, the cutoff date for the 2013 *Annual Energy Outlook*'s baseline projections of energy-related carbon dioxide (CO_2) emissions (U.S. DOE/EIA 2013b). The "2012 Policy Baseline"[1] scenario presented does not include the impacts of the President's June 2013 *Climate Action Plan* (EOP 2013a).[2] The projections of U.S. GHG emissions described here reflect national estimates considering population growth, long-term economic growth potential, historic rates of technology improvement, and normal weather patterns. They are based on anticipated trends in technology deployment and adoption, demand-side efficiency gains, fuel switching, and many of the implemented policies and measures discussed in Chapter 4.

Policies that are proposed or planned but had not been implemented as of September 2012, as well as sections of existing legislation that require implementing regulations or funds that have not been appropriated, are not included in this chapter's projections.[3] The projections include, for example, efficiency and emission standards for cars and trucks, existing appliance efficiency standards and programs, state renewable energy portfolio standards, and federal air standards for the oil and natural gas industry. They do not include additional measures from *The President's Climate Action Plan*, announced June 2013 (EOP 2013a). Projections that take into account the actions planned as a result of this announcement are contained in the U.S. *Biennial Report*, which accompanies this *Sixth National Communication*.

Projections are provided in total by gas and by sector. Gases included in this report are CO_2, methane (CH_4), nitrous oxide (N_2O), hydrofluorocarbons (HFCs), perfluorocarbons (PFCs), and sulfur hexafluoride (SF_6). Sectors reported include energy (subdivided into electric power, residential, commercial, and industrial); transportation; industrial processes; agriculture; waste; and land use, land-use change, and forestry (LULUCF). Projections for LULUCF through 2030 are presented as a range based on alternative high- and low-sequestration scenarios, while the section also describes longer-term trends in the sector.

The tables in this chapter present emission trends from 2000 through 2030. The discussion in the text focuses on the projected change in emissions between 2005 and 2020.

U.S. GREENHOUSE GAS EMISSION PROJECTIONS

The U.S. Department of Energy's (DOE's) Energy Information Administration (EIA) *Annual Energy Outlook 2013 (AEO2013)* Reference case provided the projection of energy-related CO_2 emissions in the 2012 policy baseline scenario presented in this chapter (U.S. DOE/EIA 2013b). Projected CO_2 emissions in the *AEO2013* Reference case were adjusted to match international inventory convention.[4] The U.S. Environmental Protection Agency (EPA) prepared the projections of non-energy-related CO_2 emissions and non-CO_2 emissions. The methodologies used to project non-CO_2 emissions are explained in the background document *Methodologies for U.S. Greenhouse Gas Emissions Projections: Non-CO_2 and Non-Energy CO_2 Sources* (U.S. EPA 2013b). The U.S. Department of Agriculture (USDA) prepared the estimates of carbon sequestration. Historical emissions data are drawn from the *Inventory of U.S.*

[1] The "baseline" refers to the "with measures" scenario required by the United Nations Framework Convention on Climate Change National Communications reporting guidelines (UNFCCC 2006).

[2] See the *U.S. Biennial Report* for more information on the effects of planned additional measures.

[3] Specifically, the *Annual Energy Outlook 2013*, which provides the baseline projection of energy-related CO_2 emissions, is based generally on federal, state, and local laws in effect as of the end of September 2012 (U.S. DOE/EIA 2013b).

[4] *AEO2013* estimates for CO_2 from fossil fuel combustion were adjusted for the purpose of these projections to remove emissions from bunker fuels and non-energy use of fossil fuels, and to add estimated CO_2 emissions in the U.S. territories (since these emissions are not included in *AEO2013*), consistent with international inventory convention. These changes are consistent with previous U.S. Climate Action Reports.

Greenhouse Gas Emissions and Sinks: 1990–2011 (U.S. EPA/OAP 2013). In general, the projections reflect long-run trends and do not attempt to mirror short-run departures from those trends.

All GHGs in this chapter are reported in teragrams of CO_2 equivalents (Tg CO_2e), which are equivalent to megatons. The conversions of non-CO_2 gases to CO_2e are based on the 100-year global warming potentials listed in the Intergovernmental Panel on Climate Change's (IPCC's) Second Assessment Report (IPCC 1996). Projected emissions for 2015, 2020, 2025, and 2030 are presented with historical GHG emissions from 2000 through 2011 from the 2013 U.S. GHG Inventory (U.S. EPA/OAP 2013). The base year for emission projections is 2011.

TRENDS IN TOTAL GREENHOUSE GAS EMISSIONS

Given implementation of programs and measures in place as of September 2012 and current economic projections, total gross U.S. GHG emissions are projected to be 5.3 percent lower than 2005 levels in 2020. Between 2005 and 2011, total gross U.S. GHG emissions declined significantly due to a combination of factors, including the economic downturn and fuel switching from coal to natural gas (U.S. EPA/OAP 2013). Emissions are projected to rise gradually between 2011 and 2020. However, emissions are projected to remain below the 2005 level through 2030, despite significant increases in population (26 percent) and gross domestic product (GDP) (69 percent) over that time period (Table 5-1). More rapid improvements in technologies that emit fewer GHGs, new GHG mitigation requirements, or more rapid adoption of voluntary GHG emission reduction programs could result in lower gross GHG emission levels than in the *baseline* projection.

Between 2005 and 2020, CO_2 emissions in the 2012 policy baseline projection are estimated to decline by 7.6 percent. The expected decline over this period differs from the projections presented in the *U.S. Climate Action Report* 2010 (2010 CAR) (U.S. DOS 2010). At that time,

Table 5-1 **Historical and Projected U.S. GHG Emissions Baseline, by Gas: 2000–2030** (Tg CO_2e)

Total gross U.S. GHG emissions are projected to be 5.3 percent lower than 2005 levels in 2020. CO_2 emissions are projected to decline 7.6 percent over this period.

Greenhouse Gases		Historical GHG Emissions[a]				Projected GHG Emissions			
		2000	2005	2010	2011	2015	2020	2025	2030
Carbon Dioxide[b]		5,972	6,109	5,738	5,613	5,545	5,647	5,705	5,732
Methane[c]		609	594	593	587	578	599	619	626
Nitrous Oxide[c]		359	356	344	357	343	347	359	364
Hydrofluorocarbons[c]		105	115	121	129	161	207	269	302
Perfluorocarbons[c]		13	6	6	7	6	5	6	7
Sulfur Hexafluoride[c]		19	15	10	9	10	9	10	10
International Bunker Fuels (not included in totals)		103	114	118	112	115	118	120	122
Total Gross Emissions		**7,076**	**7,195**	**6,812**	**6,702**	**6,643**	**6,815**	**6,967**	**7,041**
Sequestration Removals[d]	high sequestration	-682	-998	-889	-905	-884	-898	-917	-937
	low sequestration					-787	-614	-573	-565
Total Net Emissions	**high sequestration**	**6,395**	**6,197**	**5,923**	**5,797**	**5,759**	**5,918**	**6,050**	**6,104**
	low sequestration					**5,856**	**6,201**	**6,394**	**6,476**

[a] Historical emissions and sinks data are from U.S. EPA/OAP 2013. Bunker fuels and biomass combustion are not included in inventory calculations.

[b] Energy-related CO_2 projections are calculated from U.S. DOE/EIA 2013b, with adjustments made to remove emissions from non-energy use of fuels and international bunker fuels, and to add emissions associated with U.S. territories, which are included in totals.

[c] Non-CO_2 and non-energy CO_2 emission projections are based on U.S. EPA/OAP 2013.

[d] Sequestration removals apply only to CO_2 from the land use, land-use change, and forestry sector.

CO_2 emissions were expected to increase by 1.5 percent between 2005 and 2020, a change of about 9 percent. During the same period, CH_4 emissions are expected to grow by 1.0 percent, and N_2O emissions are expected to decline by 2.5 percent. The most rapid growth is expected in emissions of fluorinated GHGs (HFCs, PFCs, and SF_6), which are projected to increase by more than 60 percent between 2005 and 2020, driven by increasing use of HFCs as substitutes for ozone-depleting substances (ODSs).

EMISSION PROJECTIONS BY GAS

Energy-related CO_2 emission estimates are based on the *AEO2013* Reference case, with adjustments to match international inventory convention (U.S. DOE/EIA 2013b). *AEO2013* presents projections and analysis of U.S. energy supply, demand, and prices through 2040, based on results from EIA's National Energy Modeling System.[5] Key issues highlighted in *AEO2013* include the effect of eliminating the sunset provisions of such policies as Corporate Average Fuel Economy standards, appliance standards, and the production tax credit; oil and gas price and production trends; competition between coal and natural gas in electric power generation; high and low nuclear scenarios through 2040; and the impact of growth in natural gas liquids production (US DOE/EIA 2013b).

Non-CO_2 (CH_4, N_2O, HFCs, PFCs, and SF_6) and non-energy CO_2 emission projections are developed by EPA. Specific calculations to project emissions from each source category are detailed within *Methodologies for U.S. Greenhouse Gas Emissions Projections: Non-CO_2 and Non-Energy CO_2 Sources* (U.S. EPA 2013b). These projections use inventory methodologies to estimate emissions in future years based on projected changes in activity data and emission factors. Activity data used vary for each source, but include macroeconomic drivers, such as population, GDP, and energy, and source-specific activity data, such as production and use of fossil fuels and industrial production levels for iron and steel, cement, aluminum, and other products.

Carbon Dioxide Emissions

CO_2 emissions are expected to decline by 7.6 percent between 2005 and 2020. Between 2005 and 2011, emissions declined by 8.1 percent, but they are projected to increase slightly between 2011 and 2020. Energy-related CO_2 is projected to decline slightly over this time period, while non-energy CO_2 emissions (e.g., process emissions) are expected to grow between 2011 and 2020.

Projected energy-related CO_2 emissions in 2020 are 8.8 percent below their 2005 level, totaling 5,243 Tg CO_2 in 2020, assuming current policies persist. On average, energy-related CO_2 emissions decline by 0.6 percent per year from 2005 to 2020, compared with an average increase of 1.2 percent per year from 1990 to 2005. Reasons for the decline include growing use of renewable technologies and fuels; automobile efficiency improvements; slower growth in electricity demand; increased use of natural gas, which is less carbon-intensive than other fossil fuels; and an expected slow and extended recovery from the recession of 2007–2009 (U.S. DOE/EIA 2013b).

Non-energy-related CO_2 emissions are projected to increase by 12.3 percent between 2005 and 2020. Although these emissions declined between 2005 and 2011, growth in four emission sources results in overall growth: use of fossil fuels for non-energy uses (such as liquefied petroleum gas feedstock, natural gas feedstock, petrochemical feedstock, and asphalt and road oil); iron and steel production; natural gas systems; and cement production.

Methane Emissions

Between 2005 and 2020, total CH_4 emissions are estimated to increase by 1.0 percent (Table 5-2). Growth of emissions among some sources (e.g., coal mining, enteric fermentation, manure management) is largely offset by reductions among other sources (e.g., natural gas, landfills). The activities driving all of these emission sources (e.g., coal mining, livestock production, natural gas production, and waste generation) increase during this period. Emissions from many of these sources are reduced voluntarily through partnership programs. In addition, CH_4 from some natural gas activities and landfills is reduced as a co-benefit of regulations limiting volatile organic compounds (VOCs) from these sources. Increasing

---

5 This chapter presents comprehensive emissions projections through 2030. *AEO2013* covers the period 2010 through 2040.

emissions from livestock are driven by projected increases in livestock population, animal size, and an ongoing shift toward liquid waste management systems.

The quantity of methane capture-and-use projects associated with coal and landfill gas is driven in part by the prices of electricity and natural gas, which are projected to gradually increase over this period.

Nitrous Oxide Emissions

N_2O emissions are projected to decrease by 2.5 percent between 2005 and 2020. Emissions from agricultural soil management are driven by increasing crop production and the corresponding rise in nitrogen inputs to agriculture, including nitrogen fertilizer, managed manure, and crop residues. This source is estimated to account for nearly three-quarters of total N_2O emissions in 2020. N_2O emissions from stationary and mobile combustion are declining,

Table 5-2 **Select U.S. Non-CO₂ and Non-Energy CO₂ Emission Sources by Gas** (Tg CO_2e)

GHG emissions other than energy-related CO_2 include methane from natural gas, livestock, landfills, and coal; nitrous oxide from agricultural soils; and hydrofluorocarbons from the use of substitutes for ozone-depleting substances (ODS) and production of hycrochlorofluorocarbon (HCFC)-22.

Gas and Source	Historical GHG Emissions[a]				Projected GHG Emissions			
	2000	2005	2010	2011	2015	2020	2025	2030
Carbon Dioxide (CO₂)								
Non-Energy Use of Fuels	153	143	133	131	141	158	162	160
Iron and Steel Production	86	67	56	64	72	77	75	65
Natural Gas	30	30	32	32	34	37	40	42
Cement Production	40	45	31	32	42	50	53	58
Other	79	76	76	77	79	83	86	89
Methane (CH₄)								
Natural Gas	166	159	144	145	132	140	151	157
Enteric Fermentation	138	137	139	137	135	147	151	157
Landfills	112	113	107	103	102	101	99	97
Coal Mines	60	57	72	63	63	65	67	69
Manure Management	42	48	52	52	52	53	54	55
Other	91	81	79	87	95	94	96	91
Nitrous Oxide (N₂O)								
Agricultural Soil Management	227	238	245	247	250	258	265	273
Stationary and Mobile Combustion	67	57	43	40	33	28	27	27
Nitric and Adipic Acid Production	25	24	21	26	21	21	21	20
Other	39	37	35	43	39	39	46	50
Hydrofluorocarbons (HFCs)								
ODS Substitutes (HFCs)	76	99	115	122	154	200	260	291
HCFC-22 Production (HFC-23)	29	16	6	7	7	7	8	11
Semiconductors	0.3	0.2	0.4	0.3	0.3	0.4	0.4	0.5
Perfluorocarbons (PFCs)								
Aluminum	9	3	2	3	2	2	2	2
Semiconductors	5	3	4	4	4	4	4	5
Sulfur Hexafluoride (SF₆)								
Electrical Transmission and Distribution	15	11	8	7	7	7	7	8
Magnesium	3	3	1	1	1	1	1	1
Semiconductors	1	1	1	1	1	1	1	1

[a] Historical emissions and sinks data are from U.S. EPA/OAP 2013. Bunker fuels and biomass combustion are not included in inventory calculations.

largely due to improvements in emission control technologies and gradual turnover of the existing vehicle fleet (U.S. EPA/OAP 2013).

Hydrofluorocarbon, Perfluorocarbon, and Sulfur Hexafluoride Emissions

HFC emissions are estimated to increase by 80 percent between 2005 and 2020. Over the same period, PFC and SF_6 emissions are estimated to decline somewhat through increased voluntary control.

HFC emissions are increasing because of greater demand for refrigeration and air conditioning and because HFCs are predominantly used as alternatives for ODSs, such as hydrochlorofluorocarbons (HCFCs) that are being phased out under the Montreal Protocol. HFC-23 is also emitted as a by-product during the manufacture of HCFC-22. Both HFCs and HCFCs are GHGs, but HCFCs are not included here to be consistent with the United Nations Framework Convention on Climate Change (UNFCCC) guidelines (UNFCCC 2006). Growth of HFCs is anticipated to continue well beyond 2020 if left unconstrained.

Other sources of HFCs, PFCs, and SF_6 in industrial production include aluminum, magnesium, and semiconductor manufacturing and, in the case of SF_6, electricity transmission and distribution. These projections assume that voluntary emission reductions will be made in the aluminum and semiconductor industries as part of efforts to meet global voluntary reduction goals (U.S. EPA 2013b).

EMISSIONS PROJECTIONS BY SECTOR

This section presents projected GHG emissions for the following sectors: energy, transportation, industrial processes, agriculture, waste, and LULUCF (Table 5-3). These sectors largely correspond to the IPCC sector definitions used for the U.S. GHG inventory in Chapter 3 of this report. For inventory purposes, transportation is included within the energy sector and solvents are treated as a separate sector, whereas here they are included within industrial processes.

Energy

The energy sector as described in this chapter includes energy-related CO_2 emissions from electric power production and the residential, commercial, and industrial sectors. It also

Table 5-3   **Historical and Projected U.S. Greenhouse Gas Emissions Baseline, by Sector: 1990–2030** (Tg CO_2e)

Emissions from the energy, transportation, and waste sectors are projected to decline from 2005 to 2020, while emissions from the industrial processes and agriculture sectors are projected to increase, and sequestration from land use, land-use change, and forestry is projected to decline.

Sectors[b]		Historical GHG Emissions[a]				Projected GHG Emissions			
		2000	2005	2010	2011	2015	2020	2025	2030
Energy		4,258	4,321	4,104	3,981	3,936	4,038	4,141	4,207
Transportation		1,861	1,931	1,786	1,765	1,710	1,702	1,660	1,627
Industrial Processes		357	335	308	331	378	438	504	536
Agriculture		432	446	462	461	461	485	498	512
Forestry and Land Use		31	25	20	37	30	27	40	35
Waste		136	137	131	128	127	126	125	123
Total Gross Emissions		**7,076**	**7,195**	**6,812**	**6,702**	**6,643**	**6,815**	**6,967**	**7,041**
Forestry and Land Use (Sinks)[c]	high sequestration	-682	-998	-889	-905	-884	-898	-917	-937
	low sequestration					-787	-614	-573	-565
Total Net Emissions	high sequestration	**6,395**	**6,197**	**5,923**	**5,797**	5,759	5,918	6,050	6,104
	low sequestration					5,856	6,201	6,394	6,476

[a] Historical emissions and sinks data are from U.S. EPA/OAP 2013. Bunker fuels and biomass combustion are not included in inventory calculations.

[b] Sectors correspond to inventory reporting sectors, except that carbon dioxide, methane, and nitrous oxide emissions associated with mobile combustion have been moved from energy to transportation, and solvent and other product use is included within industrial processes.

[c] Sequestration is only included in the net emissions total.

includes fugitive CH_4 and non-energy CO_2 emissions from production of natural gas, oil, and coal; process emissions associated with non-energy uses of fossil fuels; and CH_4 and N_2O emissions from stationary combustion and incineration of waste for energy. Transportation-related emissions are discussed in the next section.

Under the 2012 policy baseline scenario, total energy sector emissions decline by 6.5 percent from 2005 to 2020. Energy-related CO_2 emissions decline in the electric power and residential sectors between 2005 and 2020, and increase in the industrial and commercial sectors (Table 5-4).

Electric Power

Total energy-related CO_2 from electricity production declines by 13.4 percent from 2005 to 2020, under the 2012 policy baseline scenario (Table 5-5). The growth of electricity demand (including retail sales and direct use) has slowed in each decade since the 1950s, from a 9.8 percent annual rate of growth from 1949 to 1959 to only 0.7 percent per year in the first decade of the 21st century. In the 2012 policy baseline scenario, growth in electricity demand remains relatively slow, as increasing demand for electricity services is offset by efficiency gains from new appliance standards and investments in energy-efficient equipment. Total electricity generation grows by 7 percent in the projection (0.8 percent per year) from 4,093 billion kilowatt-hours (kWh) in 2011 to 4,389 billion kWh in 2020 (U.S. DOE/EIA 2013b).

Table 5-4 **Historical and Projected U.S. Energy-Related CO_2 Emissions by Sector and Source[a]** (Tg CO_2e)

Energy-related CO_2 emissions are projected to decline in the electric power and residential sectors between 2005 and 2020, and increase in the industrial and commercial sectors.

Sector and Fuel	Historical GHG Emissions			Projected GHG Emissions	
	2005[b]	2010	2011	2020	2030
Electric Power Total	**2,402**	**2,259**	**2,166**	**2,081**	**2,224**
Petroleum	99	32	27	13	14
Natural Gas	319	399	409	446	482
Coal	1,984	1,828	1,723	1,610	1,717
Transportation[c]	**1,892**	**1,764**	**1,745**	**1,690**	**1,617**
Petroleum	1,859	1,726	1,706	1,648	1,564
Natural Gas	33	38	39	42	53
Industrial[c]	**823**	**780**	**773**	**872**	**888**
Petroleum	320	273	267	295	281
Natural Gas	389	411	416	478	499
Coal	115	96	90	99	108
Residential[c]	**358**	**335**	**329**	**317**	**299**
Petroleum	95	75	74	71	62
Natural Gas	262	259	255	245	236
Coal	1	1	1	1	1
Commercial[c]	**224**	**222**	**222**	**232**	**236**
Petroleum	51	47	47	47	45
Natural Gas	163	168	170	180	186
Coal	9	6	5	5	5
U.S. Territories	**50**	**50**	**50**	**51**	**53**
Total Energy-Related CO_2 Emissions	**5,749**	**5,409**	**5,277**	**5,243**	**5,318**

[a] U.S. DOE/EIA 2013b, with adjustments for bunker fuels, non-energy use of fossil fuels, and U.S. territories.

[b] Historical emissions data are from U.S. EPA/OAP 2013a.

[c] Sector total emissions do not include indirect emissions from electricity use.

Coal-fired power plants continue to be the largest source of electricity generation in the 2012 policy baseline scenario, but their market share declines significantly. From 42 percent in 2011, coal's share of total U.S. generation declines to 38 percent in 2020 and 37 percent in 2030 (U.S. DOE/EIA 2013b).

Most new capacity additions use natural gas or renewable energy. Natural gas-fired plants account for 44 percent of capacity additions from 2012 through 2020 in the 2012 policy baseline scenario, compared with 43 percent for renewables, 7 percent for coal, and 6 percent for nuclear. Escalating construction costs have the largest impact on capital-intensive technologies, which include nuclear, coal, and renewables. However, federal tax incentives, state energy programs, and rising prices for fossil fuels increase the competitiveness of renewable and nuclear capacity. Current federal and state environmental regulations also affect the use of fossil fuels, particularly coal. Uncertainty about future limits on GHG emissions and other possible environmental programs also reduces the competitiveness of coal-fired plants (U.S. DOE/EIA 2013b).

Residential

Total energy-related CO_2 emissions from residential energy use (excluding indirect emissions from electricity use) decline by 11.5 percent from 2005 to 2020 under the 2012 policy baseline scenario. The energy intensity of residential demand, defined as annual energy use per household, declines from 97.2 million British thermal units (Btus) in 2011 to 86.0 million Btus in 2020. The projected 12 percent decrease in intensity occurs along with a 10 percent increase in the number of homes. Residential energy intensity is affected by various factors—for example, population shifts to warmer and drier climates, improvements in the efficiency of building construction and equipment stock, and the attitudes and behavior of residents toward energy savings (U.S. DOE/EIA 2013b).

Commercial

Total energy-related CO_2 emissions from commercial energy use (excluding indirect emissions from electricity) increase by 3.7 percent from 2005 to 2020 under the 2012 policy baseline scenario. Commercial floor space grows by an average of 1.0 percent per year from 2011 to 2020, while energy consumption grows by about 0.2 percent over the same period.

Table 5-5   **Details on the Electric Power Sector**

Most new capacity additions use natural gas or renewables. Natural gas-fired plants account for 44 percent of capacity additions from 2012 through 2020 in the 2012 policy baseline scenario, compared with 43 percent for renewables, 7 percent for coal, and 6 percent for nuclear.

Electric Power by Fuel	Historical GHG Emissions						Projected GHG Emissions			
	2005		2010		2011		2020		2030	
	Emissions (Tg CO₂e)	Generation (billion kWh)	Emissions (Tg CO₂e)	Generation (billion kWh)	Emissions (Tg CO₂e)	Generation (billion kWh)	Emissions (Tg CO₂e)	Generation (billion kWh)	Emissions (Tg CO₂e)	Generation (billion kWh)
Fossil Fuels	**2,402**	**2,491**	**2,259**	**2,874**	**2,158**	**2,779**	**2,081**	**2,878**	**2,224**	**3,184**
Petroleum	99	122	32	37	27	28	13	17	14	18
Natural Gas	319	761	399	970	409	1,000	446	1,184	482	1,379
Coal	1,984	1,594	1,828	1,847	1,723	1,730	1,610	1,656	1,717	1,766
Other	8	13	12	19	11	20	11	20	11	20
Non-Fossil Fuels	**0**	**1,140**	**0**	**1,236**	**0**	**1,314**	**0**	**1,511**	**0**	**1,593**
Nuclear	0	782	0	807	0	790	0	885	0	908
Renewable	0	358	0	429	0	524	0	627	0	685
Non-Fossil % Share Generation	**31%**		**30%**		**32%**		**34%**		**33%**	
Total Generation	**3,630**		**4,110**		**4,093**		**4,389**		**4,777**	

Federal efficiency standards, which help to foster technological improvements in end uses (e.g., space heating and cooling, water heating, refrigeration, and lighting) act to limit the growth in energy consumption to less than the growth in commercial floor space (U.S. DOE/EIA 2013b).

Industrial

Total energy-related CO_2 emissions from the industrial sector (excluding indirect emissions from electricity use) increase by 5.9 percent from 2005 to 2020 under the 2012 policy baseline scenario. Despite a 31 percent increase in industrial shipments, industrial delivered energy consumption increases by only 12 percent from 2011 to 2020. The continued decline in energy intensity within the industrial sector is explained in part by a shift in the share of shipments from energy-intensive manufacturing industries (bulk chemicals, petroleum refineries, paper products, iron and steel, food products, aluminum, cement and lime, and glass) to other, less energy-intensive industries, such as plastics, computers, and transportation equipment.

Much of the growth in industrial energy consumption in the 2012 policy baseline scenario is accounted for by natural gas use, which increases by 15 percent from 2011 to 2020. With domestic natural gas production increasing sharply in the projection, natural gas prices remain relatively low. However, the mix of industrial fuels changes relatively slowly, reflecting limited capability for fuel switching in most industries (U.S. DOE/EIA 2013b).

Transportation

The transportation sector, as described in this chapter, consists of energy-related CO_2, CH_4, and N_2O from mobile source combustion. Total transportation GHG emissions decline by 11.9 percent between 2005 and 2020 under the 2012 policy baseline scenario.

CO_2 emissions from fossil fuel combustion in the transportation sector decline by 10.7 percent between 2005 and 2020. The decline occurred between 2005 and 2012, while emissions are expected to remain flat from 2012 through 2020. The growth in transportation energy consumption is flat across the projection. The transportation sector consumes 27.2 quadrillion Btus of energy in 2020, nearly the same as the level of energy demand in 2011. The projection of no growth in transportation energy demand differs markedly from the historical trend, which saw 1.1 percent average annual growth from 1975 to 2011. No growth in transportation energy demand is the result of declining energy use for light-duty vehicles (LDVs), which offsets increased energy use for heavy-duty vehicles, aircraft, marine and rail transportation, and pipelines. Higher fuel economy for LDVs more than offsets modest growth in vehicle miles traveled per driver (U.S. DOE/EIA 2013b).

N_2O emissions from mobile combustion decrease faster than energy-related CO_2 emissions, by nearly three-quarters from 2005 to 2020. Emissions from this source are declining due to improvements in emission control technologies and gradual turnover of the existing vehicle fleet (U.S. EPA/OAP 2013).

Industrial Processes

The industrial processes sector corresponds to the IPCC inventory guidelines category of the same name, plus emissions categorized as Solvent and Other Product Use (IPCC 2006). The sector includes emissions of GHGs associated with chemical transformations as part of industrial production of iron and steel, cement, nitric and adipic acid, and HCFC-22. It also includes emissions of fluorinated GHGs associated with the use of HFCs as substitutes for ODSs and other industrial uses.

Total emissions from industrial processes are projected to grow by 31 percent from 2005 to 2020 under the 2012 policy baseline scenario. From 2005 to 2011, emissions declined by 1.3 percent, but emissions are expected to grow rapidly between 2011 and 2020.

The total value of shipments from energy-intensive industries is expected to grow by an average of 1.7 percent from 2011 to 2020 in the 2012 policy baseline scenario. The iron and steel, cement, and glass industries show the greatest variability in shipments as a result of changes in economic growth assumptions. Energy efficiency improvements reduce the rate of growth of energy consumption relative to shipments. The strong growth can be explained largely by low natural gas prices that result from increased domestic production of natural gas from tight formations, as well as the continued economic recovery (U.S. DOE/EIA 2013b).

Agriculture

The agriculture sector includes CH_4 and N_2O emissions associated with livestock (e.g., enteric fermentation, manure management); crop production (e.g., agricultural soil management, rice production); and field burning of agricultural residues. CO_2 emissions and sinks associated with agricultural soils are included in the LULUCF sector. Emissions from the agriculture sector are projected to increase by 8.7 percent from 2005 to 2020 under the 2012 policy baseline scenario.

Livestock and crop production data are drawn from *USDA Agricultural Projections to 2022* (Westcott and Trostle 2013). The projections assume no domestic or external shocks that would affect global agricultural markets, normal weather, and extension of existing policies, such as the Food, Conservation, and Energy Act of 2008 (Farm Bill). Agricultural activities are extrapolated through 2030 based on their trends from 2012 through 2022 for the purpose of estimating emissions from agricultural sources.

Emissions from agricultural soil management are expected to increase by 8.8 percent between 2005 and 2020 as a result of increased crop production. Over the long run, steady global economic growth provides a foundation for continuing strong crop demand. U.S. corn-based ethanol production is projected to rebound from the 2012 decline, although the pace of further expansion slows considerably. Nonetheless, the combination of world economic growth, a depreciating dollar, and continued expansion of global biofuels production supports longer-run gains in world consumption and trade of crops (Westcott and Trostle 2013).

As a result of increased livestock production, enteric fermentation emissions are expected to rise by 7.0 percent from 2005 to 2020. Emissions from manure management rise from a combination of increased livestock populations and shift toward liquid waste management systems. High feed prices, the economic recession, and drought in the U.S. Southern Plains have combined to reduce producer returns and lower production incentives in the livestock sector over the past several years. Over the rest of the projection period, higher net returns and improved forage supplies lead to expansion of meat and poultry production (Westcott and Trostle 2013).

Waste

The waste sector includes CH_4 and N_2O emissions from landfills, wastewater treatment, and composting. Emissions from incineration of waste are included within the energy sector. Emissions from the waste sector are projected to decline by 8.1 percent between 2005 and 2020 under the 2012 policy baseline scenario.

Approximately 80 percent of emissions in the waste sector is CH_4 from landfills. Between 2005 and 2020, emissions from landfills are projected to decline, despite increasing waste disposal amounts, as a result of an increase in the amount of landfill gas collected and combusted. The quantity of recovered CH_4 that is either flared or used for energy purposes is expected to continually increase as a result of 1996 federal regulations that require large municipal solid waste landfills to collect and combust landfill gas, as well as voluntary programs that encourage CH_4 recovery and use (U.S. EPA/OAP 2013).

Land Use, Land-Use Change, and Forestry

The LULUCF sector includes net CO_2 flux from carbon (C) sequestration (such as carbon stored in trees and agricultural soils) (Table 5-6), and emissions from land-use activities (such as liming and urea fertilization of cropland and CH_4 and N_2O emissions resulting from forest fires) (Table 5-7).

LULUCF activities in 2011 resulted in a net carbon sequestration of 905.0 Tg CO_2e (246.8 Tg C). This represents an offset of 16.1 percent of total U.S. CO_2 emissions, or 13.5 percent of total U.S. GHG emissions in 2011 (U.S. EPA/OAP 2013). Forests currently account for the vast majority of net carbon sequestration among all land uses in the United States. Trends in net sequestration over the last two decades are principally the result of a positive growth-to-harvest ratio for U.S. forests nationally and small annual expansions in the area of forested land.

Table 5-6 **Projections of Net Carbon Sequestration**

In the long term, U.S. forest carbon stocks are likely to accumulate at a slower rate, and eventually may decline as a result of forestland conversion and changes in growth related to climate change and other disturbances. The timing of these changes is uncertain, represented by the range between the high- and low-sequestration scenarios.

Sources of Sequestration		Historical CO$_2$ Sink[a]				Projected CO$_2$ Sink			
		2000	2005	2010	2011	2015	2020	2025	2030
Forests[b]	high sequestration	-431	-800	-758	-762	-720	-728	-742	-755
	low sequestration					-623	-445	-397	-383
Wood Products[c]		-113	-105	-59	-72	-83	-83	-83	-83
Urban Forests		-58	-63	-68	-69	-73	-77	-82	-86
Agricultural Soils[d]		-66	-18	10	10	5	5	5	5
Landfilled Yard Trimmings and Food Scraps		-13	-12	-13	-13	-14	-15	-16	-17
Total Sequestration	**high sequestration**	**-682**	**-998**	**-889**	**-905**	**-884**	**-898**	**-917**	**-937**
	low sequestration					**-787**	**-614**	**-573**	**-565**

[a] Historical values are from U.S. EPA/OAP 2013.

[b] Estimates include carbon in above-ground and below-ground biomass, dead wood, litter, and forest soils. The high-sequestration scenario represents an extrapolation of historical inventory trends (slight annual increases in both forest land and carbon density). The low-sequestration scenario assumes that forest accumulation slows until there is no net loss or gain of forestland and carbon densities decline slightly from current rates to the historical average from 1991 through 2011. CO$_2$ emissions from forest fires are implicitly included in these estimates.

[c] Historical estimates are composed of changes in carbon held in wood products in use and in landfills, including carbon from domestically harvested wood and exported wood products (Production Accounting Approach).

[d] Includes cropland and grassland soils, while forest soils are included within forests above.

Table 5-7 **Emissions from Land Use, Land-Use Change, and Forestry** (Tg CO$_2$e)

Emissions from land use, land-use change, and forestry (LULUCF) include CO$_2$ from croplands and CH$_4$ and N$_2$O from forest fires.

Gas	Historical Emissions[a]				Projected Emissions			
	2000	2005	2010	2011	2015	2020	2025	2030
Carbon Dioxide (CO$_2$)[a]	9	9	9	9	9	9	9	9
Methane (CH$_4$)[b]	11	8	5	14	11	9	16	13
Nitrous Oxide (N$_2$O)[c]	11	8	6	13	10	9	15	13
Total	**31**	**25**	**20**	**37**	**30**	**27**	**40**	**35**

[a] CO$_2$ emissions from LULUCF include liming and urea fertilization of croplands, and peatland emissions.

[b] CH$_4$ emissions from LULUCF include emissions from forest fires.

[c] N$_2$O emissions from LULUCF include emissions from forest fires, fertilizer use in forests and settlements, and peatlands.

The amount of carbon stored in forests depends primarily on the density of carbon stored and the area of forested land. Forest carbon density can change as a forest ages and as stand dynamics change. Forest carbon density can also change as a result of forest fires, insect infestations, and other natural disturbances, as well as forest harvesting or other forest management techniques. The USDA Forest Service (USDA/FS) estimates that from 1991 to 2011, net forested area increased by an average of 0.2 percent (about 556,560 hectares [ha], or 1.4 million acres [ac]) per year (U.S. EPA/OAP 2013).

Forested areas may change when they are cleared for other land-use activities. Over time, U.S. forestland has been converted to urban/developed use, and conversions between agricultural uses and forests have also occurred. Net losses of forestland in the 1970s and 1980s, largely driven by conversion to crop uses, gave way to gains in forestland in the 1990s and 2000s, as economic returns to crops fell relative to economic returns to forests. According to USDA/FS estimates, the average carbon density of forests in the inventory increased by about 0.23 percent per year between 1991 and 2011, or about 9 percent. During the same period, annual forest sequestration of carbon amounted to 0.5 percent of the forest carbon inventory (U.S. EPA/OAP 2013).

There are indications that in the long term, U.S. forest carbon stocks are likely to accumulate at a slower rate, and eventually may decline as a result of forestland conversion and changes in forest growth related to climate change and other disturbances (Box 5-1; Haynes et al. 2007, Alig et al. 2010, Haim et al. 2011). The exact timing of these changes is uncertain, but U.S. forests are unlikely to continue historical trends of sequestering additional carbon stocks in the future under current policy conditions. While these changes may already be starting, major changes in U.S. forest inventory monitoring results are not expected in the next 5 to 10 years, partly due to lags in the time needed to collect and synthesize data for the entire nation.

For the above reasons, Table 5-6 provides two estimates for U.S. LULUCF carbon sequestration pathways to the year 2030. The high sequestration scenario (which reflects lower CO_2 emissions to the atmosphere) is an extrapolation based on recent forestland and forest carbon density accumulation rate trends (2000–2010 annual average increases of 556,560 ha [1.4 million ac] and 0.26 percent carbon density, respectively). The low sequestration scenario reflects expectations of slower accumulation of forestland and carbon density. With this scenario, forest area change declines from recent levels (accumulation of 556,560 ha [1.4 million ac] annually) and reaches a steady state of no net change in forest area in next decade. Forest carbon density declines from recent accrual rates (0.28 percent) to the 1991–2010 average (0.23 percent) by 2030.

Table 5-6 also shows CO_2 emissions or sequestration resulting from carbon stock changes in wood products, urban forests, agricultural soils, and landfilled yard trimmings and food scraps. Net CO_2 sequestration from these categories is projected to decline by 14 percent

---

Box 5-1 **2010 Resources Planning Act Assessment**

The USDA Forest Service recently published the 2010 Resources Planning Act (RPA) Assessment, which synthesizes key results of a comprehensive scientific assessment concerning the long-term outlook for the nation's forest and rangelands (USDA/FS 2012).

The RPA Assessment uses four scenarios with different assumptions about the potential rates of population growth, economic growth, land-use change, biomass energy use, and climate change over the next 50 years. This approach enables testing the sensitivity of future forest and other natural resource conditions against alternative assumptions regarding key economic, demographic, and climate variables. Viewed collectively, the RPA Assessment results highlight several long-term anthropogenic and natural forces that, absent changes in policy, demographic, or economic conditions, may act to diminish and, over time, possibly eliminate the U.S. forest carbon sink. The drivers of this anticipated decline include:

- *Aging forests:* U.S. forests are aging, and large areas of forest, particularly in the U.S. West, have reached or may reach in the next 10–20 years an age where their annual rate of growth, and thus their annual carbon sequestration rate, is expected to start declining.

- *Land-use change out of forest:* As the U.S. population increases, so too will the pressure to develop forestland for residential, commercial, and other purposes. This pressure is likely to be most acute around urban centers and in the South. All four 2010 RPA scenarios indicate a change to net losses in forestland at some point in the next 20 years.

- *Forest disturbance effects:* Climate change, wildfire, insects, disease, and other natural disturbances will continue to influence forest growth rates and mortality, leading to forest type changes under some circumstances. The combined impact of these effects can be seen in historical data on growth, age distribution, and mortality. A recent synthesis of climate change effects on forests found that area of forests affected by wildfire, invasive species, and other disturbances will increase, and that drought will lead to higher mortality and slow regeneration of some species, and altered species assemblages (Vose et al. 2012).

The forest carbon change projections from the 2010 RPA Assessment are determined by how forest area and forest growth are modified in response to changing harvest for timber products and wood energy. The carbon change projections for harvested wood products are determined primarily by how the production of solid wood products changes in response to changing U.S. and foreign demand for timber products and wood energy. Details about the 2010 RPA Assessment scenarios, the forest inventory projections, and forest sector carbon projections can be found in USDA/FS 2012, Wear 2011, and Wear et al. 2013.

from 2005 to 2020. Sequestration values for historical years are taken from U.S. EPA/OAP (2013), while projections are based on historical averages or extrapolation of historical trends over 2005–2011, depending on expected industry trends.

CO_2 sequestration in urban forests and landfilled yard trimmings and food scraps are projected to increase gradually, based on expected increases in urban land use and population. Sequestration in wood products has declined in recent years as a result of reduced home-building and wood product production during the economic downturn, but is expected to recover to the average of recent years over 2011 to 2020.

Since 2005, agricultural soils have switched from a carbon sink to a net source of CO_2 emissions. This has been driven by relatively high commodity prices since 2007, which have resulted in farmers shifting millions of hectares into crop production and an accompanying increase in CO_2 emissions from agricultural lands. According to the USDA National Agricultural Statistics Service, land area planted to crops in the United States increased by almost 2 million ha (5 million ac) between 2005 and 2012 (USDA/NASS 2013). During this same period, land enrolled in USDA's Conservation Reserve Program (CRP), which pays farmers to put environmentally sensitive cropland into conservation plantings, decreased by almost the same amount (USDA/FSA 2013). The projections for agricultural soil carbon are based on projections for cropland enrolled in the CRP, which decreases from 12.5 million ha (31.1 million ac) in 2011 to 11.3 million ha (28.5 million ac) in 2015, and then rebounds to 13 million ha (32 million ac) in 2020 (Westcott and Trostle 2013).

TOTAL EFFECT OF POLICIES AND MEASURES

Changes in Gross Emission Projections between the 2010 and 2014 Climate Action Reports
Projections of gross GHG emissions under the 2012 policy baseline scenario presented in this report are significantly lower than emission projections presented in the 2010 CAR. These differences can be traced to a combination of changes in policies, energy prices, and economic growth. The current 2012 policy baseline projection and the analogous projections from the 2010 and 2006 CARs are shown in Figure Table 5-8 and Figure 5-1 for comparison (U.S. DOS 2007 and 2010). In the 2010 CAR, emissions were projected to increase by 4.3 percent from 2005 through 2020, versus a 5.2 percent *decline* from 2005 levels projected in this report. In the 2006 CAR, the expected growth was even higher, totaling 17 percent over the same time period. Actual emissions for 2011 are significantly below those projected in past reports.

Current emissions include the effects of a number of policies that have been implemented since the analysis was completed for the 2010 CAR. These policies include the GHG emission and fuel efficiency standards for light-, medium-, and heavy-duty vehicles; various state renewable portfolio standards; the American Recovery and Reinvestment Act of 2009 (ARRA); and California Assembly Bill (AB) 32, which established the GHG emissions cap in California. CH_4 emission projections also account for GHG co-benefits from new federal air standards for the oil and natural gas industry that require controls to reduce VOC emissions. (See Chapter 4 of this report for a fuller discussion of the major regulatory changes relevant to GHG emissions.) Figure 5-2 displays the energy-related CO_2 projections contained in Reference case projections from *AEO2006* through *AEO2013*.

Top-Down Estimate of the Effects of New Policies and Measures
An analysis was conducted to disaggregate changes in emission projections due to macro-economic factors from changes resulting from policies and measures. The analysis decomposes emissions into factors representing population, per capita GDP, energy intensity, and carbon intensity of energy, referred to as a Kaya analysis (Figure 5-3). Between the 2010 and 2014 CARs, projections of population, GDP, energy use, and emissions were all adjusted (Table 5-9). By changing individual factors, the Kaya analysis can be used to associate proportions of the total change in emissions with each factor in the decomposition equation. By removing the portion of emissions change due to population and GDP changes, the remaining emissions change associated with energy and emission intensity is assumed to relate to new policies and measures and changing energy market conditions over the time period when the two sets of projections were prepared.

Table 5-8   **Comparison of 2012 Policy Baseline Projections with Previous U.S. Climate Action Reports**

In the 2010 *Climate Action Report* (CAR), emissions were projected to increase by 4.3 percent from 2005 through 2020, versus a 5.2 percent decline from 2005 levels projected in this report.

Projection	Historical GHG Emissions[a]				Projected GHG Emissions			
	2000	2005	2010	2011	2015	2020	2025	2030
2014 CAR	7,076	7,195	6,812	6,702	6,643	6,815	6,967	7,041
2010 CAR		7,109	7,074		7,233	7,416		
2006 CAR		7,550			7,942	8,330		

[a] Historical and projected years vary between CARs. For the 2014 CAR, the base year inventory is 2011; for the 2010 CAR, it was 2007; and for the 2006 CAR, it was 2004.

Figure 5-1   **Comparison of Climate Action Report Baseline "With Measures" Projections of Gross Greenhouse Gas Emissions**

Projections of gross GHG emissions under the 2012 policy baseline case presented in this report are significantly lower than emission projections presented in the 2010 *Climate Action Report* (CAR).

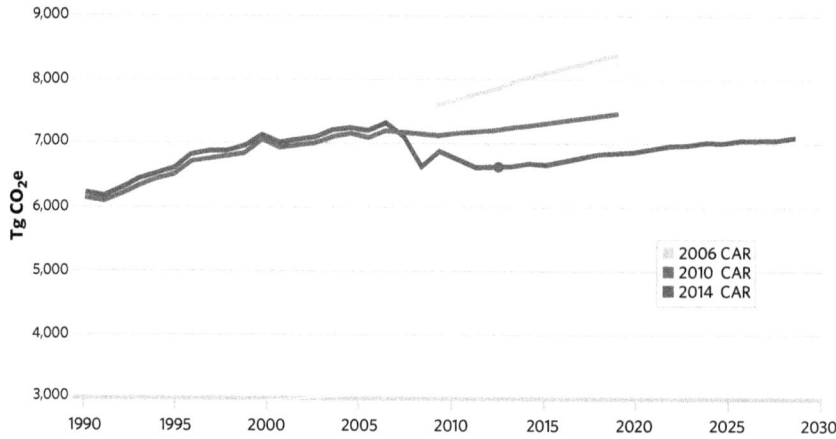

Note: Emission projections displayed are gross emissions and do not include CO_2 removals from land use, land-use change, and forestry (LULUCF). Projections from each report reflect a baseline "with measures" scenario, including the effect of policies and measures implemented at the time the projections were prepared (before 2012 in the case of the 2014 CAR), but not planned or proposed additional measures.

Figure 5-2   **Comparison of Energy-Related CO_2 Projections from *Annual Energy Outlook 2013* Reference Case Projections**

Recent projections of energy-related CO_2 emissions have declined relative to *AEO2013* projections.

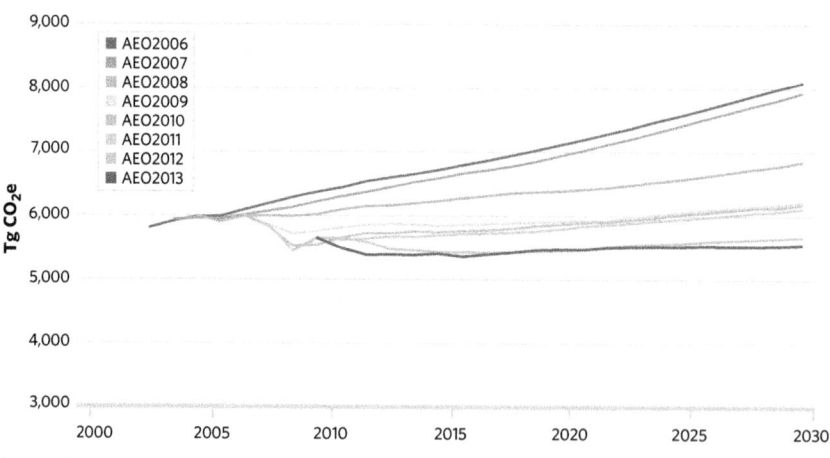

Source: U.S. DOE/EIA 2013b.

When this analysis was performed on the change in emission projections from the 2010 CAR to the 2014 CAR, about three-fifths of the total change in 2020 emission projections was found to be associated with changes in energy and emission intensity, resulting in an estimated reduction of about 350 Tg CO_2e in both 2015 and 2020 from new policies and measures implemented between 2009 and 2013 (Figure 5-4). This methodology is sensitive to various assumptions, including revisions in macroeconomic, energy, and emissions data, and cannot be used to disaggregate the effects of policy from changes due to shifts in global energy markets.

AEO2013 provided the baseline projection of energy-related CO_2 emissions (U.S. DOE/EIA 2013b). Projected CO_2 emissions in *AEO2013* were adjusted to match international inventory convention. EPA prepared the projections of non-energy-related CO_2 emissions and non-CO_2

Figure 5-3 **Normalized Kaya Identity Factors Used for Assessing the Effects of New Policies and Measures**

The analysis decomposes emissions into factors representing population, per capita gross domestic product (GDP), energy intensity, and carbon intensity of energy, and compares the figures in the 2010 *Climate Action Report* (CAR) with those in this report.

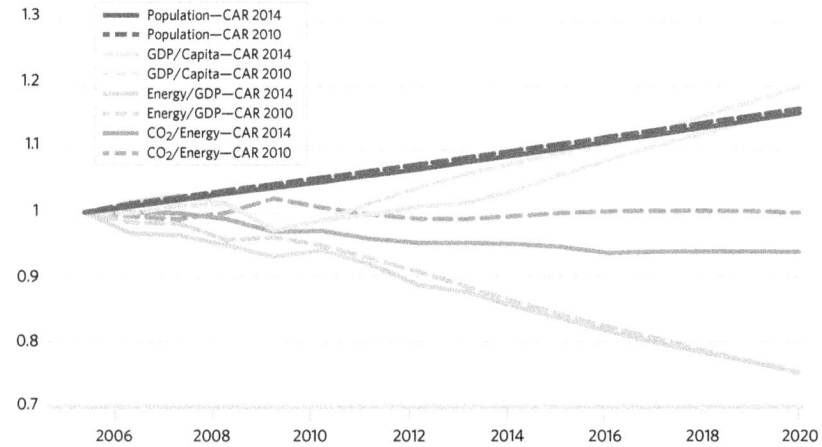

Table 5-9 **Comparison of Key Factors to Previous Climate Action Reports**

The 2014 *Climate Action Report* (CAR) reflects lower gross domestic product (GDP) and energy intensity in 2020 than was projected in the 2010 and 2006 CARs.

Factors	Assumptions for 2020		
	2006 CAR	**2010 CAR**	**2014 CAR**
Population (millions)	337	343	340
Real GDP (billion chain-weighted 2005 dollars)	19,770	17,356	16,859
Energy Intensity (Btu per 2005 chain-weighted dollar of GDP)	6,102	6,031	5,993
Light-Duty Vehicle Miles Traveled (billion miles)	3,474	3,137	2,870
Refiners Acquisition Cost of Imported Crude Oil (2005 dollars per barrel)	46.49	107.79	90.15
Wellhead Natural Gas Price (2005 dollars per thousand cubic feet)	5.06	6.39	3.48
Henry Hub (2005 dollars per thousand cubic feet)	5.45	7.23	3.75
Minemouth Coal Price (2005 dollars per ton)	20.87	25.78	43.46
Average Electricity Price (2005 cents per kilowatt-hour)	7.44	8.75	8.28
All-Sector Motor Gasoline Price (2005 dollars per gallon)	2.14	3.41	2.93
Energy Consumption	120.63	104.67	101.04

Figure 5-4 **Assessing Proportion of Change in Emission Projections**

According to an analysis of the change in emission projections from the 2010 *Climate Action Report* (CAR) to the 2014 CAR, about three-fifths of the total change in 2020 emission projections was found to be associated with changes in energy and emission intensity, resulting in an estimated reduction of about 350 Tg CO_2e in both 2015 and 2020 from new policies and measures implemented between 2009 and 2013.

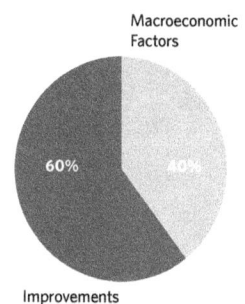

Macroeconomic Factors

60% 40%

Improvements in Energy and Emission Intensity

emissions. The methodologies used to project non-CO_2 emissions are explained in the background document *Methodologies for U.S. Greenhouse Gas Emissions Projections: Non-CO_2 and Non-Energy CO_2 Sources* (U.S. EPA 2013b). USDA and EPA prepared the estimates of carbon sequestration. Historical emissions data are drawn from the *Inventory of U.S. Greenhouse Gas Emissions and Sinks: 1990–2011* (U.S. EPA/OAP 2013). In general, the projections reflect long-run trends and do not attempt to mirror short-run departures from those trends. Information on the key factors underlying the projections is in Table 5-10.

ADJUSTMENTS

Adjustments were made to the energy-related CO_2 emissions reported in this chapter to more closely adhere to UNFCCC guidelines (UNFCCC 2006). Fuel-related emissions in U.S. territories were added based on extrapolation of historical trends because *AEO2013* does not include these emissions. Emissions of CO_2 from non-energy use of fossil fuels were subtracted from *AEO2013* projections of energy-related CO_2 and were estimated as described in the methodologies background document (U.S. EPA 2013b). Military and civilian international use of bunker fuels was subtracted from the totals and is reported separately. Emissions from fuel use in U.S. territories remain at approximately 50 Tg CO_2e from 2005 through 2020.

Bunker Fuels

Bunker fuels consist of jet fuel, residual fuel oil, and distillate fuel oil used for international aviation and marine transport. Between 2005 and 2020, GHG emissions from bunker fuels are projected to increase by 3 percent from 114 Tg CO_2 to 118 Tg CO_2. Emissions from international flights departing the United States are projected to increase by 8 percent between 2011 and 2020, while emissions from international shipping voyages are projected to increase by 2 percent over the same time period. Projections of bunker fuel emissions are subtracted from energy-related CO_2 totals from the *AEO2013* Reference case and are scaled to ensure consistent coverage as the historical GHG inventory.

Legislation and Regulations Included in the Current Projections

As discussed in Chapter 4 of this report, since the 2010 CAR the U.S. government has continued to make important progress toward reducing GHG emissions through policies and measures

Table 5-10 **Summary of Key Variables and Assumptions Used in the Projections Analysis**

Emissions are projected to remain below the 2005 level through 2030, despite significant increases in population (26 percent) and gross domestic product (GDP—69 percent) over that period.

Key Variable	Historical Values				Projected Values			
	2000	2005	2010	2011	2015	2020	2025	2030
Population (millions)	282	296	309	312	325	340	356	372
Real GDP (billion 2005 dollars)	$11,216	$12,623	$13,063	$13,299	$14,679	$16,859	$18,985	$21,355
Total Primary Energy Consumption (quadrillion Btus)	98.8	100.3	97.7	97.3	97.7	101.0	102.3	102.8
Energy Intensity (Btu per chain-weighted dollar of GDP)	8,810	7,944	7,481	7,316	6,657	5,993	5,391	4,814
Natural Gas Consumption (dry gas) (quadrillion Btus)	23.8	22.6	24.3	24.8	25.9	26.8	27.3	28.0
Petroleum Consumption (quadrillion Btus)	38.3	40.4	36.0	35.3	37.0	37.5	36.9	36.1
Coal Consumption (quadrillion Btus)	22.6	22.8	20.8	19.6	18.2	18.6	19.3	19.7
Vehicle Miles Travelled, All Vehicles (billion miles)	2,747	2,989	2,967	2,946	2,960	3,194	3,439	3,694

that promote increased investment in technologies and practices that reduce CO_2, methane, and other GHG emissions across all sectors. The projections presented in this chapter reflect this progress and include the effects of legislative and regulatory actions finalized before September 2012. In particular, the 2012 policy baseline includes regulatory and statutory changes enacted since the 2010 CAR, which relied on the *AEO2009* Reference case (U.S. DOE/EIA 2009 and U.S. DOS 2010). These regulatory and statutory changes apply to emissions in multiple sectors, including transportation, residential, commercial, and electric power.

However, the current projections of U.S. GHG emissions do not include the effects of any legislative or regulatory action that was not finalized before September 2012. For example, the 2012 policy baseline does not reflect the provisions of the American Taxpayer Relief Act of 2012, enacted on January 1, 2013, or the measures in *The President's Climate Action Plan* (U.S. Congress 2013 and EOP 2013a).

Description of NEMS and Methodology

The National Energy Modeling System (NEMS) was developed and is maintained by EIA's Office of Energy Analysis. The projections in NEMS are developed with the use of a market-based approach to energy analysis. For each fuel and consuming sector, NEMS balances energy supply and demand, accounting for economic competition among the various energy fuels and sources. The time horizon of NEMS is through 2040, approximately 25 years into the future.

NEMS is organized and implemented as a modular system. The modules represent each of the fuel supply markets, conversion sectors, and end-use consumption sectors of the energy system. NEMS also includes macroeconomic and international modules. The primary flows of information among the modules are the delivered prices of energy to end users and the quantities consumed by product, region, and sector. The delivered fuel prices encompass all the activities necessary to produce, import, and transport fuels to end users. The information flows also include other data on such areas as economic activity, domestic production, and international petroleum supply.

Each NEMS component represents the impacts and costs of existing legislation and environmental regulations that affect that sector. NEMS accounts for all combustion-related CO_2 emissions, as well as emissions of sulfur dioxide, nitrogen oxides, and mercury from the electricity generation sector. The potential impacts of pending or proposed federal and state legislation, regulations, or standards—or of sections of legislation that have been enacted but that require funds or implementing regulations that have not been provided or specified—are not reflected in NEMS.

Technology Development

The projections of U.S. GHG emissions take into consideration likely improvements in technology over time. For example, technology-based energy efficiency gains, which have contributed to reductions in U.S. energy intensity for more than 30 years, are expected to continue. However, while long-term trends in technology are often predictable, the specific areas in which significant technology improvements will occur and the specific new technologies that will become dominant in commercial markets are highly uncertain, especially over the long term.

Unexpected scientific and technical breakthroughs can cause changes in economic activities with dramatic effects on patterns of energy production and use. Such breakthroughs could enable the United States to considerably reduce future GHG emissions. While U.S. government and private support of research and development efforts can accelerate the rate of technology change, the effect of such support on specific technology developments is unpredictable.

Energy Prices

The relationship between energy prices and emissions is complex. Lower energy prices generally reduce the incentive for energy conservation and tend to encourage increased energy use and related emissions. However, a reduction in the price of natural gas relative to other fuels could encourage fuel switching that could, in turn, reduce carbon emissions. Alternatively, coal could become more competitive vis-à-vis natural gas, which could increase emissions from the power sector.

Economic Growth

Economic growth increases the future demand for energy services, such as vehicle miles traveled, amount of lighted and ventilated space, and process heat used in industrial production. However, growth also stimulates capital investment and reduces the average age of the capital stock, increasing its average energy efficiency. These two drivers work in opposing directions. However, the effect on energy service demand is the stronger of the two, so that levels of primary energy use are positively correlated with the size of the economy. The economy is projected to grow more slowly through 2020 than projected in the 2010 CAR, which is expected to slow emissions growth.

Weather and Climate

Energy use for heating and cooling is directly responsive to climate variability and change. *AEO2013* projection of CO_2 emissions account for trends in demand for heating and cooling based on the 30-year historical trend in heating- and cooling-degree days, and on state-level population projections. Therefore, the projections reflect both population migration and linear changes in heating- and cooling-degree days at a state level.

6

Vulnerability, Assessment, Climate Change Impacts, and Adaptation Measures

Human activities have dramatically altered the world's climate, oceans, land, ice cover, and ecosystems, resulting in impacts on almost every sector, including human health, agriculture, infrastructure, and natural resources. In the United States, climate change has already resulted in more frequent heat waves, extreme precipitation, larger wildfires, and water scarcity. These are serious challenges that directly affect families, communities, and jobs across the nation and all over the world. The only way to prepare and respond effectively is with a sound understanding of the changes underway and the threats and opportunities they present over time (Karl et al. 2009).

Significant progress in understanding the impacts of climate change and potential responses has been made since the publication of the *U.S. Climate Action Report* 2010 (2010 CAR) (U.S. DOS 2010), including major advances in the knowledge of Earth's past and present climate, improved capacity to project future conditions, and better understanding of vulnerabilities to the impacts of global change. The draft *Third National Climate Assessment (NCA) Report*, developed under the direction of the U.S. Global Change Research Program (USGCRP) and released for public comment in January 2013, contains expanded documentation of climate impacts and response activities across the United States. A significant change in the framing of this Third NCA Report is a focus on information that is useful for decision makers who are increasingly faced with managing climate-related risk. Unlike previous NCA reports, this report will be released electronically and will be fully searchable online, with links to the underlying data. Access will be facilitated through a number of innovative points of entry, including indicators of change and regional, sectoral, and intersectoral topics.

Like many other countries, the United States is vulnerable to current and projected climate changes. In response, the nation is increasingly emphasizing adaptation and preparedness measures to strengthen its resilience to and take advantage of potential opportunities resulting from significant change (Karl et al. 2009). Efforts are being made at multiple geographic scales to incorporate climate change into decisions at the national level (including the U.S. government), and at state, regional, and local levels (such as resource managers and policymakers within the public and private sectors) (ICCATF 2011). For example, in the fall of 2009, President Obama issued Executive Order (E.O.) 13514, *Federal Leadership in Environmental, Energy, and Economic Performance* (EOP 2009). E.O. 13514 has dramatically shifted the federal landscape of government stewardship toward sustainability and climate adaptation. In response, federal agencies completed their first set of agency-specific adaptation plans that were publicly released in February 2013.[1] These plans focus on identifying and addressing the impacts of climate change on each agency's operations, programs, and missions.

In June 2013, the President provided further direction to government agencies on reducing emissions and enhancing preparedness for climate change in his *Climate Action Plan* (EOP 2013a). Federal agencies have expanded their collaborative activities with multiple stakeholders both inside and outside of the federal government and are developing joint strategies that will address several cross-cutting issues. For example, the first national strategies for incorporating climate change into ecosystem management (NFWPCAP 2012) and managing water

[1] Develop Agency Sustainability Plans. *See* http://sustainability.performance.gov/.

supplies (ICCATF 2011) were released in 2013 and 2011, respectively. Reflecting the distributed nature of authority in the U.S. federal system as well as the need for adaptation decisions to be based on local assessments and needs, many state, local, and tribal governments have been leaders in conducting vulnerability assessments and planning and implementing adaptation activities (Bierbaum et al. 2013). These efforts are being accomplished both individually and in partnership with the federal government and with state, local, and tribal governments.

Expanding on and building from the elements in his *Climate Action Plan*, on November 1, 2013, the President issued E.O. 13653, *Preparing the United States for the Impacts of Climate Change* (EOP 2013b). This E.O. directs federal agencies to take a series of steps to enhance their efforts to build national climate preparedness and resilience and ensure the safety, health, and well-being of communities in the face of extreme weather and other impacts of climate change.

This chapter outlines, discusses, and provides examples of the following key topics:

- *Observations*: Recently observed changes in climate and the associated impacts.

- *Vulnerabilities and Impacts:* Observed and projected climate and global change vulnerabilities and impacts in the United States (regional, sectoral, and cross-cutting).

- *Research and Assessments:* Ongoing and planned research to improve the understanding of impacts, vulnerabilities, and options for response.

- *Adaptation Actions:* Ongoing adaptation measures, including examples of adaptation actions taking place at multiple scales throughout the United States.

OBSERVATIONS

Through a range of recent scientific observations, the evidence for a changing climate has strengthened considerably since the 2010 CAR. Over the past 50 years, stronger evidence coming from the scientific community indicates that human activities—primarily the burning of fossil fuels—have affected climate in unprecedented ways. Most notably, average global temperature has increased over time.

In the United States, average temperature has increased by about 1.5°F since 1900 (Karl et al. 2009). The most recent decade was the nation's warmest on record, and 2012 was the warmest single year (NOAA/NCDC 2012b). Other observations of changes in global climate include the increase in extreme weather and climate events in recent decades (NOAA/NCDC 2012a). Over the past 50 years, much of the United States has seen an increase in prolonged stretches of excessively high temperatures, a greater number of heavy downpours, and in some regions more severe droughts. Heat-trapping gases already in the atmosphere have committed us to a hotter future with more climate-related impacts over the next few decades. The magnitude of climate change beyond the next few decades depends primarily on the amount of heat-trapping gases emitted globally, now and in the future (Karl et al. 2009).

VULNERABILITIES AND IMPACTS

Many public and private efforts are analyzing the vulnerabilities of U.S. regions and sectors to the impacts of climate change. The most comprehensive, and the only official national effort, is the quadrennial NCA, which analyzes climate observations, impacts, and response options across U.S. regions and multiple sectors (NCA 2013). The Third NCA Report and many other vulnerability assessments, such as those conducted at a smaller scale across the country by the U.S. Geological Survey, document growing evidence of climate change trends and demonstrate that, like many other countries, the United States is increasingly vulnerable to current and projected changes in its climate.

However, while many effects of climate change are negative, there could be positive effects as well (Bierbaum et al. 2013), including the potential for increased agricultural productivity in northern parts of the country (Karl et al. 2009). Although potential positive effects can occur, there is extensive agreement and evidence that with current climate change mitigation policies and related sustainable development practices, global greenhouse gas (GHG) emissions will continue to grow over the next few decades, resulting in increasingly negative impacts (IPCC 2007b and Karl et al. 2009).

The upcoming Third NCA Report addresses climate impacts and vulnerabilities within some sectors individually, as well as climate-related risks and opportunities across those sectors. A common theme throughout these cross-sectoral components of the report is the connection across the sectors and how changes in one sector are amplified or attenuated through connections with other sectors. Another theme considers how decisions can influence a cascade of events that affect individual and national vulnerability and/or resilience to climate changes across multiple sectors. This "systems approach" showcases how adaptation and mitigation activities are themselves dynamic and interrelated strategies that intersect with the sectors described in this chapter. These themes also address the importance of underlying vulnerabilities and how they may influence the risks associated with climate change.

Regional Considerations

Landscapes, ecosystems, communities, and economies vary dramatically across the United States, but also share many common attributes. Each region is affected by changes in the global and national economies; each adds to the complex and multifaceted U.S. culture; each is connected to the same integrated infrastructure, such as transportation, communications, and energy systems; and they are all affected by the changing climate (Karl et al. 2009). A summary of important changes observed in each of the eight regions analyzed within the NCA is included in Table 6-1.

Table 6-1　**Regional Observations of Climate Change**

Landscapes, ecosystems, communities, and economies vary dramatically across the United States, but they also share many common attributes and are all affected by a changing climate.

Region	Observations
Northeast	Heat waves, coastal flooding due to sea level rise and storm surge, and river flooding due to more extreme precipitation events are increasingly affecting communities in the region (Horton et al. 2011).
Southeast and the Caribbean	Decreased water availability, exacerbated by population growth and land-use change, is causing increased competition for water; risks associated with extreme events, such as hurricanes, are increasing (Karl et al. 2009, Kunkel et al. 2013a).
Midwest	Longer growing seasons and rising carbon dioxide (CO_2) levels are increasing yields of some crops, although these benefits have already been offset in some instances by occurrence of extreme events, such as heat waves, droughts, and floods (Karl et al. 2009, Kunkel et al. 2013b).
Great Plains	Rising temperatures are leading to increased demand for water and energy and impacts on agricultural practices (Karl et al. 2009, Kunkel et al. 2013c).
Southwest	Drought and increased warming have fostered wildfires and increased competition for scarce water resources for people and ecosystems (Garfin et al. 2013).
Northwest	Changes in the timing of streamflow related to earlier snowmelt have already been observed and are reducing the supply of water in summer, causing far-reaching ecological and socioeconomic consequences (Karl et al. 2009, Kunkel et al. 2013d).
Alaska and the Arctic	Summer sea ice is receding rapidly, glaciers are shrinking, and permafrost is thawing, causing damage to infrastructure and major changes to ecosystems; impacts on Alaska native communities are increasing (Markon et al. 2012).
Hawaii and U.S. Affiliated Pacific Islands	Increasingly constrained freshwater supplies, coupled with rising temperatures, are stressing both people and ecosystems and decreasing food and water security (Keener et al. 2012).
Coastal Zone	Coastal lifelines, such as energy and water supply infrastructure and evacuation routes, are increasingly vulnerable to higher sea levels and storm surges, inland flooding, and other climate-related changes (Burkett and Davidson 2013).
Oceans	The oceans are currently absorbing about a quarter of human-caused CO_2 emissions to the atmosphere and more than 90 percent of the heat associated with global warming, leading to ocean acidification and the alteration of marine ecosystems (Griffis and Howard 2012).

Sectoral Considerations

Every sector of the U.S. economy is affected in some way by changes in climate, including changes in temperature, rising sea levels, and more extreme precipitation events and droughts. Such sectors as human health, water resources, agriculture, energy, and the natural environment are already experiencing the impacts of climate change at multiple scales (local, national, and international) (Karl et al. 2009). However, none of these sectors exists in isolation; each connects directly and indirectly to other sectors.

Water

The water cycle sets the stage for all life to exist, and is a driver of climate-related change through changes in precipitation, runoff, and evaporation. Water supplies and water management are also strongly affected by changes in temperature and extreme events, such as droughts and floods. Some observed impacts of climate change on the water cycle include intensified floods in some regions, summer droughts in much of the United States, and changes in seasonality of runoff (Karl et al. 2009). Water supplies are being reduced by climate change and are affecting ecosystems and livelihoods in many regions across the nation (e.g., the Southwest, the Great Plains, the Southeast, and the islands of the Caribbean and the Pacific, including the state of Hawaii).

With demand for water increasing, supplies of surface water and groundwater are already stressed. Water shortages increase the competition for water among agricultural, energy, municipal, and environmental users. Many of the expected effects of climate change on the water cycle affect human safety and health, property and infrastructure, and economy and ecology in basins across the country. Additionally, water resource managers and planners in most regions will encounter new risks, vulnerabilities, and opportunities in water management where existing practices may not be sufficient to ensure the future sustainability and safety of communities and industry (Karl et al. 2009).

Energy

The U.S. energy supply system is diverse and robust in its ability to provide a secure supply of energy with only occasional interruptions. However, current and projected impacts of climate change will shift seasonal patterns of energy use toward a reduction in heating and an increase in cooling requirements. Along with a variety of economic factors and an increase in extreme events in vulnerable areas, shifts in energy use and climate extremes pose risks to energy security. Extreme weather events and water shortages are already interrupting energy supply, and impacts are expected to increase in the future. Most vulnerabilities to and risks of interruptions in energy supply and use are created by local events, but the impacts often are national and international in scope (Wilbanks et al. 2012a, U.S. DOE and NREL 2013). Moreover, the impacts of sea level rise—in combination with storm surge and subsidence— are increasing the risks to coastal energy facilities (U.S. DOE and NREL 2013).

Transportation

The U.S. economy depends on personal and freight mobility provided by the country's transportation system. Essential products and services, such as energy, food, manufactured goods, and fuel, all depend in interrelated ways on the reliable functioning of transportation systems. There is already substantial evidence of impacts of extreme weather events on transportation systems, such as severe storms with high winds, floods, droughts (affecting barge traffic), coastal erosion, and heat waves (affecting rail systems and airports, in particular) (Figures 6-1 and 6-2). Disruptions to transportation systems related to climate change have already caused large economic as well as personal losses, and these impacts are expected to increase in response to a changing climate (Karl et al. 2009).

Agriculture

The United States produces nearly $300 billion per year in agricultural commodities, with roughly half of that coming from the production of livestock. The agriculture sector has experienced adverse impacts on crops and livestock from extreme events, and these impacts are expected to increase over the next century. Although increased carbon dioxide (CO_2) concentrations have a positive effect on some crops, agricultural productivity is expected to

Figures 6-1 and 6-2 **Effects of Extreme Events on Transportation Systems**

Essential products and services, such as energy, food, manufactured goods, and fuel, all depend on the reliable functioning of transportation systems. There is substantial evidence of impacts to U.S. transportation systems associated with severe weather, such as these photos of rail buckling under extreme heat and flooding of the Nashville MTA property in May 2010.

Photo courtesy of the U.S. Department of Transportation.   Photo courtesy of Nashville Metropolitan Transit Authority and the U.S. Department of Transportation.

decline over time in response to invasive pests and plant disease, and an increase in extreme events, such as floods, droughts, and heat waves. The locations where crops can most benefi-cially be grown are shifting northward. Climate change has the potential to affect the patterns and productivity of crop, livestock, and fishery systems at local, national, and global scales (Walthall et al. 2012).

Forestry

Forests provide numerous benefits, including wood production, clean drinking water, wildlife habitat, and recreation—and also provide carbon "sinks" that remove carbon from the atmo-sphere. Forest health decline and an increase in forest disturbances are already being ob-served and are projected to continue due to increases in the acreage burned by wildfire, the spread of insects and disease, drought, and extreme events projected as a result of climate change. At the same time, there is growing awareness that forests may play an expanded role in carbon management by storing carbon and providing resources for bioenergy production (Vose et al. 2012).

Ecosystems and Biodiversity

Climate variability and change affect humans and all living organisms through direct impacts on natural ecosystems, such as impacts on biodiversity (e.g., increased risk of extinction of species at local, regional, and national scales) and the location of species (e.g., substantial range shifts of many species of wildlife, fish, and native plants). Ecosystems provide a variety of services that are valued by society, including recreation, clean water, food, and a variety of other valued commodities. Ecosystem disruptions driven by climate change have direct im-pacts on humans, including reduced water supply availability and quality; the loss of iconic species and landscapes; and the potential for extreme events to overcome the services that ecosystems, such as coastal wetlands and barrier islands, provide in buffering the effects of severe storms (Staudinger et al. 2012).

Large-scale shifts have occurred in the ranges of species and the timing of the seasons and animal migration, and are very likely to continue. The distributions of marine fish and plankton are predominantly determined by climate, so it is not surprising that marine species in U.S. waters are moving northward and that the timing of plankton blooms is shifting. Extensive shifts in the ranges and distributions of both warmwater and coldwater species of fish have been documented (Janetos et al. 2008). For example, in the waters around Alaska, climate change already is causing significant alterations in marine ecosystems, with important impli-cations for fisheries and the people who depend on them (Karl et al. 2009). Finally, absorp-tion of more CO_2 from the air is leading to more acidic oceans, which will have broad and significant impacts on marine ecosystems, the services they provide, and the coastal econo-mies that depend on them (Box 6-1).

Box 6-1 **Ocean Acidification**

Oceans regulate climate and weather, and cycle water, carbon, and nutrients. Human activities are causing oceans to absorb increasing amounts of carbon dioxide from the air, leading to lower pH and greater acidity. When carbon dioxide reacts with seawater, it forms carbonic acid. This in turn reduces the concentration of carbonate ion, which can affect the shell formation of corals, plankton, shellfish, and other marine organisms.

Since the beginning of the Industrial Revolution, the average pH of ocean surface waters has fallen by about 0.1 units, from about 8.2 to 8.1 (total scale), resulting in an increase in acidity of approximately 30 percent (Orr et al. 2005, Feely et al. 2009). This change is at least 10 times faster than at any time over the past 50 million years.

More acidic oceans will have broad and significant impacts on marine ecosystems, the services they provide, and the coastal economies that depend on them. This more acidic environment has a dramatic effect on the growth, behavior, and survival of numerous marine organisms, including oysters, clams, urchins, corals, and calcareous plankton, which may put the marine food web at risk. Significant impacts from ocean acidification on the U.S. shellfish industry are particularly evident in the Pacific Northwest (Orr et al. 2005, Feely et al. 2009).

U.S. government agencies are participating in research efforts to increase understanding about how ocean chemistry is changing; how variable these changes are by region; what impacts they have on human and marine life, and on local, regional, and national economies; and what can be done to mitigate or adapt to ocean acidification.

Several notable state-level initiatives are also under way. For example, the Washington State Ocean Acidification Blue Ribbon Panel, convened in 2012 by Governor Gregoire, made recommendations that have led to the creation of the Washington Ocean Acidification Center at the University of Washington and other initiatives. California, Oregon, Maine, and other states are pursuing similar strategies.

The United States has also provided in-kind contributions and financial support to global efforts, such as the establishment of the new Ocean Acidification International Coordination Centre based at the International Atomic Energy Agency's Environment Laboratories in Monaco. This center will serve as an important means to develop a more comprehensive understanding of ocean acidification.

Health

Climate change threatens public safety and health in many ways, including impacts from increased extreme weather events and wildfire, decreased indoor and outdoor air quality, changes in prevalence of diseases transmitted by insects, increases in food prices, and limitations on water availability (NRC 2011). As temperatures increase, risks of heat stress, respiratory stress from poor air quality, and the spread of waterborne diseases are increasing. Absent adaptation efforts, some existing health threats will intensify, and new health threats will emerge (IWGCCH 2010).

Climate change will affect different segments of society differently because of their varying exposures and adaptive capacities. The impacts of climate change also do not affect society in isolation from other stresses. Rather, impacts can be exacerbated when climate change occurs in combination with the effects of an aging and growing population, pollution, poverty, and natural environmental fluctuations (Karl et al. 2009).

Cross-Sectoral Considerations (Linked Systems)

As noted above, climate change affects individual sectors in a variety of ways, but in managing risk and supporting adaptation decisions, it is also critical to consider cross-sectoral impacts and linkages between systems. For example, climate change affects sectors, such as water, energy, agriculture, health, and ecosystems, but also the intersections of these sectors. Some examples of recent research and observations on cross-sectoral considerations follow.

Urban Infrastructure and Vulnerability

Climate change poses a series of interrelated challenges for the country's most densely populated places: its cities. Many U.S. cities depend on aging infrastructure, such as water and sewage systems, roads, bridges, and power plants, which are in need of repair or replacement. Climate-related impacts, such as rising sea levels, storm surges, heat waves, and extreme weather events, have already compounded and will continue to compound these structural issues, stressing or even overwhelming these essential services.

In combination with the increase in coastal development, damage caused by storm surges and sea level rise is resulting in increased damage to critical infrastructure, such as roads, buildings, ports, wastewater treatment, and energy facilities. Extreme heat is another climate driver that damages transportation infrastructure, such as roads, rail lines, and airport runways (Wilbanks et al. 2012b). An example of the interdependence of infrastructure systems was observed in New York and New Jersey during Superstorm Sandy. The loss of electric power led to impacts on communications systems, which led to cascading effects in the transportation and public health sectors (Wilbanks et al. 2012b).

Land-Use and Land-Cover Change

Humans affect climate and are also vulnerable to climate impacts through land-use decisions (e.g., for land development, agriculture, or conservation of species). Adaptation options include managing vegetation to reduce heat in cities; managing landscapes to enhance environmental benefits, such as clean water supplies; restricting development in floodplains; and elevating homes to reduce vulnerability to sea level rise or flooding. Land-use and land-cover-related options for slowing the speed and intensity of climate change include expanding forests and conserving existing forest cover to pull more carbon from the atmosphere, designing cities to reduce energy use and motorized transportation demands, and altering agricultural management practices to increase carbon storage in soil (Loveland et al. 2012).

Tribal Culture, Lands, and Resources

The people, lands, and resources of indigenous communities across the United States face an array of climate change impacts and vulnerabilities that threaten many different Native communities' health, well-being, and ways of life. In parts of Alaska, Louisiana, the Pacific Islands, and other coastal locations, climate change impacts (through erosion and inundation) are so severe that some communities are already undergoing relocation from their historical homelands to which their traditions and cultural identities are tied.[2] Existing stresses on Native people's traditional food supplies, water quality and quantity, economic development, and health and safety are exacerbated by climate change (Maldonado et al. 2013, Doyle et al. 2013, Lynn et al. 2013).

Key vulnerabilities and drivers of impacts for Native communities include the loss of traditional knowledge, degradation of forests and ecosystems, lack of food security and traditional foods, water scarcity, Arctic sea ice loss, permafrost thaw, and relocation from historic homelands because of sea level rise (Hinzman et al. 2005, Dittmer 2013). In addition to the 566 federally recognized tribes and Alaska Natives, state-recognized and nonrecognized tribal groups share these vulnerabilities. Native populations are particularly vulnerable to the impacts of climate change because they depend very directly on the environment for their physical, mental, intellectual, social, and cultural well-being (Gautam et al. 2013, Cochran et al. 2013).

Water, Energy, and Land

Energy, water, and land systems interact in many ways. Energy production requires varying amounts of water (primarily for cooling) and in some cases, substantial amounts of land; water projects require energy (for treatment and delivery) and land; and land uses often depend upon availability of energy and water. Climate change impacts each of these sectors directly, but the implications of climate change on the intersections between systems are often unrecognized.

While there has been extensive study of water, energy, and land sectors individually, as well as the bilateral relationships between the sectors, there are few analyses of how multisectoral relationships are affected by a changing climate and how these relationships will influence technologies deployed in future energy systems, as well as options for reducing GHG emissions. However, the availability of energy, water, and land resources and the ways the systems interact vary across U.S. regions. Consequently the impacts, related risks, and opportunities related to climate change vary widely (Skaggs et al. 2012). Between 2003 and 2013, for example, severe weather caused an estimated 679 widespread power outages across the United States. Moreover, these and other weather-related outages during this period are estimated to have cost the U.S. economy an inflation-adjusted annual average of $18–$33 billion (COEA/DOE/OST 2013).

[2] State of Alaska Division of Community and Regional Affairs Planning and Land Management, Newtok Planning Group. 2012. See http://www.commerce.state.ak.us/dca/planning/npg/Newtok_Planning_Group.htm.

Coastal Zones

More than 50 percent of the population—approximately 164 million Americans—lives in coastal and Great Lakes watershed counties (NOAA 2011a, 2012; U.S. DOC/Census 2010). Collectively, these population centers help generate 58 percent of the national gross domestic product (NOAA 2011b). Coastal areas outside the Great Lakes region are already affected by violent storms and sea level rise, so both the lives and the livelihoods of large numbers of Americans are currently affected, with more impacts expected in the future (Burkett and Davidson 2013) (Figure 6-3). Along the shores of Great Lakes watershed counties, lake level changes are uncertain (Angel and Kunkel 2010, Milly and Dunne 2011, UGLSB 2012). However, erosion and sediment migration will be exacerbated by increased lakeside storm events, tributary flooding, and wave action due to loss of ice cover (Hayhoe et al. 2008, Uzarski et al. 2009).

Coastal and Great Lakes ecosystems are extremely vulnerable, in part because they have already been significantly altered by human activity; coastal wetlands are expected to suffer further losses of productivity and services that they provide to protect human settlements. Man-made components of coastal zones are also vulnerable to climate change, such as water supply lines, energy infrastructure, ports, tourism and fishing-based communities, and evacuation routes. As climate continues to change, repeated disruption of lives, infrastructure, and nationally and internationally important economic activities will pose challenges to populations living in coastal zones, and will aggravate existing impacts on valuable and irreplaceable natural systems (Burkett and Davidson 2013).

RESEARCH & ASSESSMENTS

As discussed above, global change is happening now and is well documented. The only way to reduce the risks of and maximize the opportunities associated with these significant changes is to enhance preparedness through a sound understanding of the changes underway, the threats and opportunities they present, and how they will change over time.

The U.S. Congress recognized this urgent need by passing the Global Change Research Act of 1990 (GCRA), which called for a federal interagency program to "assist the Nation and the world to understand, assess, predict, and respond to human-induced and natural processes of global change."[3] The U.S. Global Change Research Program (USGCRP) has been working to fulfill that mandate for the last 22 years, and is now coordinating the federal government's $2.6 billion annual investment in global change research—one of the largest such investments in the world.[4]

Figure 6-3 **Hurricane Sandy Strikes the Northeast in 2012**

More frequent and more intense storms and extreme weather events can cause widespread devastation of coastal communities, as evidenced by Hurricane Sandy's landfall on October 25, 2012.

Photo courtesy of the Federal Emergency Management Agency.

[3] Global Change Research Act of 1990. See http://globalchange.gov/about/global-change-research-act.

[4] U.S. Global Change Research Program. Budget. See http://globalchange.gov/about/budget-documents.

In addition to establishing the USGCRP, the GCRA mandates that the USGCRP develop a quadrennial report, known as the NCA. The NCA brings together the best peer-reviewed science on climate change and its impacts on the United States, leveraging research across sectors and providing a basis for future assessment and action. The current draft Third NCA Report was developed through an open and transparent process, using a broad engagement strategy that included more than 1,000 direct contributors and 240 chapter authors drawn from government scientists, academia, resource management agencies, and nongovernmental organizations (NGOs) (NCA 2013). In addition, more than 100 external organizations from the public and private sectors are now part of the NCA network, which supports the NCA activities and helps to share its findings. The Third NCA Report, due to be released in final form in the spring of 2014, is expected to become the authoritative source for information on the vulnerabilities and impacts of climate change in the United States.

In addition to this significant investment in global change research and the development of the NCA, many of the vulnerabilities discussed in the Vulnerabilities and Impacts section above are being addressed across multiple levels of government and in the private sector through programs at a variety of geographic scales for specific purposes, including management of natural resources, long-term development planning, and infrastructure investment. A goal of the previous U.S. Interagency Climate Change Adaptation Task Force, the newly created interagency Council on Climate Preparedness and Resilience (discussed in more detail below), USGCRP, and federal agencies is to bring this work together to leverage synergies and strengths among these many and varied programs. These efforts have been most recently articulated in *The President's Climate Action Plan* (EOP 2013a), released June 25, 2013, and E.O. 13653, *Preparing the United States for the Impacts of Climate Change*, issued on November 1, 2013 (EOP 2013b).

Sustained Assessment Process

A primary goal of the NCA is to help the nation anticipate, mitigate, and adapt to impacts from national and global climate change and climate variability. As the Third NCA Report was being prepared, a vision for a "sustained assessment" process took shape (NCA 2013). This includes an ongoing process of scientific evaluation and adaptive learning, improving understanding of the nation's vulnerabilities and its capacity to respond. Ongoing assessment activities, in addition to producing periodic synthesis reports as required by law, support the statutory requirements of the GCRA—to understand, predict, assess, and respond to rapid changes in the global environment. Continuous efforts to integrate new knowledge and experience can provide decision makers with more timely, concise, and useful information and permit extensive engagement with public and private partners. A well-designed and -executed sustained assessment process will also generate new insights about climate change, its impacts, and the effectiveness of societal responses. It can also help define the range of information needs of decision makers and end users relative to adaptation and mitigation, as well as the associated costs of impacts and benefits of response actions.

Indicators

Indicators are measurements or calculations that represent important features of the status, trends, or performance of a system (such as the economy, agriculture, natural ecosystems, or changes in Arctic sea ice cover). Indicators are used to identify and communicate changing conditions to inform both research and management decisions. Part of the vision for the sustained NCA process described above is a system of physical, ecological, and societal indicators that communicate key aspects of physical climate changes, climate impacts, vulnerabilities, and preparedness for the purpose of informing both decision makers and the public with scientifically valid information. Ideally, this system would be scalable for multiple geographic levels of use, would augment and expand on existing agency efforts when possible, and would include indicators to measure adaptive capacity or the effectiveness of adaptation actions (Janetos et al. 2012).

A robust public–private working group is currently dedicated to developing such indicators, which are expected to include both *current* indicators (which describe what is happening now and what happened in the past) and *leading* indicators (which represent potential future states of the system). Most of these indicators focus on the United States, but some include global trends to provide context or a basis for comparison.

Scenarios and Regional Climate Information

Scenarios used for the draft Third NCA Report included information on global and regional climate, sea level rise, and land-use and socioeconomic conditions (NCA 2013). Major new reports were developed for each of the eight regions of the United States, documenting historic climate trends, as well as producing standardized projections under the Intergovernmental Panel on Climate Change Special Report on Emissions Scenarios A2 and B1 (relatively high-end and low-end emission scenarios, respectively) (Nakićenović et al. 2000). In addition, an interagency report was developed on the current state of knowledge of global sea level rise, including four potential scenarios for the year 2100 resulting from climate-related processes, such as thermal expansion of the oceans, melting of ice sheets, and other factors. These scenarios were provided to NCA authors to help them build internally consistent views of future impacts across sectors and regions.[5]

Transparency and Review

The Third NCA Report process has made major strides toward maximizing the transparency of data and sources that underlie the report's key conclusions (NCA 2013). This NCA report is one of the first major U.S. government reports that will be delivered electronically, facilitating access via Internet links to all of the underlying data and publications. The findings will provide a foundation for a new comprehensive Web-based system for providing shared data and analytic capabilities, known as the Global Change Information System, currently being developed. In addition, "traceable accounts" have been developed for all of the key findings in the chapters. These accounts document the authors' process for coming to their conclusions, including an itemization of remaining uncertainties.

Engagement and Communications

Partnerships, two-way communication, and ongoing and meaningful engagement are critical to promoting understanding of and action toward addressing climate change. The USGCRP's 2012 Strategic Plan and the five National Research Council (NRC) *America's Climate Choices* reports underscore the importance of engagement and communications to informing decisions and achieving meaningful action (USGCRP 2012, NRC 2011). Partnerships and engagement strategies among federal and nonfederal participants are needed to (1) communicate effectively about climate vulnerabilities, impacts, risks, and opportunities; (2) enhance the relevance of actionable information; (3) encourage capacity building; (4) create opportunities for meaningful engagement of end users and public and private decision makers to inform the substance of the assessment; and (5) offer opportunities for input, evaluation, review, and feedback. To this end, an important component of the NCA is NCAnet: the "network of networks" that will help to build the content of the assessment and communicate the NCA process and products to a broader audience. *The President's Climate Action Plan* emphasizes the importance of partnerships across all levels of government and with the private sector to build national climate resilience (EOP 2013a).

SAMPLE U.S. ADAPTATION ACTIONS

Over the last decade, all levels of the U.S. government have increased efforts to plan for and implement climate adaptation (to address and prepare for impacts) and mitigation (to reduce emissions). In the past four years, the United States has made major strides toward increasing climate preparedness and resilience, initially through implementation of E.O. 13514 (EOP 2009) and the Interagency Climate Change Adaptation Task Force, and followed by *The President's Climate Action Plan* (EOP 2013a) and E.O. 13653, *Preparing the United States for the Impacts of Climate Change* (EOP 2013b). Given the federal system of government and the need for decisions and actions at the local level, adaptation, resilience, and preparedness activities necessarily take place at all levels of government and across the public and private sectors.

The federal government itself has made substantial progress in incorporating adaptation activities across the country, though they are not widely known or recognized by the public. Several of the most significant efforts are documented in the following section. In many cases,

[5] National Climate Assessment and Development Advisory Committee. Scenarios for Climate Assessment and Adaptation. January 9, 2013. See http://scenarios.globalchange.gov/content/scenarios.

even more significant progress has been made within U.S. regions, states, and cities, with cities in particular making major strides toward resilience and sustainability goals. Even with this progress, however, the nation must do more to avoid or adapt to serious impacts of climate change that have large social, environmental, and economic consequences.

The sample sector- and region-specific impact summaries and adaptation projects included in this section demonstrate the variety and scale of adaptation efforts in progress within the United States. The examples are illustrative and are not a comprehensive listing of all efforts across the nation.

Federal Government Adaptation Actions

In the spring of 2009, the Obama administration convened the Interagency Climate Change Adaptation Task Force, co-chaired by the White House Council on Environmental Quality (CEQ), the White House Office of Science and Technology Policy, and the National Oceanic and Atmospheric Administration (NOAA), and including representatives from more than 20 federal agencies. In October 2009, President Obama signed E.O. 13514, which directed federal agencies to reduce GHG pollution, eliminate waste, improve energy and water performance, and leverage federal purchasing power to support clean energy technologies and environmentally responsible products (EOP 2009). In addition, E.O. 13514 required all federal agencies to assess their vulnerabilities to the impacts of climate change and directed the Adaptation Task Force to develop a report with recommendations for how the federal government could strengthen policies and programs to better prepare the nation to adapt to a changing climate.

In its October 2010 progress report to the President, the Adaptation Task Force articulated a set of policy goals and recommendations that called for collaborative approaches within the federal government to address key cross-cutting issues related to climate change adaptation (ICCATF 2010). Specifically, the Adaptation Task Force recommended that the federal government:

- **Encourage and mainstream adaptation planning across the federal government**, including through adaptation planning within federal agencies.

- **Improve integration of science into decision making, including** through prioritizing activities that address science gaps important to adaptation decisions, building science translation capacity to improve the communication and application of science to meet the needs of decision makers, and developing an online data and information clearinghouse for adaptation.

- **Improve water resource management in a changing climate**, including through strengthening data and information systems for understanding climate change impacts on water and developing a national action plan to strengthen climate change adaptation for freshwater resources.

- **Protect human health by addressing climate change in public health activities**, including through enhancing the ability of federal decision makers to incorporate health considerations into adaptation planning and building integrated public health surveillance and early-warning systems to improve detection of health risks from climate change.

- **Facilitate the incorporation of climate change risks into insurance mechanisms**, including through exploration of a public–private partnership to produce an open-source risk assessment model.

- **Develop a strategic action plan focused on strengthening the resilience of coastal, ocean, and Great Lakes communities and ecosystems to climate change.**

- **Develop a strategy for reducing the impacts of climate change on the nation's fish, wildlife, and plant resources and their habitats.**

- **Enhance efforts to support international adaptation**, for example, by developing a government-wide strategy to support multilateral and bilateral adaptation activities and integrate adaptation into relevant U.S. foreign assistance programs.

- **Coordinate capabilities of the federal government to support adaptation at all levels**, including through partnerships addressing local, state, and tribal needs.

In October 2011, the Adaptation Task Force released a second progress report that outlined the federal government's progress in expanding and strengthening the nation's capacity to better understand, prepare for, and respond to extreme events and other climate change impacts (ICCATF 2011). The 2011 report also provided an update on actions in key areas of federal adaptation efforts, including building resilience in communities; safeguarding critical natural resources, such as freshwater; and providing accessible climate information and tools to help decision makers manage climate risks. The next such report is anticipated in 2014.

The President's Climate Action Plan

In June 2013, the Obama Administration released the nation's first comprehensive *Climate Action Plan* (EOP 2013a). The plan outlines actions the federal government will take to cut carbon pollution, prepare the United States for the impacts of climate change, and work with the international community to significantly reduce emissions and forge a truly global solution to this global challenge. The plan acknowledges that even as the nation takes steps to cut carbon pollution, it must also prepare for the impacts of a changing climate that are already being felt across the country. Building on the progress noted above, the plan:

* Directs federal agencies to support local climate-resilient investment by removing barriers or counterproductive policies and modernizing programs, and establishes a short-term task force of state, local, and tribal officials to advise on key actions the federal government can take to help strengthen communities on the ground.

* Highlights innovative strategies in the Hurricane Sandy-affected U.S. Northeast to strengthen communities against future extreme weather and other climate impacts. For example, building on a new, consistent flood-risk reduction standard established for the Sandy-affected region, agencies will update flood-risk reduction standards for all federally funded projects.

* Launches an effort to create sustainable and resilient hospitals in the face of climate change through a public–private partnership with the healthcare industry.

* Maintains agricultural productivity by delivering tailored, science-based knowledge to farmers, ranchers, and landowners; and helps communities prepare for drought and wildfire by launching a National Drought Resilience Partnership and by expanding and prioritizing forest and rangeland restoration efforts to make areas less vulnerable to catastrophic fire.

* Pledges to continue identifying innovative ways to help America's most vulnerable communities prepare for and recover from the impacts of climate change through annual federal agency "Environmental Justice Progress Reports."

* Commits to the development of actionable climate science; the production of the NCA report and vulnerability assessments within economic sectors (including energy, health, transportation, food supply, oceans, and coastal communities); and the development of climate preparedness tools and information needed by state, local, and private-sector leaders through a centralized "toolkit" and a new Climate Data Initiative.

To build on this progress, President Obama signed E.O. 13653, *Preparing the United States for the Impacts of Climate Change*, on November 1, 2013 (EOP 2013b). In particular, this E.O. directs federal agencies to:

* **Modernize federal programs to support climate-resilient investments:** Agencies will examine their policies and programs and find ways to make it easier for cities and towns to build smarter and stronger. Agencies will identify and remove any barriers to resilience-focused actions and investments—for example, policies that encourage communities to rebuild to past standards after disasters instead of to stronger standards—including through agency grants, technical assistance, and other programs in sectors from transportation and water management to conservation and disaster relief.

* **Manage lands and waters for climate preparedness and resilience:** America's natural resources are critical to its economy, health, and quality of life. E.O. 13653 directs agencies to identify changes that must be made to land- and water-related policies, programs, and regulations to strengthen the climate resilience of U.S. watersheds, natural resources, and ecosystems, and the communities and economies that depend on them. Federal agencies

will also evaluate how to better promote natural storm barriers, such as dunes and wet-lands, as well as how to protect the carbon sequestration benefits of forests and lands to help reduce the carbon pollution that causes climate change (EOP 2013b).

- **Provide information, data, and tools for climate change preparedness and resilience:** Scientific data and insights are essential to help communities and businesses better under-stand and manage the risks associated with extreme weather and other impacts of climate change. E.O. 13653 instructs federal agencies to work together and with information users to develop new climate preparedness tools and information that state, local, and private-sector leaders need to make smart decisions. In keeping with the President's Open Data Initiatives project, agencies will also make extensive federal climate data accessible to the public through an easy-to-use online portal.[6]

- **Plan for climate change-related risk:** Recognizing the threat that climate change poses to federal facilities, operations, and programs, E.O. 13653 builds on the first-ever set of federal agency adaptation plans released early in 2013 and directs federal agencies to develop and implement strategies to evaluate and address their most significant climate change-related risks (EOP 2013b).

To implement these actions, E.O. 13653 establishes an interagency Council on Climate Pre-paredness and Resilience, chaired by the White House and composed of more than 25 agen-cies, which will succeed the Adaptation Task Force established in 2009. Because state, local, and tribal leaders across the country are already contending with more frequent or severe heat waves, droughts, wildfires, storms and floods, and other impacts of climate change, the E.O. also establishes a State, Local, and Tribal Leaders Task Force on Climate Preparedness and Resilience. This new task force will provide recommendations to the President on remov-ing barriers to investments aimed at strengthening resiliency; modernizing federal grant and loan programs to better support state, local, and tribal efforts; and developing the information and tools that communities need to prepare for climate change (EOP 2013b).

National Cross-Cutting Adaptation Strategies

At the recommendation of the previous Adaptation Task Force, the National Ocean Council (NOC), members of Congress, and external groups, such as the National Research Council, beginning in 2009, federal agencies prioritized an initial set of issues for consideration and developed a series of cross-cutting strategies to reduce the impacts of climate change on the nation's natural resources. The first of these, the October 2011 *National Action Plan: Priorities for Managing Freshwater Resources in a Changing Climate*, was developed by federal agencies working with stakeholders to plan for adequate water supplies in a changing climate, while protecting water quality, human health, property, and aquatic ecosystems (ICCTF 2011). Federal agencies also partnered with state and tribal representatives to develop a *National Fish, Wildlife and Plants Climate Adaptation Strategy* to address the impacts climate change is having on U.S. natural resources and the people and economies that depend on them (NFWPCAP 2012); the final strategy was released in March 2013. In addition, as part of President Obama's *National Policy for the Stewardship of the Ocean, Our Coasts, and the Great Lakes* (EOP 2010), in April 2013 NOC released the final *National Ocean Policy Implementation Plan* (NOC 2013), which includes a series of actions to address "Resiliency and Adaptation to Climate Change and Ocean Acidification," one of nine priority objectives identified by the National Ocean Policy.

Agency Adaptation Plans

In response to the directive given to federal agencies in E.O. 13514 (EOP 2009), in March 2011, CEQ issued guidance on how agencies should integrate climate change adaptation into their planning, operations, policies, and programs (EOP/CEQ 2011). In response, Agency Adaptation Plans were submitted to CEQ as part of an annual sustainability planning process in June 2012 and were released for public review in February 2013.[7] These plans integrate ad-aptation planning into the operations, policies, and programs of all federal agencies. For example:

- The U.S. Department of Transportation's (DOT's) plan describes how increased flooding would affect the transportation sector and notes that the Federal Highway Administration will develop guidance for incorporating climate change considerations into the planning

[6] See http://www.whitehouse.gov/innovationfellows/open-data-initiatives.

[7] See http://www.whitehouse.gov/administration/eop/ceq/Press_Releases/February_07_2013.

and design of projects in coastal areas. DOT has explicitly authorized use of its state transportation funds for adaptation activities (U.S. DOT 2013).

- The U.S. Department of Homeland Security (DHS) is working to ensure the nation's resilience to more frequent or extreme natural disasters, including the need to ensure safety and stability in the Arctic,[8] and prepare for changing conditions along the nation's borders (U.S. DHS 2012). DHS has developed planning scenarios that include consideration of a series of cascading impacts associated with increased intensity of hurricanes and a nearly ice-free Arctic in summer with thinner ice cover in winter.

- The U.S. Environmental Protection Agency (EPA) identified potential climate-related risks to air quality and the availability and quality of water resources as critical topics. EPA is currently conducting regional assessments to identify areas of greatest priority, including identifying the most vulnerable populations and developing plans to address these priorities (U.S. EPA 2012).

These plans are meant to be living documents. Moreover, E.O. 13653, *Preparing the United States for the Impacts of Climate Change* (EOP 2013b), requires that each federal agency update its Agency Adaptation Plan to include:

- Identification and assessment of climate change-related impacts on and risks to the agency's ability to accomplish its missions, operations, and programs.

- A description of programs, policies, and plans the agency has already put in place, as well as additional actions the agency will take, to manage climate risks in the near term and build resilience in the short and long terms.

- A description of how any climate change-related risk identified in the plan that is deemed so significant that it impairs an agency's statutory mission or operation will be addressed, including through the agency's existing reporting requirements.

- A description of how the agency will consider the need to improve climate adaptation and resilience, including the costs and benefits of such improvement, with respect to agency suppliers, supply chain, real property investments, and capital equipment purchases, such as updating agency policies for leasing, building upgrades, relocation of existing facilities and equipment, and construction of new facilities.

- A description of how the agency will contribute to coordinated interagency efforts to support climate preparedness and resilience at all levels of government, including collaborative work across agencies' regional offices and hubs, and through coordinated development of information, data, and tools.

The federal government will also be working to bring agencies together to address many of the common challenges that the plans identified. These challenges include the need to provide better, more locally relevant information on climate change impacts; to ensure coordination of federal action to support adaptation efforts at the local level; to better integrate climate considerations into planning and investment decisions to ensure they are viable over the long term; and to protect federal facilities and personnel from extreme events and other impacts. These interactions will be facilitated by a "community of practice" developed across federal agencies to address adaptation-related issues. This community includes individuals from more than 55 agencies and subagencies who are responsible for adaptation planning, demonstrating that more federal employees are now integrating climate adaptation planning into their day-to-day activities. The community shares best practices and is building a "knowledge network" to support adaptation activities.

Selected Examples of Interagency and Agency-Specific National Adaptation-Related Initiatives

Managing Drought—In 2006, the National Integrated Drought Information System (NIDIS) was established by Congress to help support a more proactive response to drought.[9] The Web-based U.S. Drought Portal provides public access to NIDIS, which includes decision-support tools like the Drought Early Warning System.[10] The NIDIS implementation team also conducts workshops and meetings at federal, state, and local levels to facilitate and inform

---

[8] The waters of the Arctic are gradually opening up, not only to new resource development, but also to new shipping routes that may reshape the global transport system and affect U.S. national security interests. While these developments offer opportunities for growth, they are potential sources of competition and conflict for access and natural resources.

[9] See http://www.drought.gov/drought/.

[10] See http://www.drought.gov/drought/content/regional-programs/regional-drought-early-warning-system.

stakeholders. NIDIS is made possible by the collaboration of 16 different federal agencies, as well as state, local, and tribal partners.

Coordinating Disaster Response—The Federal Emergency Management Agency (FEMA) has found that every dollar it spends on hazard mitigation provides the nation with about four dollars in future benefits (U.S. DHS/FEMA 2011, MMC 2005). FEMA, the U.S. Army Corps of Engineers (USACE), and state agencies are helping to address flood risks through the Silver Jackets program, which creates interagency teams to simplify access to critical flood risk mitigation and planning resources and provides communities with a single point of contact to the federal government on these issues.[11]

Incorporating Adaptation into Disaster Recovery during Superstorm Sandy—Recognizing the need to better publicize existing data sets and the development of climate-related, decision-support tools, the federal government introduced a suite of future flood risk tools to ensure that investments minimize risk to the greatest degree possible. FEMA, CEQ, USGCRP, NOAA, and USACE came together to combine various data sets and sources of expertise to produce tools accessible to local decision makers (Box 6-2).

Managing Wildfire—In response to requirements of the Federal Land Assistance, Management, and Enhancement Act of 2009, the Wildland Fire Leadership Council directed the development of the National Cohesive Wildland Fire Management Strategy.[12] This strategy is a collaborative process with active involvement from all levels of government and NGOs, as well as the public, to seek national, all-lands solutions to wildland fire management issues.

Addressing Sea Level Rise—The U.S. Geological Survey, National Park Service (NPS), U.S. Fish and Wildlife Service (FWS), and private and nonprofit partner organizations have been engaged in a research project aimed at assessing the vulnerability of Assateague National Seashore to sea level rise and increased erosion along the North Atlantic Seaboard, and identifying adaptation actions to ensure that the resources of the seashore remain resilient. Findings will be used to inform a coast-wide assessment of threats from sea level rise and related habitat conservation recommendations.

Conserving Biodiversity—The Bureau of Land Management (BLM), the U.S. Forest Service (USFS), and Colorado State University are currently assessing and inventorying limber pines in Wyoming, Montana, and Colorado, to evaluate stand structure and the extent of mountain

---

Box 6-2 **Hurricane Sandy Rebuilding Task Force**

Hurricane Sandy hit the U.S. Northeast in late October 2012. Sandy was the deadliest hurricane of the season, and the second-costliest hurricane in U.S. history. Many links have been made between Hurricane Sandy and climate-related global changes, such as warming oceans, greater atmospheric moisture, and sea level rise.

In December 2012, President Obama signed Executive Order (E.O.) 13632, *Establishing the Hurricane Sandy Rebuilding Task Force* (EOP 2012). The E.O. directed the Sandy Task Force to "ensure that the Federal Government continues to provide appropriate resources to support affected State, local, and tribal communities to improve the region's resilience, health, and prosperity by building for the future," including in the face of climate change. The Sandy Task Force built on lessons learned during previous disasters, where experience has shown that planning for long-term rebuilding must begin, even as the response is ongoing.

Working within the National Disaster Recovery Framework, the Sandy Task Force partnered with federal, state, and local officials, as well as the private sector and nonprofit, community, and philanthropic organizations to promote recovery in a unified and coordinated manner and to incorporate adaptation principles. The Sandy Task Force also provided decision makers with information on potential impacts of climate change in the region, in user-friendly and useful formats or products, so that they can recover and rebuild in a way that increases their resilience to future weather events.

On August 19, 2013, the Sandy Task Force released its final strategy for rebuilding the affected region (HSRTF 2013). This strategy will ensure that families, small businesses, and communities are stronger, more economically competitive, and better able to withstand future storms, and will serve as a model for communities across the country.

[11] See http://www.nfrmp.us/state/.

[12] See http://www.forestsandrangelands.gov/strategy/.

pine beetle and white pine blister rust infestations. Also, BLM, USFS, NPS, FWS, the Bureau of Indian Affairs, states, universities, and several nonprofit partners have conducted the first-ever range-wide genetics survey of ponderosa pine. These research studies will increase understanding of the ability of these species to adapt to climate change and to identify genetically unique populations as priorities for conservation.

Building Understanding of Climate Change Impacts—One example of federal efforts in this area is NOAA's Regional Integrated Science and Assessments (RISA) program, which supports research teams that help expand and build the nation's capacity to prepare for and adapt to climate variability and change.[13] RISA teams work with public and private user communities to advance understanding; develop knowledge on impacts, vulnerabilities, and response options; develop products and tools to enhance the use of science in decision making; and test governance structures for managing scientific research. In addition, NOAA's Coastal Services Center provides technology, information, and management strategies for local, state, and national organizations to address challenges associated with flooding, hurricanes, sea level rise, and other coastal hazards.[14]

Protecting Human Health—The U.S. Department of Health and Human Services' Centers for Disease Control and Prevention's (CDC's) Climate and Health Program developed the Climate-Ready States and Cities Initiative to help state and city health departments plan and prepare for the potential health effects of climate change.[15] The initiative is currently working with eight states and two cities to assess, plan, and implement health-related climate change adaptation programs. Strategy development relies on the Building Resilience Against Climate Effects (BRACE) framework, which is a five-step sequential process for developing successful human health-related climate change adaptation. The framework includes vulnerability assessment, projection of disease burden, identification of adaptation options, implementation, and evaluation.

Managing Natural Resources—The U.S. Department of the Interior (DOI) is developing information and tools to manage U.S. natural resources and support state and local efforts to prepare for climate change. DOI's WaterSMART program helps states deal with rapid population growth, climate change, aging infrastructure, and land-use changes.[16] BLM is currently conducting 15 Rapid Ecoregional Assessments across the U.S. West and Alaska to promote cross-boundary collaboration and informed decision making, and to facilitate collaborative development and prioritization of regional conservation, restoration, and climate adaptation strategies and actions.[17] Additionally, DOI's Bureau of Reclamation recently completed a study defining current and future imbalances in water supply and demand in the Colorado River Basin and adjacent areas (U.S. DOI/BR 2012).

Supporting the Agricultural Sector—The U.S. Department of Agriculture (USDA) has integrated climate change objectives into its strategic plans. USDA is expanding its focus on climate-related research and delivery capacity across its agencies to provide climate services to rural and agricultural stakeholders through existing programs, including the Cooperative Extension Service, the USDA Service Centers, and the Forest Service Climate Change Resource Center (USDA 2010). In June 2013, USDA announced plans to develop Regional Climate Hubs that will provide climate-related scientific and technical support, assessments, outreach, and education for the agriculture sector.[18] In addition, USDA is working with farmers in the Environmental Quality Incentives Program to improve water-use efficiency through measures that allow farmers to grow more crops with less water.[19] The USDA *Climate Change Science Plan*, developed by an interagency USDA team, provides farmers, ranchers, foresters, landowners, resource managers, policymakers, and federal agencies with science-based knowledge to manage the risks, challenges, and opportunities of climate change and position themselves for the future (USDA/GCTF 2010).

Building a More Resilient Transportation Sector—To better understand potential climate change impacts on transportation infrastructure and identify adaptation strategies, DOT is conducting a comprehensive study of climate change impacts in the Mobile Bay region, with the intention of developing methods and tools that can be used nationwide.[20] In addition, the Federal Transit Administration (FTA) is providing public transportation officials across the

---

[13] *See* http://cpo.noaa.gov/ClimatePrograms/ClimateandSocietalInteractions/RISAProgram/AboutRISA.aspx.

[14] *See* http://www.csc.noaa.gov/.

[15] *See* http://www.cdc.gov/climateandhealth/climate_ready.htm.

[16] *See* http://www.doi.gov/watersmart/html/index.php.

[17] *See* http://www.blm.gov/wo/st/en/prog/more/Landscape_Approach/reas.html.

[18] *See* http://www.usda.gov/oce/climate_change/regional_hubs.htm.

[19] *See* http://www.nrcs.usda.gov/wps/portal/nrcs/detailfull/national/programs/financial/eqip/?cid=nrcs143_008334.

[20] *See* http://www.fhwa.dot.gov/environment/climate_change/adaptation/ongoing_and_current_research/gulf_coast_study/.

country with information on transit use during emergency response and on building the resilience of public transportation assets and services to weather and climate risks. FTA has also established a new Emergency Relief Program that incentivizes incorporating actions to build climate resilience into disaster recovery efforts.[21] Additionally, the Federal Aviation Administration is analyzing aviation facility, service, and equipment profile data for vulnerability to a combination of potential storm surge impacts caused by climate change (U.S. DOT 2013).

Preparing for Future Energy Needs—The U.S. Department of Energy (DOE) recently conducted an assessment of climate change impacts on the U.S. energy sector and opportunities to make the energy system more resilient to climate-related risks (U.S. DOE and NREL 2013). DOE is also contributing to enhanced climate preparedness and resilience by facilitating basic scientific discovery; enhancing research, development, demonstration, and deployment of more climate-resilient energy technologies; convening and partnering with stakeholders, including industry and federal, state, and local leaders; and providing technical information and assistance. These efforts include research and development programs to reduce the energy and water intensity of electricity generation and use, and transportation fuels production; to expand and modernize the electric grid; and to enhance energy efficiency and reduce energy demand for buildings, appliances, and vehicles.

In addition, DOE is developing information and tools that will help local and regional planners anticipate climate change effects on the energy system and adaptation needs. DOE is providing technical assistance and guidance for state and local energy assurance planning, as well as support and assistance to help communities prepare for climate impacts and to address challenges, such as simultaneous restoration of electricity and fuel supply. Many of these programs will have co-benefits of both increasing climate preparedness and resilience and reducing carbon pollution to slow the effects of climate change.

Developing Tools to Support Local Decisions—EPA is supporting local decision makers through a variety of programs and online tools, including the Climate Ready Estuaries (CRE) program[22] and the Climate Ready Water Utilities Working Group.[23] EPA's CRE program has supported more than 30 coastal adaptation projects in collaboration with 19 National Estuary Programs from Charlotte Harbor, Florida, to Puget Sound, Washington. EPA's Water/Wastewater Agency Response Network helps water utility managers respond to and recover from emergencies that affect water system integrity and can lead to health risks from sewer system failures.[24] These projects have used the best available science for the development of climate change vulnerability assessments and have developed ecosystem-based adaptation strategies. Finally, EPA has developed a National Stormwater Calculator, a desktop application that estimates the annual amount of rainwater and frequency of runoff from a specific site anywhere in the United States (including Puerto Rico). Estimates are based on local soil conditions, land cover, and historic rainfall records, and the calculator accesses several national databases that provide soil, topography, rainfall, and evaporation information for the chosen site.[25]

The USFS is creating similar decision-support tools for natural resource managers. The Template for Assessing Climate Change Impacts and Management Options generates reports capturing and organizing information for specific locations and natural resource issues by synchronizing climate change literature with mapping tools and climate models.[26] Another tool, ForWarn, is a satellite-based forest disturbance monitoring system for assessing change.[27] It offers tools to attribute forest changes to insects, disease, wildfire, storms, human development, or unusual weather. Archived data allow ForWarn users to track, compare, and monitor forest disturbances that have occurred across the conterminous United States since 2000. Finally, iTree is a software suite for urban and community forestry monitoring, analysis, and benefits assessment.[28] iTree quantifies urban forest structure, environmental effects, and values.

Supporting Community-Level Resilience—The U.S. Department of Housing and Urban Development's (HUD's) Office of Policy Development and Research is helping to develop a toolkit of HUD initiatives that will provide new resources to communities to address the challenges resulting from climate change and growth patterns at the local level. In addition, HUD

---

[21] *See* http://www.fta.dot.gov/map21_15025.html.

[22] *See* http://www.epa.gov/CRE/.

[23] *See* http://water.epa.gov/infrastructure/watersecurity/climate/.

[24] *See* http://www.awwa.org/resources-tools/water-knowledge/emergency-preparedness/water-wastewater-agency-response-network.aspx.

[25] *See* http://www.epa.gov/nrmrl/wswrd/wq/models/swc/.

[26] *See* http://www.forestthreats.org/research/tools/taccimo.

[27] *See* http://www.fs.fed.us/ccrc/tools/forwarn.shtml.

[28] *See* http://www.itreetools.org/.

Sustainable Communities Regional Planning Grants encourage grant recipients to integrate climate adaptation into their regional housing, land use, and transportation planning.[29] The Regional Plan Association (RPA) of New York City is one of a number of HUD grantees incorporating climate information to enhance resilience of critical infrastructure to severe storms and coastal flooding. The RPA will also assess the urban design implications of flood protection standards to develop new example standards, codes, and regulations for municipalities that will better equip them to adapt to extreme climate conditions.

Protecting Government Facilities—The National Aeronautics and Space Administration (NASA) has created an integrated effort between its Earth Science Division and Office of Infrastructure to look at the long-term effects of climate change for NASA's facilities, many of which are in climate-sensitive areas, and to enable more informed future planning for its facilities and resource management.[30] In addition, through the Prediction of Worldwide Energy Resource project and Web portal, NASA provides user-friendly weather and solar data that help the energy, building, and agricultural industries plan for climate impacts.[31]

Designing Infrastructure for the Future—In 2011, USACE issued new guidance on how its projects, systems, and programs can respond to future changes in sea level (USACE 2011). In the long term, USACE will use this information to incorporate climate change considerations into existing and new civil works infrastructure and ecosystem restoration projects in coastal areas to improve safety and resilience.

Regional, State, Local, and Tribal Adaptation Initiatives

The federal government recognizes that state and local action is essential to ensuring that the nation is prepared for the impacts of climate change. Across the country, communities are taking steps to protect themselves and invest in lasting, resilient infrastructure. Through E.O. 13653, the President has directed federal agencies to take action to support these communities in their efforts to increase climate preparedness and resilience, including forging new partnerships with state and local governments to improve the preparedness and resilience of cities and towns and to ensure that taxpayer dollars are used efficiently to promote stronger, safer communities (EOP 2013b). Tables 6-2 and 6-3 highlight selected examples of state, regional, local, and tribal adaptation efforts, which are in many cases accomplished with federal support or in coordination with multiple federal agencies.

International Adaptation Activities

Presidential Policy Directive on Global Development and Global Climate Change Initiative
In September 2010, President Obama issued the Presidential Policy Directive on Global Development (PPD).[32] The PPD calls for elevating development as a core pillar of American foreign policy and for addressing global climate change as a key development initiative. Adaptation to climate change is specifically identified as a central component of the PPD and is one of the three pillars of the Obama administration's Global Climate Change Initiative (GCCI).[33]

As part of the GCCI, the United States is helping countries prepare for potentially severe climate change impacts (U.S. DOS 2012). For example, glacier retreat could have a devastating impact on water supply in Andean nations, India, Nepal, Bangladesh, Afghanistan, Pakistan, and Central Asia. The United States is building capacity for water resource management and supporting research on hydrological cycles, glacier dynamics, and adaption for downstream communities, as well as building climate resilience in least-developed countries (LDCs) and small-island developing states that are most vulnerable to extreme weather and other climate impacts. Support to the multilateral Pilot Program for Climate Resilience has leveraged $285 million in contributions from other developed country governments to help vulnerable developing countries, including several LDCs, pilot and demonstrate approaches for incorporating climate risk and resilience into development policies and planning.[34]

The President's Climate Action Plan reiterates U.S. support of international adaptation actions through historic investments in bolstering the capacity of countries to respond to climate change, including through the GCCI (EOP 2013a). The plan outlines efforts that expand bilateral cooperation with major emerging economies; strengthen government and local

[29] See http://portal.hud.gov/hudportal/HUD?src=/program_offices/sustainable_housing_communities/sustainable_communities_regional_planning_grants.

[30] See https://c3.nasa.gov/nex/projects/?tag=CASI%20-%20Climate%20Adaptation%20Science%20Investigators.

[31] See http://power.larc.nasa.gov/.

[32] Presidential Policy Directive on Global Development, September 2012. See http://foreignassistance.gov/Initiative_GCC_2012.aspx?FY=2012#ObjAnchor.

[33] President Obama's Development Policy and the Global Climate Change Initiative. See http://www.whitehouse.gov/sites/default/files/Climate_Fact_Sheet.pdf.

[34] *Climate Investment Funds: Response to Call for Inputs from the Technology Executive Committee. See* http://unfccc.int/ttclear/sunsetcms/storage/contents/stored-file-20130422151608983/CIF_EE.pdf.

Table 6-2 **Examples of State-Level Adaptation Activities**

Several states are taking action to address the preparedness and resilience of their cities and towns and to ensure that taxpayer dollars are used efficiently to promote stronger, safer communities.

State	Adaptation Action
Alaska	The Alaska Climate Change Impact Mitigation Program provides funds for hazard impact assessments to evaluate climate change-related impacts, such as coastal erosion and thawing permafrost.[a]
California	California is implementing building standards mandating energy and water efficiency savings, advancing both adaptation and mitigation. The State Adaptation Plan calls for a 20 percent reduction in per-capita water use.[b]
Florida	Florida legislators have passed a law supporting low-water-use landscaping techniques and have established state zoning statutes that allow regional authorities to establish adaptation zones in preparation for sea level rise in projected impact areas.[c]
Hawaii	Hawaii has adopted a water code that calls for integrated management, preservation, and enhancement of natural systems (Keener et al. 2012).
Kentucky	The *Action Plan to Respond to Climate Change in Kentucky: A Strategy of Resilience* identifies six goals to protect ecosystems and species in a changing climate (KDFWR 2010).
Louisiana	The 2012 *Comprehensive Master Plan for a Sustainable Coast* includes both protection and restoration activities addressing land loss from sea level rise, subsidence, and other factors over the next 50 years (CPRAL 2012).
Maine	Maine's *Coastal Sand Dune Rules* require that structures greater than 2,500 square feet be set back at a distance that is calculated based on the future shoreline position and considering 0.6 meters (m) (or 2 feet [ft]) of sea level rise over the next 100 years (MDEP 2012).
Maryland	Maryland legislators passed the *Living Shorelines Act* to reduce hardened shorelines throughout the state. The state government also created the "Building Resilience to Climate Change" policy, which establishes practices and procedures related to facility siting and design, new land investments, habitat restoration, government operations, research and monitoring, resource planning, and advocacy.[d]
Massachusetts	In Massachusetts, each school district has a designated school that acts as an evacuation site in the event of an emergency. After identifying a need for infrastructure to protect vulnerable citizens during a heat event, the Massachusetts Health Department—in partnership with the Center for Disease Control and Prevention's (CDC's) Climate-Ready States and Cities Initiative—is working with the state's Department of Education to secure funds to install air conditioning in these schools so they can be used as cooling shelters during extreme heat events.[e]
Montana	Montana maintains a statewide climate change Web site to help stakeholders access relevant and timely climate information, tools, and resources (Bierbaum et al. 2013).
New Mexico	New Mexico's Active Water Resource Management program allows for temporary water rights changes in real time in case of drought (Propst 2012).
North Carolina	In partnership with CDC's Climate-Ready States and Cities Initiative, North Carolina has mapped storm surge predictions against the location of critical infrastructure of public health significance. Using inundation estimates at 0.5, 1, and 2 m (1.6, 3.3, and 6.6 ft) the health department has been able to determine vulnerable drinking water sources and drinking water and wastewater treatment facilities that would be adversely affected, and has begun planning to mitigate these risks.[f]
Pennsylvania	The state government established polices to encourage the use of green infrastructure and ecosystem-based approaches for managing stormwater and flooding (Solecki et al. 2012).
Rhode Island	Rhode Island requires that public agencies considering land-use applications accommodate a 0.9-1.5-m (3-5-ft) rate of sea level rise (Bierbaum et al. 2013).
Texas	Texas coordinated the response to the 2011 drought through the National Integrated Drought Information System, Regional Integrated Science and Assessments (Southern Climate Impacts Planning Program and the Climate Assessment for the Southwest), and state and private-sector partners based on previously completed anticipatory planning and preparedness efforts (SCIPP 2010).

[a] *See* http://www.climatechange.alaska.gov/docs/iaw_accimp_27aug08.pdf.

[b] *See* http://www.climatechange.ca.gov/adaptation/water.html.

[c] *See* http://scholarship.law.wm.edu/cgi/viewcontent.cgi?article=1003&context=wmelpr.

[d] *See* http://www.cakex.org/case-studies/2829.

[e] Massachusetts Health and Human Services. Climate Change. *See* http://www.mass.gov/eohhs/gov/departments/dph/programs/environmental-health/exposure-topics/public-health-implications-of-climate-change.html.

[f] North Carolina Department of Health and Human Services. Occupational and Environmental Epidemiology: Climate and Health. *See* http://epi.publichealth.nc.gov/oee/programs/climate.html.

Table 6-3 **Examples of Regional and Local Adaptation Activities**

Across the country, communities are taking steps to protect themselves and invest in lasting, resilient infrastructure.

Local or Regional Government	Adaptation Action
Satellite Beach, FL	Collaboration with the Indian River Lagoon National Estuary Program led to the incorporation of sea level rise projections and policies into the city's comprehensive growth management plan (Gregg et al. 2011).
Portland, OR	Portland updated its city code to require on-site stormwater management for new development and redevelopment, and provides a downspout disconnection program to help promote on-site stormwater management.[a]
Lewes, DE	In partnership with Delaware Sea Grant, ICLEI-Local Governments for Sustainability, the University of Delaware, and state and regional partners, the City of Lewes undertook a stakeholder-driven process to understand how climate adaptation could be integrated into its hazard mitigation planning process. Recommendations for integration and operational changes were adopted by the City Council and are currently being implemented (Lewes 2011).
Groton, CT	Groton partnered with federal, state, regional, local, nongovernmental, and academic partners through EPA's Climate Ready Estuaries program to assess vulnerability to and devise solutions for sea level rise (Stults and Pagach 2011).
San Diego Bay, CA	Five municipalities partnered with the Port of San Diego, the airport, and more than 30 organizations with direct interests in the future of San Diego Bay to develop the San Diego Bay Sea Level Rise Adaptation Strategy. The strategy identified key vulnerabilities for the bay and adaptation actions that can be taken by individual agencies, as well as through regional collaboration (Solecki et al. 2012).
Chicago, IL	Through a number of development projects, the city has added 55 acres of permeable surfaces since 2008 and has more than four million square feet of green roofs planned or completed (Bierbaum et al. 2013).
King County, WA	King County created the King County Flood Control District in 2007 to address increased impacts from flooding through such activities as maintaining and repairing levees and revetments, acquiring repetitive loss properties, and improving countywide flood warnings (Bierbaum et al. 2013).
New York City, NY	Through a partnership with FEMA, the city has updated FEMA Flood Insurance Rate Maps based on more precise elevation data. The new maps will help stakeholders better understand their current and future flood risks and allow the city to more effectively plan for climate change (NPCC2 2013).
	In partnership with CDC's Climate-Ready States and Cities Initiative, New York City also used climate models to develop a more sensitive and customized heat-warning system to better protect New Yorkers during heat waves. This was achieved by studying retrospective hospitalization and mortality data, projections for relevant climate conditions (such as temperature and humidity), and localized modeling of the urban heat island effect (NPCC2 2010).
Southeast Florida Regional Climate Compact	Broward, Miami-Dade, Palm Beach, and Monroe counties have jointly committed to partner in reducing greenhouse gas emissions and adapting to climate impacts.[b] They have already made significant progress in regional planning to address sea level rise.
Phoenix, AZ; Boston, MA; Philadelphia, PA; and New York, NY	Climate change impacts are being integrated into public health planning and implementation activities that include creating more community cooling centers, neighborhood watch programs, and reductions in the urban heat island effect (Vogel et al. 2011, Horton et al. 2011, White-Newsome et al. 2011).
Boulder, CO; New York, NY; and Seattle, WA	Water utilities in these communities are using climate information to assess vulnerability and inform decision making (Vogel and Smith 2010).
Philadelphia, PA	In 2006, the Philadelphia Water Department began a program to develop green stormwater infrastructure, intended to convert more than one-third of the city's impervious land cover to "Greened Acres," which include green facilities, green streets, green open spaces, green homes, and stream corridor restoration and preservation (Wilbanks et al. 2012b).

[a] *See* http://www.cnt.org/repository/Portland.pdf.

[b] Southeast Florida Regional Climate Compact. See http://southeastfloridaclimatecompact.org/pdf/compact.pdf.

community planning and response capacities, such as by increasing water storage and water use efficiency to cope with the increased variability in water supply; develop innovative financial risk management tools, such as index insurance to help smallholder farmers and pastoralists manage risks associated with changing rainfall patterns and drought; and distribute drought-resistant seeds and promote management practices that increase farmers' ability to cope with climate impacts (EOP 2013a).

USAID Programs

The U.S. Agency for International Development (USAID) is investing in the scientific capacity of partner countries, and improving access to and use of climate information to help societies identify vulnerabilities and evaluate potential adaptation strategies (U.S. DOS 2012). The following programs are examples of USAID's work to provide access to timely and user-driven information and to help communities adapt to climate variability and change.[35]

SERVIR—A collaborative effort between USAID and NASA, the SERVIR program provides 10 countries in Central America and the Caribbean, 18 countries in East Africa, and 6 countries in the Hindu Kush-Himalaya region with satellite imagery and user-friendly weather and climate information, informing decision making in health, environmental management, disaster preparedness, and other areas.[36] SERVIR supports national governments, universities, NGOs, and the private sector.

Climate Services Partnership (CSP)—The CSP was formed at the first International Conference on Climate Services in 2011 to improve understanding and application of climate services among decision makers and practitioners in developing countries. The CSP draws from a broad membership to promote the matching of the best information with those who need to use it in decision making. In doing so, the CSP supports the Global Framework for Climate Services, a formal international system that facilitates the coordinated support of climate services worldwide. The CSP is also building the capacity of national weather services to deliver climate information products to stakeholders in government ministries and the private sector.

High Mountain Adaptation Partnership—Created in 2010, the High Mountain Adaptation Partnership grew out of the Adaptation Partnership, which was founded by the United States, Spain, and Costa Rica to facilitate enhanced action on adaptation. The partnership also built on a series of activities that USAID and the National Science Foundation organized in glacier-dependent areas. The partnership has created a community of practice that brings together physical and social scientists, development practitioners, policymakers, and planners, with the aim of improving knowledge, fostering South–South information exchange, and mobilizing resources for applied research and multi-stakeholder-based adaptation projects in the Hindu Kush-Himalaya, Andes, Central Asia, and other high mountain regions. The program has pioneered new rapid assessment techniques for studying the risks of glacier lakes.

The Mountain Institute—Women are disproportionately vulnerable to climate change impacts, but often have high levels of skill in leading and supporting adaptation actions. USAID aims to make its adaptation efforts inclusive and gender sensitive and to demonstrate ways to effectively integrate this perspective into adaptation programs. For example, in Peru in 2010, USAID supported The Mountain Institute in conducting a series of community workshops to analyze climate vulnerability and test ways to integrate a gender approach into adaptation. Women identified the need to conserve local ecosystems, such as high Andean wetlands and grasslands, which are critical for water regulation, especially in the context of melting glaciers. The project also provided leadership and climate change adaptation training to women serving on municipal councils.

U.S. Department of State

Also in support of international adaptation efforts, the U.S. Department of State focuses on development and implementation of effective international adaptation policies and programs and promotes the integration of adaptation considerations into diplomatic and development initiatives in sectors that will be affected by climate change, such as agriculture, water, and disaster risk management.[37]

[35] USAID. Global Climate Change Adaptation Activities. *See* http://www.usaid.gov/what-we-do/environment-and-global-climate-change/global-climate-change-adaptation/global-climate-change-adaptation-activities.

[36] SERVIR is a Spanish language acronym for Regional Visualization and Monitoring System.

[37] U.S. Department of State, Global Climate Change. *See* http://www.state.gov/e/oes/climate/.

BUILDING ON PROGRESS

In the last several years, major progress has been made on adaptation planning and implementation across all levels of government in the United States, including a focus on research, assessments, and adaptation. At the national level, the most recent Third NCA Report is a major step forward in building both scientific understanding and important partnerships focused on reducing risk, and the new "sustained assessments" approach is explicitly designed to support adaptation decisions (NCA 2013). The previous Adaptation Task Force has produced a large number of adaptation initiatives and has overseen the development of adaptation and sustainability plans for every federal agency. The interagency Council on Climate Preparedness and Resilience, informed by the recommendations of the new State, Local, and Tribal Leaders Task Force, will continue and build on this work.

Some states, including California, have taken significant steps toward increasing energy efficiency, reducing emissions, and increasing preparedness. Many other states have joined regional efforts to curb emissions. New York, Philadelphia, Chicago, San Francisco, and many smaller cities and towns have made impressive progress in reducing their vulnerability to climate-related impacts. In addition, Native American tribes in Alaska, the Pacific Northwest, and elsewhere have engaged in comprehensive climate change adaptation planning. At the same time, U.S. investments in adaptation efforts internationally have substantially expanded, and they are reducing the vulnerability of many developing countries to climate change.

Although much more work needs to be done both domestically and internationally, the United States has made major progress since publishing the 2010 CAR. The most dramatic evidence to date of the U.S. commitment to managing emissions, increasing preparedness, and providing leadership in the domestic and international arenas can be found in the President's June 2013 *Climate Action Plan* (EOP 2013a) and the November 1, 2013, E.O. 13653, *Preparing the United States for the Impacts of Climate Change* (EOP 2013b).

7

Financial Resources and Transfer of Technology

The United States is committed to assisting developing countries in their efforts to mitigate and adapt to climate change. Since the period covered by the *U.S. Climate Action Report 2010* (2010 CAR) (U.S. DOS 2010), the United States has significantly ramped up its provision of climate finance. Climate change has become a major thrust of U.S. diplomatic and development assistance efforts and has been integrated into the core operations of all major U.S. foreign assistance agencies.

The United States is using the full range of institutions—bilateral, multilateral, development finance, and export credit—to mobilize private finance and invest strategically in building lasting resilience to unavoidable climate impacts; to reduce emissions from deforestation and land degradation; and to support low-carbon development strategies and the transition to a sustainable, clean energy economy. The United States is working to ensure that its capacity-building and investment support is efficient, effective, innovative, based on country-owned plans, and focused on achieving measurable results with a long-term view toward economic and environmental sustainability.

Climate change has become a major focus of U.S. diplomatic and development objectives through a series of significant policy directives. The 2010 Presidential Policy Directive on Global Development[1] identified the Global Climate Change Initiative (GCCI) as one of three priority U.S. development initiatives.[2] GCCI provides a platform upon which the United States builds climate change considerations into its foreign assistance operations. The 2010 U.S. *First Quadrennial Diplomacy and Development Review* also identified climate change as one of the main pillars of U.S. diplomacy and international development (U.S. DOS and USAID 2010). The 2012 U.S. Agency for International Development (USAID) *Climate Change and Development Strategy* sets out principles, objectives, and priorities for USAID climate change assistance from 2012 through 2016 (USAID 2012). This strategy prioritizes not only clean energy, sustainable landscapes, and adaptation, but also integration: factoring climate change knowledge and practice into all USAID programs to ensure all sector portfolios are climate resilient and, where possible, reduce greenhouse gas (GHG) emissions.

In addition, the Overseas Private Investment Corporation (OPIC) has adjusted its policies to shift its international investments into climate-friendly activities. As the U.S. government's development finance institution, OPIC mobilizes private capital toward development challenges, and in doing so contributes to U.S. development and foreign policy objectives. OPIC has pledged to reduce GHG emissions associated with its investments by 30 percent by 2018 and by 50 percent by 2023, and to promote clean energy and energy efficiency investments. OPIC has dramatically expanded its commitments to renewable resources, up 30-fold since 2007. OPIC has also introduced new tools for developing-country investors, such as direct financing for energy efficiency improvements; insurance against regulatory changes, such as cuts in renewable energy feed-in tariffs; and protection against government interference in the use of carbon credits.

The United States remains committed to supporting multilateral climate change and environment funds, including the Climate Investment Funds (CIFs) and the Global Environment Facility (GEF). The United States has pledged $2 billion to the CIFs, and to date has contributed

---

[1] Fact Sheet: U.S. Global Development Policy. *See* http://www.whitehouse.gov/the-press-office/2010/09/22/fact-sheet-us-global-development-policy.

[2] Foreign Assistance Initiatives. *See* http://foreignassistance.gov/InitiativeLanding.aspx.

$1.137 billion. For the GEF's fifth replenishment (GEF-5) for fiscal years (FYs) 2011–2014, the United States has pledged $575 million, an increase of more than 50 percent from the U.S. GEF-4 pledge.

In FY 2010, the United States made its first contributions to the Least Developed Countries Fund (LDCF) and the Special Climate Change Fund (SCCF). The United States is now one of the largest donors to these multilateral adaptation funds, having contributed $120 million between FYs 2010 and 2012. The United States has supported the development of the Green Climate Fund (GCF) since the concept was first proposed, has actively participated on the Transitional Committee that negotiated the GCF Governing Instrument, and remains committed to helping operationalize an effective and efficient GCF as a member of its Board.

At the 15th Conference of the Parties (COP-15) to the United Nations Framework Convention on Climate Change (UNFCCC) in Copenhagen, the United States committed to working with other developed countries to collectively provide resources approaching $30 billion in "fast start" finance (FSF) during the period 2010–2012 to support developing countries in their mitigation and adaptation efforts. In conjunction with other developed country Parties to the UNFCCC, the United States also agreed to the goal of collectively mobilizing $100 billion per year in climate finance by 2020, from a wide variety of public and private sources, to address the needs of developing countries in the context of meaningful mitigation actions and transparency on implementation.

As noted in Decision 1 of COP-18 in Doha, developed country Parties successfully achieved the FSF goal (UNFCCC 2013). U.S. climate finance was $7.5 billion[3] from FYs 2010 through 2012, and reached more than 120 countries through bilateral and multilateral channels, meeting the President's commitment to provide America's fair share of the collective pledge.[4] This $7.5 billion consists of more than $4.7 billion of congressionally appropriated assistance, more than $1.9 billion of development finance, and $749 million of export credit. The $4.7 billion in appropriated assistance represents a fourfold increase in annual climate assistance since 2009, with a ninefold increase in adaptation assistance.

This chapter provides details on U.S. climate finance by channels and instruments, thematic pillar, and region; describes U.S. efforts to mobilize private climate finance; and illustrates examples of U.S. contributions to capacity building and transfer of technology.

CHANNELS AND INSTRUMENTS

U.S. climate finance is provided through several different channels that can broadly be grouped into three categories: (1) congressionally appropriated finance, delivered through both bilateral and multilateral channels; (2) development finance, delivered through OPIC; and (3) export credit, delivered through the U.S. Export-Import Bank (Ex-Im).

Congressionally Appropriated Assistance

The United States provides congressionally appropriated, climate change-dedicated, grant-based assistance via the GCCI, as well as additional congressionally appropriated grant-based assistance that delivers climate co-benefits. This assistance is delivered through both bilateral and multilateral channels.

Bilateral Climate Finance

Grant-based U.S. bilateral climate assistance is programmed directly through bilateral, regional, and global programs. These programs are principally supported by USAID, and also through the U.S. Department of State (DOS), Millennium Challenge Corporation (MCC), and other U.S. government agencies.[5] Allocation decisions for each program are made by the administering U.S. government agency. Dedicated U.S. climate assistance is targeted to help the most vulnerable countries adapt to climate change impacts, and countries with significant opportunities to mitigate their GHG emissions (Box 7-1).

Multilateral Climate Finance

Multilateral climate change funds feature institutional structures governed jointly by developed and developing countries, and play an important role in promoting a coordinated, global response to climate change. U.S. contributions to multilateral climate funds—channeled through the U.S. Department of the Treasury and DOS—leverage funding from other

---

[3] The totals reported here reflect slight revisions to previously reported levels, based on updated information received since the release of the November 2012 Fast Start Finance (FSF) report (U.S. DOS 2012).

[4] While the U.S. FSF reports use the term "provided" to describe U.S. support, the term "committed" is used in this report to be consistent with the new Biennial Report Common Tabular Format guidelines, and to be consistent with the terminology used in the *Biennial Report* and the *Sixth National Communication*. For further information related to U.S. methodologies, see http://www.state.gov/e/oes/rls/rpts/car6/index.htm.

[5] In counting and aggregating climate finance, the United States includes programs that have a primary mitigation and/or adaptation purpose, as well as activities with significant climate co-benefits (e.g., relevant biodiversity and food security activities). In the case of programs for which only part of the activity is targeted toward a climate objective, only the relevant financial support is counted, rather than the entire program budget. (For more information, see the *Biennial Report* and associated documentation at http://www.state.gov/e/oes/rls/rpts/car6/index.htm.

Box 7-1 **Millennium Challenge Corporation**

The Millennium Challenge Corporation (MCC) was founded in 2004 with a focused mandate to reduce poverty through economic growth. Two of MCC's founding principles are country ownership and a focus on results. These principles lead MCC to support investments that reflect countries' own priorities for poverty reduction, and offer the most promise for returns in terms of increased incomes.

The United States recognizes that people's livelihoods and well-being depend on reliable and equitable access to natural resources. Toward this end, the United States will help partner countries strengthen their capacity to preserve and enhance ecosystem functions and natural wealth that are vital to achieving long-term poverty reduction and development outcomes, and will help communities build resilience to environmental stressors, such as climate change, water scarcity, and natural disasters. Among other approaches, these goals are achieved by incorporating cost-effective, technically, and economically viable, measures into projects that can promote energy efficiency, improve water resource management, support less carbon-intensive land-use practices, improve institutional capacity for environmental management, and help protect worker and public health and safety.

For example, in an effort to increase the incomes of Indonesia's poor in targeted districts, the MCC-funded $332.5 million Green Prosperity Project will provide commercial and grant financing to help mobilize greater private-sector investment in renewable energy and sustainable land-use practices. This project will also provide technical assistance to support project preparation, improve land-use planning, and strengthen local and regional capacity to pursue low-carbon development.

governments, development partners, and the private sector to enable large-scale infrastructure investments with a range of tailored financial products across a wide range of countries. As with bilateral finance, U.S. contributions to multilateral climate funds are allocated to adaptation, clean energy, and sustainable landscape activities.

During FY 2010–2012, U.S. multilateral climate change finance amounted to $1.2 billion. This total includes the CIFs (which include the Clean Technology Fund, the Forest Investment Program, the Pilot Program for Climate Resilience, and the Scaling-Up Renewable Energy Program in Low-Income Countries), the GEF, the LDCF, the SCCF, and the Forest Carbon Partnership Facility.

Development Finance and Export Credit

OPIC and Ex-Im play a critical role by using public funds to mobilize much larger sums of private investment directed at mitigation through loans, loan guarantees, and insurance in developing countries.

Table 7-1 summarizes U.S. climate finance by channel. Tables 7-3 through 7-6 at the end of this chapter present climate-related U.S. financial contributions to the GEF, overall

Table 7-1 **U.S. Climate Finance by Channel** (in US$ millions)[a]

U.S. climate finance was $7.5 billion during fiscal years 2010, 2011, and 2012, and reached more than 120 countries through bilateral and multilateral channels. The $4.7 billion in appropriated assistance represents a fourfold increase in annual climate assistance since 2009, and a ninefold increase in adaptation assistance.

Channel	2010	2011	2012	Total
Congressionally Appropriated Assistance (USAID, State, Treasury, MCC, and other U.S. agencies)	$1,587.9	$1,884.1	$1,261.7	$4,733.7
Development Finance (OPIC)[b]	$155.1	$1,114.8	$721.6	$1,991.5
Export Credit (Ex-Im)	$253.2	$194.7	$301.2	$749.1
Total	**$1,996.2**	**$3,193.6**	**$2,284.5**	**$7,474.3**

[a] These numbers do not include private investment leveraged.

[b] These figures include only OPIC projects related to climate change, and are therefore counted under fast start finance (FSF). However, OPIC's renewable resources portfolio (renewable energy, sustainable water, and agriculture) totals exceed the FSF-eligible totals being reported here. OPIC figures in this document reflect commitments made in the specified year and do not take into account any cancellations that may occur in subsequent years.

Note: Ex-Im = Export-Import Bank of the United States; GHG = greenhouse gas; MCC = Millennium Challenge Corporation; OPIC = Overseas Private Investment Corporation; USAID = U.S. Agency for International Development contributions to multilateral institutions, and bilateral and regional contributions related to the implementation of the UNFCCC.

CLIMATE FINANCE BY THEMATIC PILLAR

U.S. climate finance falls under three thematic pillars: adaptation, clean energy, and sustainable landscapes, the last of which focuses largely on helping countries to slow, halt, and reverse defor- estation and related GHG emissions (primarily through reducing emissions from deforestation and forest degradation, or REDD+). The latter two pillars are often described jointly as mitigation.

Adaptation—Promoting Climate Resilience

For adaptation, dedicated U.S. climate assistance prioritizes countries, regions, and popula- tions that are highly vulnerable to climate change impacts. By increasing resilience in key sec- tors, such as food and water security, coastal management, and public health, U.S. programs help vulnerable countries prepare for and respond to increasing climate- and weather-related risks. Assistance identifies and disseminates adaptive strategies, makes accessible the best available projected climate change impact and weather data to counterparts, and builds the capacity of partner governments and civil society partners to respond to climate change risks.

Sample Activities: Adaptation

SERVIR[6]—Globally, USAID and the National Aeronautics and Space Administration (NASA) have provided more than $41 million from FY 2010 through 2013, to increase the application of satellite data, ground-based observations, and forecasts directly tailored to the needs of decision makers to help them avoid climate-related hazards and improve development outcomes. SERVIR partners with international institutions in Central America, Eastern and Southern Africa, and the Hindu Kush-Himalaya region to reach governmental and other key decision makers. It also provides a Web-based platform to improve open access to satellite information, imagery, and other decision-support tools to inform agriculture, water, energy, health, forest and land planning and management, ecotourism, and disaster preparedness and response, among other areas. SERVIR has leveraged approximately $1 million in private-sector resources and services, including hardware, software, and wireless services from partners, including Cable and Wireless, ESRI, and Google.

FEWS NET—USAID, working with the U.S. Geological Survey (USGS), NASA, the National Oceanic and Atmospheric Administration (NOAA), and the U.S. Department of Agriculture (USDA), is investing more than $13 million annually for FYs 2010–2013 to support the Famine Early Warning Systems Network. FEWS NET provides information and early warning on sea- sonal climate patterns and challenges to food and water security in communities vulnerable to climate variability and change; monitors agriculture, climate, and market data; and helps decision makers anticipate and respond to food insecurity. This and other efforts are trans- forming the ability of developing countries to use science to improve their decision-making processes and strategies.

R4 Rural Resilience Initiative—USAID is piloting new approaches to insurance to help poor farmers manage weather risks. In Senegal, for example, USAID is investing $8 million in the R4 Rural Resilience Initiative, which will overcome cash constraints by enabling the poorest farmers to pay for their insurance with their labor by working extra days on community risk reduction projects, such as improved irrigation or soil management. USAID is also supporting the expansion of an index-based livestock insurance program from Kenya to Ethiopia to help protect herding families from losses due to severe drought. This initiative has leveraged $1.2 million in private investment and expertise from global re-insurer Swiss Re.

C-CAP—In the Pacific Islands region, USAID is supporting a five-year, $23.6 million Coastal Community Adaptation Program (C-CAP) to help reduce the vulnerability of coastal com- munities to the impacts of climate change. C-CAP is building local capacity for disaster risk reduction and preparedness, and integrating climate-resilient policies and practices into long- term land-use plans and building standards. The program is expected to benefit approximate- ly 90 communities in up to 12 Pacific Island nations.

PPCR—During FYs 2010–2012, the United States contributed $84 million to the Pilot Program for Climate Resilience (PPCR), which works to increase resilience and protect vulnerable

---

[6] SERVIR is a Spanish language acronym for Regional Visualization and Monitoring System.

populations in 18 countries. The PPCR is providing funds to help six Caribbean countries improve disaster management in response to devastating hurricanes and flooding. PPCR funding will help save thousands of lives and avoid billions of dollars in economic losses through improved planning and weather forecasting.

Mitigation—Accelerating Growth and Supporting Transitions to Low-Carbon Economies

Clean Energy

For clean energy, dedicated U.S. climate assistance focuses on countries and sectors offering significant emission reduction potential over the long term, as well as countries that offer the potential to demonstrate leadership in sustained, large-scale deployment of clean energy. In terms of sector coverage, clean energy includes renewable energy and energy efficiency and excludes natural gas and other fossil fuel power plant retrofits. The United States also supports regional energy programs that improve the enabling environments for regional energy grids to distribute clean energy, as well as global programs that focus chiefly on information sharing and building coalitions for action on clean energy technologies and practices.

Although climate finance generally refers to investing in low-carbon infrastructure, it is equally important from a climate impact point of view to address financing for high-carbon forms of energy. In June 2013, President Obama called for an end to U.S. government support for public financing of new coal power plants overseas, except for (1) the most efficient coal technology available in the world's poorest countries in cases where no other economically feasible alternatives exist, or (2) facilities deploying carbon capture and sequestration technologies (EOP 2013a). As part of this new commitment, the United States is working to secure the agreement of other countries, export credit agencies, development finance institutions, and multilateral development banks to adopt similar policies as soon as possible.

In September 2013, the leaders of Denmark, Finland, Iceland, Norway, and Sweden joined the United States in ending public financing for new coal-fired power plants overseas, except in rare circumstances, and the United Kingdom announced a similar commitment in November 2013. The United States also welcomes the decisions made by the World Bank and the European Investment Bank to adopt similar policies. Furthermore, the United States remains committed to phasing out subsidies that encourage wasteful consumption of fossil fuels. President Obama is calling for the elimination of U.S. fossil fuel tax subsidies in his FY 2014 budget, and the United States will continue to collaborate with partners around the world toward this goal (EOP 2013a).

Sample Initiatives: Clean Energy

AIP—During FYs 2010–2012, USAID invested more than $15 million in the Africa Infrastructure Program (AIP) to provide clean energy capacity-building and transaction advisory assistance across sub-Saharan Africa. AIP is helping partner governments and agencies in African countries to plan and implement the key institutional, legal, commercial, and regulatory reforms that are needed to attract private investment in clean energy. AIP also provides specific technical assistance and advisory services to support governments in evaluating and negotiating clean energy projects.

Ex-Im Support—Ex-Im committed $749.1 million to support renewable energy exports to developing countries during FYs 2010–2012. These authorizations were made in the form of loans, financial guarantees, and export credit insurance policies. This financing will establish more than 850 megawatts (MW) of clean electricity generation capacity, mainly from new solar power plants and wind energy farms. For example, Ex-Im provided a $48.6 million loan to support the Novo Gramacho biogas project in Brazil. The funding will support the export of proprietary biogas cleaning technology. Additionally, Ex-Im has provided substantial support for solar energy in India. Estimates are that Ex-Im financed more than 30 percent of the projects allocated under National Solar Mission in India, under Phase 1, which recently concluded.

OPIC Support—During FYs 2010–2012, OPIC committed $1,991.5 million in climate change financing support, predominately for clean energy projects. The wide variety of clean energy projects OPIC supported in 2012 illustrate the breadth of its work, which covers a range of project sizes and structures. OPIC's FY 2012 projects include a $16.7 million loan to develop a new 12-MW biomass power plant in Pakistan, which will be the first renewable energy biomass plant to supply power to the national grid, and $250 million in financing to support the construction of a solar power plant in an underdeveloped region of South Africa.

SEAD, CESC, Global LEAP—As part of the Clean Energy Ministerial process, the U.S. Department of Energy (DOE) is implementing a range of programs aimed at expanding the use of energy efficiency and renewable energy technologies.

The Super-Efficient Equipment and Appliance Deployment (SEAD) initiative supports the acceleration of global energy efficiency gains for internationally traded equipment and appliances by pulling super-efficient appliances and equipment into the market through cooperation on incentives, procurement, awards, and research and development (R&D) investments, and by bolstering national or regional minimum efficiency standards.

The Clean Energy Solutions Center (CESC) is a Web-based, knowledge-sharing platform that aims to help governments design and adopt policies and programs that support the deployment of low-carbon technologies.

The Global Lighting and Energy Access Partnership (Global LEAP, formerly known as the Solar and LED Energy Access initiative, or SLED) is developing a global quality assurance program for off-grid lighting products and small solar kits for rural electrification. Global LEAP also is supporting the expansion of the Lighting Africa activities spearheaded by the World Bank Group to new regions, including India. At COP-15 in Copenhagen, the United States announced its intent to contribute $35 million over five years to these programs as part of the Climate Renewables and Efficiency Deployment Initiative.

Power Africa —Power Africa is a new initiative to double access to power in sub-Saharan Africa. More than two-thirds of the population of sub-Saharan Africa is without electricity, and more than 85 percent of people living in rural areas lack access. Power Africa will build on Africa's enormous power potential, including the potential to develop clean geothermal, hydro, wind, and solar energy. This initiative will help countries develop newly discovered resources responsibly, build out power generation and transmission, and expand the reach of mini-grid and off-grid solutions.

CTF—The United States contributed $714.6 million during FYs 2010–2012 to support the critical work of the Clean Technology Fund. The CTF catalyzes clean energy investments in emerging economies with rapidly growing emissions by helping countries achieve access to renewable energy, green growth, and energy efficiency in transport, industry, and agriculture. The CTF is working with 18 countries on projects, such as wind power in Egypt, sustainable urban transportation in the Philippines, and energy efficiency in Turkey. The funds are channeled toward projects that focus on scaling up proven technologies, thereby promoting new markets for maximum impact. To date, the CTF has approved 41 projects for a total of $2.3 billion. These funds have leveraged $18.8 billion in co-financing, including $5.8 billion from the multilateral development banks and $13 billion from other sources, and have contributed to reducing 525 teragrams of carbon dioxide equivalent (Tg CO_2e) emissions—the equivalent of taking 99 million cars off the road for a year.

SREP—During FYs 2010–2012, the United States contributed $28 million to the Scaling-up Renewable Energy Program (SREP), which is working to expand energy access in eight countries. To date, approved projects in Kenya, Nepal, and Honduras are using $46 million in SREP funds to leverage $562 million in co-financing and build 250 MW of sustainable energy capacity. The Maldives will use SREP funds to increase renewable energy production from 1 percent of power generated to 16 percent. The SREP projects will supply energy that is cleaner and 10–20 percent cheaper than diesel-generated power, and help the Maldives government save at least $7 million in fuel subsidies per year.

ENERGY STAR—The U.S. Environmental Protection Agency's ENERGY STAR program has arrangements with agencies in several other countries, allowing them to implement ENERGY STAR for a variety of products and building types. These bilateral agreements on products delineate program responsibilities to promote, monitor, and enforce ENERGY STAR in their markets. Most of these product partnerships are limited to office equipment because of the global nature of the products. All of these international efforts allow ENERGY STAR to work closely with other government agencies and stakeholders to harmonize test procedures and specification levels, where appropriate.

PACE—Launched in 2009, the U.S.-India Partnership to Advance Clean Energy (PACE) focuses on spurring low-carbon inclusive development by supporting R&D of clean energy. Since PACE's launch, the U.S. government has mobilized about $2 billion in public and private resources for clean energy projects in India. In addition, the United States and India have launched a $125 million Joint Clean Energy Research and Development Center, which includes pledges of $25 million from the U.S. and Indian governments and an additional $75 million in matching private funds.

Sustainable Landscapes

For activities related to land-use-related mitigation (or "sustainable landscapes"), including REDD+, dedicated U.S. climate change assistance works to combat unsustainable forest clearing (for example, for agriculture and illegal logging), and is helping ensure good governance at local and national levels to support the sustainable management of forests. U.S. support prioritizes mitigation potential; countries with the political will to implement large-scale efforts to reduce emissions from deforestation, forest degradation, and other land-use activities; and potential for investments in monitoring, reporting, and verification of forest cover and GHG emission reductions. The United States also provides multilateral funding to support all three phases of REDD+, from readiness (Phase 1) through strategy implementation (Phase 2), to payment for results (Phase 3).

Sample Initiatives: Land-Use-Related Mitigation

FCPF, FIP—The United States funds the Readiness Fund of the Forest Carbon Partnership Facility (FCPF), which supports 36 developing countries in preparing strategies and programs, as well as engaging stakeholders, to advance REDD+. The United States also funds the Forest Investment Program (FIP), which supports efforts to strengthen forest governance and institutional capacity, as well as measures to reduce drivers of deforestation outside the forest sector in eight countries. U.S. funding for the FCPF Carbon Fund helps pilot an international results-based system that will reward progress made in reducing deforestation and the associated emissions. Together the FCPF and FIP have contributed to advancing global knowledge and technical approaches to REDD+, as well as supporting the strategies and programs that will lead to increased forest protection, reduced GHG emissions, and the many other benefits provided by healthy, intact tropical forests.

SilvaCarbon—The interagency SilvaCarbon program is an effort to build the capacity of selected countries in Africa, South America, and Southeast Asia to use forest and terrestrial carbon measurement and monitoring tools and technologies, and demonstrate and compare related methodologies. The program is supported by $8 million from DOS and $12 million from USAID, as well as funding from the participating technical agencies.

CARPE—USAID's landmark Central Africa Regional Program for the Environment (CARPE) is now transitioning into its third phase with a $13.6 million investment from USAID. The third phase of CARPE will include two major components: the Central Africa Forest Ecosystems Conservation (CAFEC) program and the Environmental Monitoring and Policy Support (EMAPS) program. CAFEC promotes responsible management of tropical forests. EMAPS strengthens central African nations' capacity to better govern their natural resources, develop new scientific methods to monitor changes to forests, and manage natural resources in a way that strengthens biodiversity and reduces landscape-related GHG emissions.

Forging International Partnerships

The United States is a strong supporter of partnerships and coalitions focused on practical action to address the drivers of climate change (Box 7-2).

Sample Initiatives: Forging International Partnerships

GMI—Formerly known as the Methane to Markets Partnership, the Global Methane Initiative (GMI) aims to reduce methane emissions and advance the abatement, recovery, and use of

Box 7-2 **Climate and Clean Air Coalition**

DOS invested $12.5 million in the Climate and Clean Air Coalition. Launched in 2012, this voluntary, collaborative global partnership unites governments, intergovernmental organizations, the private sector, and civil society to quickly reduce short-lived climate pollutants (SLCPs), such as methane, black carbon, and many hydrofluorocarbons (HFCs). According to a United Nations Environment Programme/World Meteorological Organization study aggressive action on these pollutants could avert 0.5°C (0.9°F) of warming by 2050, while preventing more than two million premature deaths each year and avoiding more than 30 million tons of annual crop losses by 2030 (UNEP and WMO 2011).

The Coalition focuses high-level attention on this issue to help catalyze major reductions of SLCPs. These actions can be undertaken now using current technologies. Major efforts include reducing methane and black carbon from waste and landfills; avoiding methane leakage, venting, and flaring from oil and gas production; phasing down HFCs through new technologies; and addressing black carbon from brick kilns, cookstoves, and diesel engines.

Since its launch in February 2012, the Coalition has rapidly grown from six country partners to 32, and has brought on leading international organizations, including UNEP, the World Bank, and the United Nations Development Programme, with more than 60 total international partners. In less than 18 months, the Coalition has attracted more than $40 million in funding support and has launched nine action-oriented initiatives to reduce SLCPs.

methane as a valuable clean energy source. GMI achieves this by creating an international network to build capacity, develop strategies and markets, and remove barriers to methane reduction project development in partner countries.

The United States has been a strong leader of GMI. U.S. contributions of $74.4 million through FY 2012 have mobilized more than $465 million in investment from other partner countries, development banks, the private sector, and members of the GMI Project Network. Under the GMI, the United States has cumulatively provided technical, financial, or capacity-building support to several hundred global projects. U.S. activities contributed to the reduction of methane emissions by approximately 30 Tg CO_2e in 2011 alone; cumulative emission reductions exceed 160 Tg CO_2e.

LEDS GP—DOS is investing $2 million in the Low Emission Development Strategies Global Partnership (LEDS GP). Through workshops and collaboration on a wide range of topics, the LEDS GP has brought together more than 100 countries, more than 100 institutions, and more than 700 LEDS practitioners to engage in peer learning and training on low-emission development. The partnership operates three regional platforms for cooperation, one each in Asia, Latin America, and Africa. In 2013, the LEDS GP will focus on building capacity on financing LEDS, connecting LEDS experts, and developing tools to make the case for low-emission development (Box 7-3).

TFA 2020—Tropical Forest Alliance (TFA) 2020 is a public–private-sector alliance launched in 2012 by the United States and the Consumer Goods Forum, a business network of more than 400 global retailers and producers from 70 countries with over $3 trillion in annual sales. Other TFA 2020 partners include the Netherlands, Norway, the United Kingdom (UK), Conservation International (CI), the Dutch Sustainable Trade Initiative, and World Resources Institute (WRI). All TFA 2020 partners agree to take voluntary actions to reduce the tropical deforestation associated with global commodities, such as palm oil, soy, beef, and paper and pulp. TFA 2020 is a whole-of-U.S. government effort, engaging a full range of expertise across U.S. government agencies.

The Alliance is open to new government, business, and civil society partners who agree to undertake specific voluntary actions to address commodity-driven tropical deforestation. On July 1, 2013, USAID announced that it will contribute $5.5 million to a new public–private partnership that will mobilize an additional $17.2 million from financial and in-kind contributions for an innovative tropical forest monitoring tool called Global Forest Watch (GFW) 2.0. Partners include WRI, which will develop the tool, as well as Google, the Government of Norway, the University of Maryland, and Staples, among others. GFW 2.0 will support TFA 2020 efforts to reduce commodity-driven tropical deforestation by bringing together satellite imagery and monitoring systems, mobile technology, and multiple overlay maps and tree

Box 7-3 **Enhancing Capacity for Low-Emission Development Strategies**

As an organizing framework for much of its climate change mitigation assistance, the United States supports a cross-cutting objective—building national capacity for low-emission development strategies. During the fast start finance (FSF) period, the United States launched the Enhancing Capacity for Low-Emission Development Strategies (EC-LEDS) program. EC-LEDS supports developing countries' efforts to pursue low-emission, climate-resilient economic development and growth. The program now has official partnerships with more than 20 countries.

The EC-LEDS program supports the development and implementation of country-driven LEDS by providing targeted technical assistance for efforts, such as GHG inventories, economic and emissions modeling and analysis, and landscape and clean energy-related interventions. Going forward, the EC-LEDS program will continue to support partner governments in both the development and the implementation of their LEDS, using a country's own strategy to guide U.S. investments in actionable projects and programs that reduce long-term emission trajectories.

* In Colombia, the United States supported the development of "marginal abatement cost" curves to identify and prioritize emission reduction opportunities in five key sectors—energy, transport, agriculture, housing, and waste. This has led to several specific mitigation opportunities being identified and further developed by Colombian Ministry experts.

* In partnership with the Philippines Climate Change Commission, U.S. experts are supporting the preparation of the next Philippines GHG inventory. This work is enhancing institutional arrangements and coordination around climate change, and resulting in a more robust data collection and archiving system for long-term planning.

* In Bangladesh, the United States is working closely with the government to assess Bangladesh's coastal wind power potential, paving the way for private investment. By delivering high-quality data on wind resource characteristics, the project helps private companies decide whether and where to invest in wind energy.

cover loss alert systems to provide detailed, near-real-time information on tropical forests. USAID will support all aspects of development, including working with developing country partners to ensure they have the capacity to access and use GFW 2.0.

CEM—The Clean Energy Ministerial (CEM) is a high-level global forum to promote policies and programs that advance clean energy technology, share lessons learned and best practices, and encourage the transition to a global clean energy economy. DOE played a crucial role in launching the CEM and hosted the first meeting of ministers in Washington, D.C., in June 2010. There are 23 developed and developing country governments voluntarily participating in the CEM; together they represent 90 percent of global clean energy investment and 80 percent of global GHG emissions.

The CEM is organized around a three-part strategy: high-level policy dialogue, technical cooperation, and engagement with the private sector and other stakeholders. The technical cooperation takes place through 13 wide-ranging initiatives. CEM's low-cost, high-impact technical work facilitates international coordination that amplifies each government's clean energy deployment efforts and helps nations reduce carbon emissions, improve energy security, provide energy access, and sustain economic growth. The United States leads or co-leads eight of those initiatives, including SEAD and Global LEAP.

CERC—In November 2009, President Obama directed DOE and President Hu directed China's Ministry of Science and Technology and National Energy Administration to explore a new model for bilateral cooperation in clean energy research. The U.S.-China Clean Energy Research Center (CERC), launched shortly thereafter, is a $150-million joint R&D program carried out by three U.S. CERC consortia (one each for energy-efficient buildings, clean vehicles, and advanced coal) and their counterparts in China, with 50/50 division of funding costs between the United States and China, and with $75 million provided by private sources (UNEP and WMO 2011).

BREADTH OF SUPPORT AND PRIORITY REGIONS

U.S. climate finance is notable for its geographic breadth: more than 120 countries received U.S. climate finance in the period 2010–2012 across all regions.

The United States prioritizes its assistance to different countries and regions, depending on their relative thematic importance. U.S. clean energy programs prioritize today's major emerging economies and tomorrow's potentially large GHG emitters. U.S. sustainable landscapes programming focuses on globally important tropical forests, such as those in Central Africa, the Amazon, and Southeast Asia. For adaptation assistance, the United States prioritizes its support to the most vulnerable developing countries, such as the least developed countries (LDCs), small-island developing states (SIDS), and Africa, in line with the commitments made in the Copenhagen Accord. In FY 2012, the United States provided nearly 80 percent of its country-specific adaptation funding to LDCs, SIDS, or Africa.

Figure 7-1 shows the regional distribution of U.S. FSF for programs that can be attributed to a particular country or region. (The figure does not include global or multiregional programs.)

New and Additional Climate Finance

International assistance for climate change continues to be a major priority for the United States. The U.S. administration seeks new funding from Congress on an annual basis. Since ratifying the Convention, which is where the term "new and additional" was first used, U.S. international climate finance increased from virtually zero in 1992 to an average of $2.5 billion per year during the FSF period (2010 to 2012). During the FSF period, average annual appropriated climate assistance increased fourfold compared with 2009 funding levels. U.S. climate assistance has increased in the context of an overall increasing foreign assistance budget.

Mobilizing Private Climate Finance

While maintaining a strong core of public climate finance is essential, the United States also recognizes that private finance must play a key role in mitigation and adaptation in developing countries. The reasons are abundant. First, private investors manage resources that dwarf available public resources, and these resources can often be distributed more quickly and efficiently than public-sector resources. Second, because of the scale of the climate problem, public funds alone will never be sufficient to adequately address climate change. Further, more efficient leveraging of private investment can enable the nation to use the available public resources in areas and sectors where the private sector is unlikely to invest enough on its own, particularly in areas like adaptation for the most vulnerable and least developed countries. Finally, a large share of mitigation-related investments can deliver a financial return and, therefore, lend themselves to private investment. As a result, private finance has been and will continue to be the dominant force driving economic growth in most economies. How it is channeled will determine whether that growth is low in carbon and resilient to changes in climate.

Toward that end, the United States is actively working to combine its significant, but finite, public contributions with targeted, smart policies to mobilize maximum private investment in climate-friendly activities in developing countries. The U.S. government is looking to use public funds where they are catalytic—where a targeted and timely injection of public finance creates new markets and opportunities for low-carbon investment that would not otherwise occur. Continuing to execute this vision will be especially important as developed countries, including the United States, work toward a collective goal of mobilizing $100 billion per year in public and private climate finance for developing countries by 2020, in the context of meaningful mitigation actions and transparency on implementation.

The United States is laying the foundation for larger-scale investments (1) by encouraging OPIC's development finance and Ex-Im Bank's export credit authorities to invest in clean energy technologies and create new products tailored toward climate change solutions; and (2) by leveraging significant private-sector investments across all three pillars through bilateral and multilateral programs. The United States will continue to place special emphasis on working with developing countries to develop strong regulatory frameworks and national policies to attract international capital flows, mobilize domestic flows, and create the right institutional framework for domestic action.

The United States has also been working with its developed country partners to collectively develop and coordinate strategies for scaling up climate-friendly investment in developing countries. In April 2013, the United States held an inaugural meeting of climate ministers and senior officials from development and finance ministries to explore ways to coordinate more

Figure 7-1

Regional Distribution of Country-Specific Congressionally Appropriated Funds for FY 2010–2012

U.S. clean energy programs prioritize today's major emerging economies and tomorrow's potentially large greenhouse gas emitters. U.S. sustainable landscapes programming focuses on globally important tropical forests, such as those in Central Africa, the Amazon, and Southeast Asia. For adaptation, dedicated U.S. climate assistance prioritizes countries, regions, and populations that are highly vulnerable to the impacts of climate change.

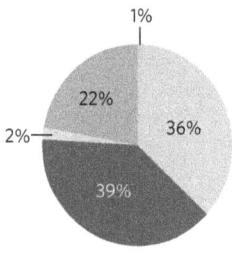

- Middle East
- Africa
- Asia
- Europe and Eurasia
- Latin America and Carribean

Note: Shares of funds depicted here do not include global, multilateral, or multiregional funds, for which country allocations are not currently available.

closely on using public resources and policies to mobilize the maximum amount of total investment in climate action. The developed countries in attendance agreed to focus on strengthening and augmenting key tools that are provided through existing public finance institutions that operate at the nexus with the private sector: development finance institutions, multilateral development banks, key multilateral climate change funds, and export credit agencies. The United States will continue to play an active role internationally to help coordinate this work going forward.

Sample Initiatives: Mobilizing Private Climate Finance

ACEF—Launched in 2012, the Africa Clean Energy Finance (ACEF) Initiative is an example of innovative U.S. government approaches to mobilizing private-sector financial resources to address climate change. ACEF seeks to address sub-Saharan Africa's acute energy needs by mobilizing private investment in clean energy projects, ranging from household-level solar energy to utility-scale power plants. ACEF represents a new way of doing business that harnesses the best of the U.S. government's technical and financial expertise. By combining $20 million in grant-based financing from DOS, project planning expertise from the U.S. Trade and Development Agency, and financing and risk mitigation tools from OPIC, ACEF will catalyze hundreds of millions of dollars in financing from OPIC, which will then leverage hundreds of millions of dollars in private investment. ACEF demonstrates how a very limited amount of grant-based public resources—when surgically applied—can catalyze a much larger pool of finance that can bring climate projects to fruition at scale.

USAID-India Clean Energy—USAID announced in June 2013 that it will facilitate a new private–public investment of $100 million in India's clean energy sector via Nereus Capital, an alternative asset manager investing in industries undergoing transformative change. This investment, announced during the fourth annual U.S.-India Strategic Dialogue, will be mobilized by USAID's Development Credit Authority in partnership with the U.S.-based institutional investor Northern Lights Capital Group.

CTI PFAN—As of the end of 2012, the Climate Technology Initiative Private Financing Advisory Network (CTI PFAN) has successfully mobilized about $300 million in private investment to implement clean energy projects in developing countries. PFAN financial professionals work with project developers and other project proponents to structure the project and develop a business plan, with supporting investor pitch, so that the merits of the project can be presented to the international private financial community with the goal of securing debt and/or equity investment for implementation. In addition, USAID is investing $1 million in the PFAN-Asia program to expand investment in clean energy in developing countries in Asia. Activities will link private-sector financiers with clean energy project developers to increase access to private financing for clean energy. Participating countries are expected to include Cambodia, Indonesia, Laos, Malaysia, Philippines, Thailand, and Vietnam.

OPIC Clean Energy—As a result of making the renewable resources sector an agency-wide priority in 2007, OPIC increased its total clean energy financing from $50 million in 2007 to an average of $663.8 million annually over the period 2010–2012. This support is expected to leverage an estimated $2.7 billion in additional private investment.

Technology Development and Transfer

Since 2009, the United States has engaged in a wide range of activities with developing countries and economies in transition, with the primary goal of promoting the development and deployment of climate-friendly technologies and practices. The United States promotes its technology development and transfer activities bilaterally, plurilaterally, and multilaterally.

At all levels of activity, the principal U.S. focus is to help support the development of the policies and regulations and overall institutional scaffolding that is required to facilitate technology transfer actions. For example, the United States works bilaterally with individual countries on capacity-building activities on appliance efficiency standards, renewable energy policies, and smart-grid regulatory schemes. Plurilaterally, the United States works with other countries on regional initiatives to transform market structures that will expedite the technology flows. Finally, on the multilateral level, the United States contributes to such global technology transfer institutions as the UNFCCC's Technology Executive Committee and Climate Technology Center and Network.

The United States has also worked extensively on the CTI, a multilateral initiative originally established at the first Conference of the Parties to the UNFCCC in 1995 to foster international cooperation for accelerated development and diffusion of climate-friendly technologies and practices. Since July 2003, CTI has been operating under an implementing agreement of the International Energy Agency that includes the United States, Australia, Austria, Canada, Finland, Germany, Japan, Norway, Republic of Korea, Sweden, and the UK. Through a variety of capacity-building activities, CTI has promoted meaningful technology transfer to and among developing countries and countries in transition. Specific activities include technology needs assessments, seminars and symposia, implementation activities, training courses, information dissemination, and support activities. In addition to their current and future environmental benefits, these efforts are promoting near- and long-term global economic and social stability through creation of jobs and associated strengthening of local and regional infrastructure.

For the most part, U.S. assistance is dedicated to "soft" technology transfer, as "soft" technology often needs to be in place before "hard" technology can be installed. However, much of OPIC's and Ex-Im's activities, which do finance hard technologies on the ground, such as wind turbines and solar panels, can be characterized as "hard" technology transfer. Table 7-2 presents specific examples of U.S. involvement in technology development and transfer activities. Please note that this table does not represent an exhaustive list of these activities.

Additionally, several U.S. government agencies have helped U.S.-based companies access international markets, thus providing clean energy and climate-friendly technologies around the world. For example, In FY 2013 the U.S. Department of Commerce (DOC) welcomed delegates from 105 countries to clean energy-focused trade shows in the United States and organized related trade missions to several key markets. Since the launch of the interagency Civil Nuclear Trade Initiative, the Renewable Energy and Energy Efficiency Export Initiative, and the Environmental Exports Initiative, DOC officials have led U.S. climate-friendly technology exporters to China, India, Japan, Indonesia, the Philippines, Mexico, Chile, Brazil, Turkey, Vietnam, the Middle East, and Central/Eastern Europe, with more visits to occur in 2014 and beyond. U.S. government agencies have also played a key role in helping foreign governments establish regulations and incentives that support the deployment of clean energy.

Table 7-2 **Examples of U.S. Technology Development and Transfer Activities**

For the most part, U.S. assistance is dedicated to "soft" technology transfer activities, as "soft" technologies often need to be in place before "hard" technologies can be installed. However, much of OPIC's and Ex-Im's activities, which do finance hard technologies on the ground, such as wind turbines and solar panels, can be characterized as "hard" technology transfer. This table presents specific examples of U.S. involvement in technology development and transfer activities.

Purpose	Description	Recipient	Sector	U.S. Funding	Public or Private Sector	Factors Enabling Project's Success	Technology Transferred	Impact on GHG Emissions/ Sinks
Global Methane Initiative								
Reduce methane emissions and advance the abatement, recovery, and use of methane as a valuable clean energy source.	Focuses on an international network to build capacity, develop strategies and markets, and remove barriers to methane reduction project development in partner countries.	Several hundred global projects and activities.	Agriculture, coal mine methane, municipal solid waste, oil and gas systems, wastewater.	$38.4 million (FY 2009-2012). $74.4 million total since inception in 2005.	Public and private	High-quality emission data, technical capability, availability of financing, policy incentives, valuable use for gas, capacity training.	Best practices/technologies for evaluating and measuring methane emissions from target sectors; mitigation technologies/best practices, such as coal mine and landfill methane capture systems, biodigester, and technologies for reducing oil and gas sector methane emissions.	Reduced methane emissions by approximately 23 Tg CO_2e in 2012 alone; cumulative emission reductions exceed 150 Tg CO_2e.

Table 7-2 (Continued)　　**Examples of U.S. Technology Development and Transfer Activities**

Purpose	Description	Recipient	Sector	U.S. Funding	Public or Private Sector	Factors Enabling Project's Success	Technology Transferred	Impact on GHG Emissions/ Sinks
Super-efficient Equipment and Appliance Deployment (SEAD)								
Advance global market transformation of energy-efficient equipment and appliances.	Provides peer community, research, data, and tools to help turn knowledge into action to accelerate the transition to a clean energy future through effective appliance and equipment energy efficiency programs.	16 governments participate in the SEAD initiative of the Clean Energy Ministerial (CEM). Non-CEM countries engage on a case-by-case basis.	Electricity	$11.45 million (FY 2009–2012).	Public and private	Peer-to-peer exchange among technical and policy experts from participating governments; existence of complementary activities that develop clear, broadly accepted test procedures for products; and collaborating with industry to ensure their participation in promoting a transition to energy-efficient products.	SEAD data and analysis inform regional appliance standards processes, international test procedure harmonization activities, and capacity building for test laboratories.	Employing current best practices in SEAD economies can by 2030 reduce annual electricity demand by over 2000 billion kilowatt-hours. These measures would decrease CO_2 emissions over the next two decades by 11 billion tons (1,000 Tg CO_2e).
Global Lighting and Energy Access Partnership (Global LEAP)								
Advance global market transformation toward higher-performing, higher-efficiency solar-powered lanterns and direct current (DC)-powered appliances designed for off-grid markets to advance energy access.	Supports quality assurance activities for solar-powered lanterns for off-grid lighting, a global competition in two categories (lights and televisions) to identify the best DC-powered products in the market for use in an off-grid context, and efforts to advance commercially viable mini-grid solutions for rural energy access.	DOE, in coordination with other donor governments and development partners, including Italy, Japan, UK, the World Bank, International Finance Corporation, UNDP, and the UN Foundation. Global LEAP is a CEM initiative.	Off-grid electricity	$2.15 million (FY 2009–2012).	Public and private	Close coordination and collaboration with World Bank group partners to leverage comparative strengths; strong stakeholder engagement efforts; market analysis to select appropriate products for competition; broadly accepted test procedures; collaboration to give off-grid customers greater choice and information about available products.	Over 40 solar-powered lighting devices have been certified through the Global LEAP-supported quality assurance framework, used by the World Bank Group's Lighting Africa program, and now adopted by the IEC, an international standards-setting body. The Global LEAP competitions identify the top DC-powered televisions and DC-powered light-emitting diode (LED) lights (used with off-grid solar home systems); winners to be announced in spring 2014.	An estimated 138,600 metric tons of CO_2e (0.1386 Tg CO_2e) have been avoided. The climate benefits are even more significant when the black carbon implications of kerosene lighting are considered.

Table 7-2 (Continued)   **Examples of U.S. Technology Development and Transfer Activities**

Purpose	Description	Recipient	Sector	U.S. Funding	Public or Private Sector	Factors Enabling Project's Success	Technology Transferred	Impact on GHG Emissions/ Sinks
SERVIR								
Increased capacity to utilize geospatial information.	USAID and NASA collaboration to build capacity of regional institutions in developing countries to improve environmental management and climate change resilience through the application of geospatial information in decision making.	Regional Center for Mapping Resources for Development and member country governments in East Africa, International Center for Integrated Mountain Development and member country governments in the Himalaya Hindu-Kush Region, Water Center for the Humid Tropics of Latin America and the Caribbean, and member country governments in Central America.	Water, agriculture, energy, land cover, climate, disasters, biodiversity.	$41.7 million over FY 2010–2013.	Public	Science backstopping from NASA, user engagement support from USAID, partnership with regional institutions.	Geographic information system (GIS), remote sensing, land cover classification, hydrologic modeling.	Decision support will aid land and forest management, monitoring, emission estimations, and policy improvement leading to emission reductions.
Famine Early Warning System Network (FEWS NET)								
Establish more effective, sustainable networks that reduce vulnerability to food insecurity.	Assesses short- to long-term vulnerability to food insecurity with environmental information from satellites and agricultural and socio-economic information from field representatives. Conducts vulnerability assessments and contingency and response planning, aimed at strengthening host country food security networks.	Afghanistan, Burkina Faso, Chad, Djibouti, Eritrea, Ethiopia, Guatemala, Haiti, Honduras, Kenya, Malawi, Mali, Mauritania, Mozambique, Nicaragua, Niger, Rwanda, Somalia, Sudan, Uganda, Zambia, Zimbabwe.	Adaptation	Average $13 million per year.	Public	The combined U.S. environmental monitoring expertise of NASA, NOAA, and USGS; implementation by host country field staff.	Information networks: remote sensing, data acquisition, processing, and analysis; GIS analytical skills. Equipment to facilitate adaptation: GIS hardware and software.	N/A

Table 7-2 (Continued)   **Examples of U.S. Technology Development and Transfer Activities**

Purpose	Description	Recipient	Sector	U.S. Funding	Public or Private Sector	Factors Enabling Project's Success	Technology Transferred	Impact on GHG Emissions/ Sinks
SilvaCarbon								
Build capacity and provide tools for improved measurement and monitoring of forest carbon.	A multi-agency U.S. government effort to improve developing country capacity for forest and other terrestrial carbon measurement and monitoring, through coordinated support for tool and methodology development and training to use appropriate methods for building and implementing forest carbon monitoring systems.	Bilateral programs with the governments of Colombia, Peru, Ecuador, Vietnam, and Gabon. Regional training activities in South and Central America, Congo Basin, and Southeast Asia.	Forests and other sectors impacting land use, including agriculture watershed management, protected areas.	Approximately $20 million (FY 2010–2012).	Public	Focus on agency coordination and very close coordination with recipient country government technical agencies.	Remote sensing, geospatial analysis methods, forest inventory design, and field collection tools.	Providing countries with improved capacity to measure and report on current carbon stocks and emissions and use information together with other natural resource management data to reduce emissions from future deforestation.

Notes: This table does not represent an exhaustive list of these activities. CO$_2$e = carbon dioxide equivalent; DOE = United States Department of Energy; FY = fiscal year; N/A = not applicable; NASA = National Aeronautics and Space Administration; NOAA = National Oceanic and Atmospheric Administration; Tg = teragram; UNDP = United Nations Development Programme; USAID = United States Agency for International Development; USGS = United States Geological Survey.

Table 7-3   **U.S. Financial Contributions to the Global Environment Facility for Climate Change Activities** (in US$ millions)

During fiscal years 2010–2012, the United States allocated $149 million for Global Environment Facility programs related to climate change.

Multilateral Institution	2010	2011	2012
Global Environment Facility	44	45	60

Table 7-4 **Annual U.S. Financial Contributions to Multilateral Institutions** (in US$ millions)

The U.S. government provides direct funding to multilateral institutions and programs in support of sustainable economic development and poverty alleviation. Although in many cases a portion of this funding supports climate change activities, in almost all cases it is not currently possible to identify that amount. Therefore, this table represents total U.S. government contributions to these multilateral development institutions and funds, including amounts not directly attributable to climate change activities.

Institutions, Funds, and Programs	2010	2011	2012
Poverty Reduction and Economic Growth (Multilateral Development Banks)			
International Bank for Reconstruction and Development	–	–	117.36
International Development Association	1,262.50	1,352.53	1,325.00
Inter-American Development Bank	204.00	–	81.20
Enterprise for the America Multilateral Investment Fund	25.00	24.95	25.00
Inter-American Investment Corporation	4.67	20.96	4.66
Asian Development Bank	–	211.37	106.59
Asian Development Fund	–	–	100.00
African Development Bank	–	–	32.42
African Development Fund	155.00	65.83	223.95
Multilateral Debt Relief for International Development Association	–	–	167.00
Multilateral Debt Relief for African Development Fund	–	–	7.50
Food Security			
Global Agriculture and Food Security Program	66.60	99.80	160.00
International Fund for Agricultural Development	30.00	37.44	30.00
Environmental Trust Funds			
Clean Technology Fund	300.00	184.63	229.63
Forest Investment Program	20.00	30.00	37.50
Pilot Program for Climate Resilience	55.00	10.00	18.70
Scaling-Up Renewable Energy Program in Low-Income Countries	–	10.00	18.70
Global Environment Facility[c]	86.50	89.82	119.82
Least Developed Countries Fund	30.0	25.0	25.0
Special Climate Change Fund	20.0	10.0	10.0
Forest Carbon Partnership Facility	10.0	8.0	–
Partnership for Market Readiness	5.0	–	2.5
Other Multilateral Institutions, Funds, and Programs			
United Nations Development Programme[b]	100.50	84.78	82.00
United Nations Environment Programme[a, b]	11.50	7.70	7.70
OAS Development Assistance Programs[a, b]	5.00	4.75	3.50
UN Women[b, d]	9.00	6.00	7.50
World Trade Organization Technical Assistance[a, b]	1.05	1.20	1.15
International Civil Aviation Organization[a, b]	0.95	0.95	0.95
Montreal Protocol Multilateral Fund[b]	35.30	35.50	36.45
Intergovernmental Panel on Climate Change/UNFCCC[b]	13.00	10.00	10.00
International Contributions for Scientific, Educational, and Cultural Activities[a, b]	1.00	1.85	–
World Meteorological Organization Voluntary Co-operation Programme[a, b]	2.05	2.09	2.09
UN Human Settlements Program (UN HABITAT)[b]	2.05	2.00	1.90

[a] These international organizations also receive assessed contributions through the Contributions to International Organizations account.
[b] Voluntary contributions from International Organizations and Programs account.
[c] These numbers reflect fiscal year funding—i.e. "2005" funding is FY 2005 funding. The U.S. fiscal year begins October 1st of the preceding year and ends on September 30th.
[d] 2010 was the last year there was a breakout between the UN Development Fund for Women ($6 million) and UNIFEM Trust Fund ($3 million) accounts. For 2011 and 2012, the line items were merged.

Note: OAS = Organization of American States; UN = United Nations; UNFCCC = United Nations Framework Convention on Climate Change; UNIFEM = United Nations Development Fund for Women.

Table 7-5   **2010 Bilateral and Regional Contributions Related to the Implementation of the UNFCCC**
               (in US$ millions)

Fiscal year 2010 bilateral and regional contributions related to the implementation of the United Nations Framework Convention on Climate Change amounted to almost $2,000 million. This includes grant-based assistance, development finance, and export credit. In the case of grant-based assistance, some funding covers multiple countries and/or regions. As a result of enhanced data collection methodologies and improvements made to data collection over time, some data in this table may vary slightly from data reported separately.

Recipient Country/Region	Energy	Forestry and Agriculture	Adaptation	Total
Grant-Based Assistance	**915.3**	**242.4**	**430.3**	**1,587.9**
Multiple Regions, Multiple Countries	467.9	120.1	301.1	889.4
Africa				
Africa—Multiple Countries	9.6	15.9	12.0	37.5
Angola	0.0	0.0	0.4	0.4
Democratic Republic of the Congo	2.3	7.9	0.3	10.4
Ethiopia	0.0	0.0	5.0	5.0
Ghana	1.0	0.0	0.0	1.0
Kenya	1.5	1.0	4.2	6.7
Liberia	1.0	0.4	0.0	1.4
Malawi	138.8	2.0	0.0	140.8
Mali	0.0	0.2	2.0	2.2
Mozambique	2.0	1.0	1.5	4.5
Nigeria	1.5	0.0	0.0	1.5
Rwanda	0.0	0.0	2.3	2.3
Senegal	0.0	1.5	0.0	1.5
Tanzania	0.0	3.3	2.2	5.5
Uganda	1.5	1.5	0.0	3.0
Zambia	0.0	1.5	0.0	1.5

No regional total is provided because "multiple region" funds also go to this region.

Asia				
Asia—Multiple Countries	12.0	9.3	22.4	43.7
Afghanistan	48.6	0.0	0.0	48.6
Bangladesh	0.0	0.0	3.0	3.0
Cambodia	0.0	3.0	1.0	4.0
China	2.0	0.0	0.0	2.0
India	11.3	5.0	4.0	20.3
Indonesia	5.0	17.5	0.0	22.5
Kazakhstan	0.4	0.0	0.0	0.4
Kyrgyzstan	1.5	0.0	0.0	1.5
Maldives	0.0	0.0	3.0	3.0
Marshall Islands	0.0	0.0	0.0	0.1
Mongolia	48.7	0.0	0.0	48.7
Nepal	0.0	0.0	3.0	3.0
Pakistan	63.8	0.0	0.0	63.8
Philippines	4.0	0.0	0.3	4.3
Tajikistan	0.9	0.0	0.0	0.9

No regional total is provided because "multiple region" funds also go to this region.

Europe & Eurasia				
Albania	0.0	0.0	1.5	1.5

Table 7-5 (Continued) **2010 Bilateral and Regional Contributions Related to the Implementation of the UNFCCC** (in US$ millions)

Recipient Country/Region	Energy	Forestry and Agriculture	Adaptation	Total
Armenia	1.3	0.0	0.0	1.3
Georgia	2.4	0.0	0.0	2.4
Macedonia	2.0	0.0	0.0	2.0
Moldova	2.0	0.0	2.0	4.0
Ukraine	5.0	0.0	0.0	5.0
No regional total is provided because "multiple region" funds also go to this region.				
Latin America & Caribbean				
Latin America & Caribbean—Multiple Countries	16.0	28.0	10.3	54.3
Brazil	1.0	6.0	0.0	7.0
Colombia	2.0	1.3	0.0	3.3
Dominican Republic	0.0	0.0	2.0	2.0
Ecuador	0.0	1.0	1.4	2.4
El Salvador	0.0	1.0	0.0	1.0
Guatemala	0.0	3.0	1.4	4.4
Guyana	0.0	1.0	0.0	1.0
Haiti	43.0	0.0	3.0	46.0
Jamaica	0.0	0.0	1.0	1.0
Mexico	2.2	3.0	0.0	5.2
Panama	0.0	2.5	0.0	2.5
Peru	0.0	4.5	0.0	4.5
No regional total is provided because "multiple region" funds also go to this region.				
Middle East				
Jordan	12.8	0.0	0.0	12.8
Other Operating Units	0.0	0.0	40.0	40.0
No regional total is provided because "multiple region" funds also go to this region.				
Development Finance	**155.1**	**0.0**	**0.0**	**155.1**
Afghanistan	7.6	0.0	0.0	7.6
India	35.4	0.0	0.0	35.4
Mexico	20.3	0.0	0.0	20.3
Nigeria	69.8	0.0	0.0	69.8
Ukraine	22.0	0.0	0.0	22.0
Export Credit	**253.2**	**0.0**	**0.0**	**253.2**
Chile	0.0	0.0	0.0	0.0
Honduras	158.6	0.0	0.0	158.6
India	6.0	0.0	0.0	6.0
Jamaica	0.1	0.0	0.0	0.1
Kenya	6.8	0.0	0.0	6.8
Mexico	81.2	0.0	0.0	81.2
South Africa	0.4	0.0	0.0	0.4
4 countries <$500,000 (Bangladesh, Chile, Namibia, Uganda)	0.1	0.0	0.0	0.1
COMBINED TOTAL	**1,323.5**	**242.4**	**430.3**	**1,996.2**

Table 7-6   **2011 Bilateral and Regional Contributions Related to the Implementation of the UNFCCC** (in US$ millions)

Fiscal year 2011 bilateral and regional contributions related to the implementation of the United Nations Framework Convention on Climate Change amounted to $3,137.6 million. This includes grant-based assistance, development finance, and export credit. In the case of grant-based assistance, some funding covers multiple countries and/or regions.

Recipient Country/Region	Energy	Forestry and Agriculture	Adaptation	Total
Grant-Based Assistance	**962.4**	**361.5**	**560.2**	**1,884.1**
Multiple Regions, Multiple Countries	332.6	132.8	351.7	817.1
Africa				
Africa—Multiple Countries	12.6	26.2	13.9	52.6
Ethiopia	0.0	7.0	16.1	23.1
Ghana	0.6	4.0	0.0	4.6
Kenya	4.6	0.1	5.4	10.0
Malawi	141.1	5.9	3.0	150.0
Mali	0.0	0.0	3.0	3.0
Mozambique	0.0	0.0	4.0	4.0
Nigeria	2.8	0.0	3.5	6.3
Rwanda	0.0	1.0	4.8	5.8
Senegal	0.0	0.0	3.0	3.0
South Africa	4.9	0.0	0.0	4.9
Tanzania	0.0	0.7	3.2	3.9
Uganda	0.0	0.0	3.0	3.0
Zambia	0.0	5.0	0.8	5.8
No regional total is provided because "multiple region" funds also go to this region.				
Asia				
Asia—Multiple Countries	15.2	13.4	20.6	49.1
Afghanistan	73.5	0.0	0.0	73.5
Bangladesh	0.0	0.0	20.1	20.1
Cambodia	0.0	5.0	2.0	7.0
China	3.8	0.0	0.0	3.8
India	7.5	4.0	3.4	14.9
Indonesia	266.8	83.9	10.2	360.9
Kyrgyz Republic	0.3	0.0	0.0	0.3
Maldives	0.0	0.0	3.0	3.0
Nepal	0.0	3.0	4.4	7.4
Pakistan	42.0	0.0	0.0	42.0
Philippines	5.6	3.0	4.0	12.6
Tajikistan	0.6	0.0	0.0	0.6
Timor-Leste	0.0	0.0	3.0	3.0
Vietnam	4.0	4.0	3.0	11.0
No regional total is provided because "multiple region" funds also go to this region.				
Europe & Eurasia				
Europe & Eurasia—Multiple Countries	9.1	1.0	1.0	11.1
Albania	0.4	0.0	0.0	0.4
Armenia	0.4	0.0	1.1	1.5
Bosnia and Herzegovina	0.8	0.0	0.0	0.8
Georgia	2.0	0.5	1.0	3.5
Macedonia	0.5	0.0	0.0	0.5
Moldova	0.3	0.0	0.0	0.3
Ukraine	6.0	0.0	0.0	6.0
No regional total is provided because "multiple region" funds also go to this region.				

Table 7-6 (Continued)   **2011 Bilateral and Regional Contributions Related to the Implementation of the UNFCCC** (in US$ millions)

Recipient Country/Region	Energy	Forestry and Agriculture	Adaptation	Total
Latin America & Caribbean				
Latin America & Caribbean—Multiple Countries	5.0	17.4	9.3	31.7
Barbados	0.0	0.0	2.3	2.3
Bolivia	0.0	0.9	0.0	0.9
Brazil	4.2	3.8	0.0	8.0
Chile	0.2	0.0	0.0	0.2
Colombia	4.5	2.0	2.0	8.5
Dominican Republic	0.0	0.0	3.0	3.0
Ecuador	0.0	5.9	0.0	5.9
El Salvador	0.3	0.0	0.1	0.4
Guatemala	0.0	7.1	3.5	10.6
Haiti	1.8	0.0	1.5	3.3
Honduras	0.0	2.0	0.0	2.0
Jamaica	0.0	0.0	3.0	3.0
Mexico	6.2	8.0	0.0	14.2
Peru	0.0	14.0	2.0	16.0
No regional total is provided because "multiple region" funds also go to this region.				
Middle East				
Egypt	0.5	0.0	0.0	0.5
Morocco	1.8	0.0	2.5	4.3
Other Operating Units	0.0	0.0	39.0	39.0
No regional total is provided because "multiple region" funds also go to this region.				
Development Finance	**1,113.9**	**0.9**	**0.0**	**1,114.8**
Multiple countries	50.0	0.0	0.0	50.0
Cambodia	0.0	0.9	0.0	0.9
Georgia	58.0	0.0	0.0	58.0
India	213.8	0.0	0.0	213.8
Jordan	3.0	0.0	0.0	3.0
Kenya	310.0	0.0	0.0	310.0
Liberia	90.0	0.0	0.0	90.0
Peru	123.0	0.0	0.0	123.0
St. Kitts and Nevis	16.1	0.0	0.0	16.1
Thailand	250.0	0.0	0.0	250.0
Export Credit	**194.7**	**0.0**	**0.0**	**194.7**
Multiple Regions, Multiple Countries	5.0	0.0	0.0	5.0
Brazil	0.1	0.0	0.0	0.1
Chile	2.2	0.0	0.0	2.2
Guatemala	4.6	0.0	0.0	4.6
India	180.0	0.0	0.0	180.0
Jamaica	0.4	0.0	0.0	0.4
Mexico	2.3	0.0	0.0	2.3
Namibia	0.1	0.0	0.0	0.1
COMBINED TOTAL	**2,271.0**	**362.4**	**560.2**	**3,193.6**

Table 7-7   **2012 Bilateral and Regional Contributions Related to the Implementation of the UNFCCC** (in US$ millions)

Fiscal year 2012 bilateral and regional contributions related to the implementation of the United Nations Framework Convention on Climate Change amounted to $2,278.0 million. This includes grant-based assistance, development finance, and export credit. In the case of grant-based assistance, some funding covers multiple countries and/or regions.

Recipient Country/Region	Energy	Forestry and Agriculture	Adaptation	Total
Grant-Based Assistance	**585.9**	**277.5**	**398.2**	**1,261.7**
Multiple Regions, Multiple Countries	382.7	141.0	180.4	704.1
Africa				
Africa—Multiple Countries	11.7	17.2	16.9	45.7
Burkina Faso	1.8	0.0	0.0	1.8
Cape Verde	0.0	0.0	41.0	41.0
Democratic Republic of the Congo	0.0	2.2	0.0	2.2
Ethiopia	0.0	0.0	22.9	22.9
Gabon	0.0	0.2	0.0	0.2
Kenya	4.0	1.0	3.5	8.5
Liberia	5.5	4.4	1.8	11.7
Malawi	0.0	3.0	5.0	8.0
Mozambique	0.0	0.0	4.7	4.7
Nigeria	3.4	0.0	1.7	5.1
Rwanda	0.0	0.0	3.5	3.5
Senegal	0.0	0.0	2.0	2.0
South Africa	3.1	0.0	0.0	3.1
Tanzania	0.0	0.2	5.9	6.1
Uganda	0.0	0.0	3.0	3.0
Zambia	0.0	5.0	0.8	5.8

No regional total is provided because "multiple region" funds also go to this region.

	Energy	Forestry and Agriculture	Adaptation	Total
Asia				
Asia—Multiple Countries	5.4	8.5	17.6	31.5
Afghanistan	79.6	0.0	0.0	79.6
Bangladesh	4.5	2.0	9.0	15.5
Cambodia	0.0	3.6	4.0	7.5
China	2.2	0.0	0.0	2.2
India	4.6	4.0	2.0	10.6
Indonesia	3.0	8.4	4.1	15.6
Kazakhstan	2.0	0.0	0.0	2.0
Kyrgyz Republic	0.7	0.0	0.0	0.7
Maldives	0.0	0.0	2.0	2.0
Nepal	0.0	4.5	4.8	9.3
Pakistan	31.8	0.0	0.0	31.8
Papua New Guinea	0.0	2.0	0.0	2.0
Philippines	3.0	5.8	2.8	11.6
Timor-Leste	0.0	0.0	2.0	2.0
Vietnam	2.0	1.9	3.0	6.9

No regional total is provided because "multiple region" funds also go to this region.

Table 7-7 (Continued)   **2012 Bilateral and Regional Contributions Related to the Implementation of the UNFCCC**
(in US$ millions)

Recipient Country/Region	Energy	Forestry and Agriculture	Adaptation	Total
Europe & Eurasia				
Europe & Eurasia—Multiple Countries	3.4	0.0	0.0	3.4
Albania	0.4	0.0	0.0	0.4
Armenia	1.6	0.0	0.0	1.6
Bosnia and Herzegovina	0.6	0.0	0.0	0.6
Georgia	4.0	0.8	0.1	4.8
Macedonia	0.8	0.0	0.2	1.0
Ukraine	7.1	0.0	0.0	7.1
No regional total is provided because "multiple region" funds also go to this region.				
Latin America & Caribbean				
Latin America & Caribbean—Multiple Countries	6.4	18.0	7.0	31.4
Barbados	0.0	0.0	1.5	1.5
Brazil	0.0	8.7	0.0	8.7
Colombia	4.0	4.5	3.0	11.5
Dominican Republic	0.0	0.0	3.0	3.0
Ecuador	0.0	2.8	2.0	4.8
El Salvador	0.7	0.0	0.1	0.8
Guatemala	0.0	4.5	3.1	7.6
Haiti	0.0	0.0	3.5	3.5
Honduras	0.1	1.3	4.0	5.3
Jamaica	0.0	1.0	2.0	3.0
Mexico	5.4	10.4	0.0	15.8
Peru	0.0	10.7	2.6	13.4
No regional total is provided because "multiple region" funds also go to this region.				
Middle East				
Jordan	0.5	0.0	0.0	0.5
Morocco	0.7	0.0	0.0	0.7
Other Operating Units	0.0	0.0	22.0	22.0
No regional total is provided because "multiple region" funds also go to this region.				
Development Finance	**721.6**	**0.0**	**0.0**	**721.6**
India	261.9	0.0	0.0	261.9
Pakistan	16.7	0.0	0.0	16.7
Peru	193.0	0.0	0.0	193.0
South Africa	250.0	0.0	0.0	250.0
Export Credit	**301.2**	**0.0**	**0.0**	**301.2**
Multiple Regions, Multiple Countries	11.5	0.0	0.0	11.5
Barbados	6.4	0.0	0.0	6.4
Brazil	80.7	0.0	0.0	80.7
India	201.6	0.0	0.0	201.6
Mexico	1.0	0.0	0.0	1.0
COMBINED TOTAL	**1,608.7**	**277.5**	**398.2**	**2,284.5**

8

Research and Systematic Observations

OVERVIEW

The United States is committed to understanding the issues driving global change and to conducting the energy research that will lead to global emission reductions over the long run. The United States is providing global leadership in developing the fundamental scientific and technological foundation for understanding the causes and consequences of climate and global change, reducing scientific uncertainties, and supporting adaptation and mitigation actions to manage risks and produce benefits at local, regional, and global scales.

The United States places a high priority on research and development (R&D) needed to understand, observe, and respond to global change. Major U.S. investment in climate and related global change science over the past few decades has greatly increased understanding of global climate change, including its attribution to human influences. Now, as the effects on people's well-being are already being felt in the form of more heat waves, alterations in rainfall patterns on which agriculture depends, and coastal communities increasingly at risk from rising seas, scientific knowledge of the integrated Earth system is even more critical as the foundation for responding effectively.

During the last few years, U.S. government agencies have put forward a coordinated set of investments in global change science to gain new theoretical knowledge of Earth system processes; to maintain and enhance a mix of atmospheric, oceanic, land, and space-based observing systems; to advance predictive capabilities through the next generation of numerical models; to promote advances in computational capabilities, data management, and information sharing; and to further develop an expert scientific workforce in the United States and worldwide.

For the period 2009–2013, the United States has invested roughly $12.5 billion in global change science (USGCRP 2011, 2012b, 2013). Additional investments under the American Recovery and Reinvestment Act (ARRA) have contributed to enhancing research infrastructure, building next-generation cyberinfrastructure assets, and awarding many new research grants and graduate fellowships.

The U.S. government is also making major investments in R&D to support clean energy and climate change mitigation technologies. The United States has committed to accelerating the development and deployment of technologies to reduce greenhouse gas (GHG) emissions, while increasing energy end-use efficiency. These steps would enable the nation to greatly reduce GHG emissions and stabilize GHG atmospheric concentrations at a level that avoids dangerous human interference with the climate system.

To address these challenges, the Obama administration and Congress have continued to build on these efforts, such as with the creation of the Advanced Research Projects Agency-Energy (ARPA-E), to spur a revolution in clean energy technologies. Overall, ARRA has provided more than $25 billion in additional funding for R&D activities across a broad portfolio of GHG mitigation options, including high-performance buildings; efficient manufacturing; advanced vehicles; clean biofuels; wind, solar, geothermal, hydropower, and nuclear fusion;

[1] See http://www.globalchange.gov/about/global-change-research-act.

[2] Additionally, the Defense Meteorological Satellites Program (DMSP) supplies some data sets that are useful for climate science.

carbon capture and sequestration; advanced energy storage; a more intelligent electric grid; and techniques for reducing emissions and/or increasing uptake of carbon dioxide (CO_2) in agriculture and forestry.

This chapter is divided in three major parts. The first two sections discuss how the United States pursues research and observations of global change, while the third section focuses on U.S. energy research and technology. Collectively, these commitments to research, observations, and technology demonstrate continuing U.S. leadership in understanding and responding to climate and global change.

Research on Global Change

As the essential capacities for research and observations are widely distributed across U.S. government agencies, they are brought together into a single interagency program. Created by the Global Change Research Act (GCRA) of 1990,[1] the U.S. Global Change Research Program (USGCRP) advances the legislative mandate to deepen basic scientific understanding while providing information and tools to support the nation's and the world's preparation for, and response to, global change. The United States is fostering greater coordination across its agencies and the international scientific community than ever before, in areas that include Earth observations, model development and use, assessments of climate change and impacts in the United States and worldwide, and data and information sharing.

At its core, global change is an issue that requires a coordinated, international response. Over the past three years, the United States has enhanced coordination with other nations and international organizations on global change research activities, promoted increased international access to scientific data and information, and fostered increased participation in international global change research by developing nations. In partnership with the International Council for Science (ICSU), the International Social Science Council (ISSC), and the Belmont Forum, the United States, is helping to shape the future of international global change research coordination.

In addition, during the last three years, the United States and international scientific communities have embarked upon, and are on the verge of completing, the Fifth Assessment of the Intergovernmental Panel on Climate Change (AR5) (IPCC 2013; the Working Group I report has now been released, with the remaining reports to follow soon). U.S. researchers are playing critical and wide-ranging roles in the assessment, serving as working group co-chairs, co-ordinating lead authors, lead authors, contributing authors, review editors, and reviewers. The U.S. government also directly supports the IPCC Task Force on National Greenhouse Gas Inventories, as well as the IPCC Working Group II Technical Support Unit (which also supports U.S. authors and contributors to all IPCC Working Groups).

Observing Systems

All of these research and assessment activities depend on the existence of a comprehensive, continuous, integrated, and sustained set of physical, chemical, biological, and societal observations of global change and its impacts. The current portfolio upon which the U.S. and international research enterprise relies includes satellite, airborne, ground-based, and ocean-based missions, platforms, and networks that provide measurements of the Earth system variables important for understanding global change.

The United States supports a large number of civilian remote-sensing satellites that supply climate-related information. These satellites are operated by the National Aeronautics and Space Administration (NASA), the National Oceanic and Atmospheric Administration (NOAA), and the U.S. Geological Survey (USGS) Earth-observing satellites.[2] The United States also supports extensive nonsatellite observational capabilities across multiple federal agencies, providing the backbone for many global observing networks. For example, the United States sponsors half of the platforms deployed in the global ocean (3,860 of 7,723), with 72 other countries providing the remainder.

The United States achieved new milestones with the launch of critical new satellite observing systems, including the Suomi National Polar-orbiting Partnership (NPP), Landsat-8, and Aquarius (in partnership with the Space Agency of Argentina). New surface-based networks, such as the National Ecological Observatory Network (NEON) and Ocean Observatories Initiative (OOI) are well on their way to operation, creating a next generation of *in situ* observing capabilities. And the U.S. Department of Energy (DOE) Atmospheric Radiation Measurement (ARM) Climate Research Facility received $60 million in ARRA funding to build its next-generation facility for climate research, deploying an expansive array of new instruments, as well as the cyberinfrastructure needed to support the increased data volume and distribution requirements.

Energy Research and Technology

To address the challenge of transitioning the U.S. energy portfolio in the face of climate change, the Obama administration and Congress are working to spur a revolution in clean energy technologies. The research and innovation activities in this arena, which span multiple federal agencies, are organized around such goals as reducing emissions from energy supply, end use, and infrastructure; capturing and sequestering CO_2 and reducing emissions of other GHGs; measuring and monitoring emissions; and bolstering the contributions of basic science to innovation.

This section describes how these technology research and innovation activities are organized around these goals and are achieved through such mechanisms as the new Bioenergy Research Centers (BRCs), Energy Frontier Research Centers (EFRCs), and the multidisciplinary DOE Energy Innovation Hubs. These investments build on the $400 million in ARRA funds for establishing ARPA-E within DOE to help overcome the long-term and high-risk technological barriers to the development of clean energy options.

Furthermore, the United States believes that well-designed multilateral collaborations focused on achieving practical results can accelerate development and commercialization of new technologies. Thus, the United States has initiated or joined a number of technology collaborations in hydrogen, carbon sequestration, nuclear energy, and fusion that address many energy-related concerns, including climate change. These include the Carbon Sequestration Leadership Forum (CSLF), the Generation IV International Forum (GIF), and the ITER international fusion experiment.

RESEARCH ON GLOBAL CHANGE

Global change is happening now. Increases in population, industrialization, and human activities have altered the world's climate, oceans, land, ice cover, and ecosystems. Decision makers at every level of government, across every geographic region, and in every economic sector are demanding clear information about global change in order to plan, prepare, adapt, and respond. Responding effectively depends on a sound understanding of the changes underway, the threats and opportunities they present, and how they will evolve over time.

The U.S. Congress recognized this urgent need in 1990 by mandating USGCRP to "assist the Nation and the world to understand, assess, predict, and respond to human-induced and natural processes of global change."[3] USGCRP is designed to fulfill that mandate by coordinating the federal government's $2.7 billion annual investment in global change research—the largest such investment in the world. The science portfolio managed by the USGCRP federal agencies spans scales from atoms, to ecosystems, to the entire planet, and includes changes being wrought by human behaviors as well as by natural forces. It encompasses laboratory experiments, field research, computer modeling, scientific assessment, and observations of Earth from land, air, sea, and space.

This vast body of work is carried out by 13 federal agencies, each with its own mission and areas of expertise. Since USGCRP's founding in 1990, these federal agencies have coordinated their investments and activities in global change science to create and maintain a mix of atmospheric, oceanic, land-, and space-based observing systems; gain new theoretical knowledge of Earth system processes and the causes and consequences of global change; advance Earth system understanding and predictive capabilities through numerical modeling; promote advances in computational capabilities, data management, and information sharing; and develop an expert scientific workforce. These activities have proven critical to improving scientific understanding of the rich interconnections and feedbacks within the Earth system; the

[3] *See* http://www.globalchange.gov/about/global-change-research-act.

significant role of human activities in climate and related global change; and the current and potential future rates, magnitudes, and impacts of this change.

These investments stand today as the foundation of current understanding, in the United States and worldwide. Today, USGCRP continues to advance fundamental scientific understanding of global change. However, recognizing that global change and its consequences are happening already, USGCRP is also focusing on a new priority: ensuring that its science is as immediately decision-relevant as possible.

A New 10-Year Strategic Plan for USGCRP

As mandated by the GCRA, USGCRP is required to develop a National Global Change Research Plan every 10 years. In April 2012, USGCRP released a new research plan that describes in detail how it will fulfill this role and its congressional mandate over the next decade. Entitled *The National Global Change Research Plan 2012–2021* (USGCRP 2012b), the plan lays out specific goals and objectives to generate fundamental new scientific knowledge and to disseminate this knowledge in readily available and directly useful ways to decision makers and citizens.

This 10-year strategic plan—which reflects recommendations from multiple reports of the U.S. National Academies, dozens of listening sessions with stakeholders around the country, public comments on a draft plan, and collaborative planning among the USGCRP agencies— charts a course that will advance USGCRP's legislative mandate to deepen basic scientific understanding, while providing information and tools to support the nation's and the world's preparation for, and response to, global change. This includes strengthening and expanding fundamental understanding of climate change and its interactions with other critical drivers of global change, more effective collaboration among researchers in the natural and social sciences, increased interagency cooperation to sustain ongoing assessments of global change impacts, and robust dialogues with diverse audiences to enhance communication of scientific knowledge.

Under the new strategic plan, USGCRP will coordinate federal research efforts through the following four strategic goals:

- *Goal 1: Advance Science*—Advance scientific knowledge of the integrated natural and human components of the Earth system.

- *Goal 2: Inform Decisions*—Provide the scientific basis to inform and enable timely decisions on adaptation and mitigation.

- *Goal 3: Conduct Sustained Assessments*—Build sustained assessment capacity that improves the nation's ability to understand, anticipate, and respond to global change impacts and vulnerabilities.

- *Goal 4: Communicate and Educate*—Advance communications and education to broaden public understanding of global change and develop the scientific workforce of the future.

In particular, the plan calls for greater coordination than ever before across U.S. agencies and the international scientific community in a number of critical areas, including (1) observations of Earth, including both satellite and *in situ* observations for monitoring global change and understanding its key processes; (2) development, testing, and application of sophisticated models, the principal tools used to anticipate future changes and understand the possibility of tipping points in the Earth system; (3) assessments of climate change and impacts in the United States, synthesizing across peer-reviewed scientific literature and other credible sources; (4) sharing of information to support adaptation and mitigation response needs; and (5) communication of scientific findings to diverse audiences, including the public, Congress, and the global scientific community.

A substantial amount of work is underway to achieve this vision, building from the foundation in fundamental global change research and research infrastructure over the last two-plus decades. Achieving these goals will continue to depend on integrating observations of all essential Earth system components and processes, which is essential for developing theories and explanations of the causes and consequences of global change. These theoretical advances must in turn be captured and tested in integrated modeling systems for further advancing fundamental scientific understanding and informing decision making about responding to global change. Finally, success in all of these areas will need to build on continuing advances in information management and data sharing to aid scientific progress and to communicate with and inform society.

The following section discusses in more detail the principles for advancing global change science embodied in the 2012–2021 USGCRP strategic plan (as per Goal 1 above), as well as examples of major research accomplishments in the last three years. Chapter 6 of this report provides a detailed description of actions by U.S. government agencies to deliver credible, timely, and relevant information grounded in the best available science, as well as to advance an inclusive, broad-based, and sustained process for assessing and communicating scientific knowledge of the impacts, risks, and vulnerabilities associated with climate change, in support of decision making across the United States (as per Goals 2 and 3 above). Chapter 9 provides a detailed description of actions by U.S. federal agencies to support national global change-related communication and education efforts (as per Goal 4 above), including gaining greater understanding of the public's science and information needs through engagement and dialogue.

Advancing Global Change Science

Scientific knowledge of the integrated Earth system is the foundation for responding effectively to global change. The USGCRP agencies define a research program that acknowledges the complexity of global change as both a scientific and a societal challenge. To meet this challenge, USGCRP embraces multiple forms of integration: across the components of the Earth system (including humans), across observations and modeling, across space and time, across scientific disciplines, across domestic and international partnerships, and across the capabilities of science and the needs of decision makers.

As articulated in the new USGCRP strategic plan, these aims are being accomplished through the pursuit of five objectives:

* *Earth System Understanding*—Advance fundamental understanding of the physical, chemical, biological, and human components of the Earth system, and the interactions among them, to improve knowledge of the causes and consequences of global change.

* *Science for Adaptation and Mitigation*—Advance understanding of the vulnerability and resilience of integrated human–natural systems, and enhance the usability of scientific knowledge in supporting responses to global change.

* *Integrated Observations*—Advance capabilities to observe the physical, chemical, biological, and human components of the Earth system over multiple space and time scales, to gain fundamental scientific understanding and monitor important variations and trends.

* *Integrated Modeling*—Improve and develop advanced models that integrate across the physical, chemical, biological, and human components of the Earth system, including the feedbacks among them, to represent more comprehensively and predict more realistically global change processes.

* *Information Management and Sharing*—Advance the capability to collect, store, access, visualize, and share data and information about the integrated Earth system, the vulnerabilities of integrated human–natural systems to global change, and the responses to these vulnerabilities.

Although these five objectives are defined distinctly and discussed separately, they describe one integrated body of knowledge and practice: seeking answers to fundamental scientific questions about the integrated Earth system, and harnessing that improved scientific understanding to support the development of actions in response to global change. Areas of increased emphasis in USGCRP under its new strategic plan include:

* Fostering new research at the interface between the study of the physical climate system and the biological sciences.

- Improving integration of the social, behavioral, and economic sciences within the larger global change research enterprise.

- Recognizing the interplay between climate change and other dimensions of global change, such as land-use change, alteration of biogeochemical cycles, pollution, and biodiversity loss.

- Improving understanding of climate system extremes, thresholds, and tipping points.

- Assessing the vulnerability of sectors, regions, and populations, and supporting iterative risk management of these vulnerabilities through adaptation and mitigation responses.

These efforts are complemented by the ongoing efforts of the U.S. Carbon Cycle Science Program, which finished its planning for carbon cycle research in the upcoming decade with its 2011 release of its *U.S. Carbon Cycle Science Plan* (Michalak et al. 2011). This plan outlines a strategy for refocusing U.S. carbon cycle research based on the current state of the science, and provides funding agencies with community-recommended research priorities over the next decade. Global in scale and recognizing a strong need for international cooperation and collaboration, the plan is organized around how natural processes and human actions affect the carbon cycle on land, in the atmosphere, and in the oceans; how policy and management decisions affect the levels of the primary carbon-containing gases in the atmosphere; and how ecosystems, species, and natural resources are affected by increasing GHG concentrations, the associated changes in climate, and carbon management decisions. In addition to reaffirming the need for basic research and for continuing the current areas of research in carbon cycle science and successful efforts, such as the North American Carbon Program (NACP), the 2011 plan outlines specific recommendations for new priorities, such as the consequences of carbon management activities, the direct impacts of CO_2 on ecosystems, and the need to coordinate researchers from the natural and social sciences to address societal concerns.

Observing Systems

All of this research depends on the existence of a comprehensive, continuous, integrated, and sustained set of physical, chemical, biological, and societal observations of global change and its impacts. These are essential for improving the understanding of the components and processes of the Earth system and the causes and consequences of global change. As will be discussed in more detail in the Systematic Observations section of this chapter, the current observational portfolio upon which the U.S. and international global change research enterprise relies includes satellite, airborne, ground-based, and ocean-based missions, platforms, and networks that provide measurements of the Earth system variables important for understanding global change.

Understanding the complexity of the global, integrated Earth system requires simultaneous recording of diverse observations, maintained over long time periods. Effective Earth system observation requires both remotely sensed and *in situ* observations from all domains—atmosphere, ocean, land, and ice—that are then transformed into products, information, and knowledge through analysis and integration in both time and space. For most measurements, no single approach can provide all the needed observations of sufficient quantity and quality, requiring coordination across platforms and instruments. In addition, such observations should be sustained in a well-calibrated state for decades (over multiple generations of observing systems) to separate long-term trends from short-term variability, and should have global coverage at sufficient spatial resolution to account for variability across a wide range of scales.

For example, two new NASA efforts—the Aquarius satellite mission and the Salinity Processes in the Upper Ocean Regional Study (SPURS) field campaign—will complement the information about sea surface salinity that, for more than a century, has been collected only from ships, surface buoys, and profiling floats. These unprecedented new ocean observations will enhance this sometimes-sparse data record of complex interactions between evaporation, precipitation, and ocean circulation worldwide. These observations are important because regional variations in ocean salinity can influence the ocean's ability to absorb, transport, and store heat, freshwater, and CO_2, and, therefore, drive further changes in atmospheric circulation and the hydrologic cycle.

Other efforts to integrate observations to improve fundamental understanding of Earth system processes include recent USGS work to assess the amount of carbon stored on and within the U.S. land surface, and future plans, under NEON, to combine site-based data with remotely sensed data to document and understand changes in the nation's ecosystems. This sustained, long-term measurement of the climate system is complemented by process-based research to document the Earth system's response to global change over broad space and time scales.

Modeling Capabilities

Integration across Major Classes of Models—In addition, this research depends on the development, use, and, increasingly, integration of three classes of models to improve understanding of the causes and consequences of global change: Earth system models (ESMs); integrated assessment models (IAMs); and impact, adaptation, and vulnerability (IAV) models. Of these, ESMs have the most comprehensive representations of physical and biological systems and their interactions; thus, they are essential tools for exploring Earth system complexities predicting the behavior of the climate system, and interpreting observed changes in climate and weather.

New and enhanced models are expected to make important contributions toward advancing fundamental understanding of climate change, as well as informing future policymaking, planning, and decision support for sectors, such as energy, natural resources, food, and water, and national security. Used in conjunction with climate and ESMs, so-called IAV models are designed for assessments of potential climate change impacts, critical vulnerabilities, and effective adaptation strategies in such sectors as agriculture, coastal systems, energy, transportation, health, forestry, fisheries, and ecosystem services. These IAV models also assist in the development of more informative and comprehensive scenarios of drivers of future climate forcing, socioeconomic vulnerability, and adaptive capacity.

Also, IAMs combine the drivers and consequences of climate change within a consistent modeling framework. At the center of IAMs are representations of present and possible future human activities (e.g., changes in emissions, land, or water uses) and their potential influence on the Earth system.

Enhanced Modeling Capabilities—The major U.S. modeling centers—NOAA Geophysical Fluid Dynamics Laboratory and National Centers for Environmental Prediction, NASA Goddard Institute for Space Studies and Global Modeling and Assimilation Office, and National Science Foundation (NSF)/U.S. Department of Energy (DOE) National Center for Atmospheric Research—continue to lead in developing, evaluating, and applying ESMs and other modeling systems, as well increasing the accessibility of model output to user communities.

Under the auspices of USGCRP, the climate and global change modeling community has taken advantage of rapidly advancing computing resources to work toward a number of goals. To provide regional-scale information for planning purposes, the resolution at which models are being run has continued to increase as ESMs aim to provide information at scales that are relevant to decision makers. New numerical methods, grids, and parameterizations have been introduced to meet the challenges of running these models at unprecedentedly fine resolutions.

Coupled Model Intercomparison Project—These modeling centers, along with other USGCRP agencies, such as USGS, are playing a critical role in Phase 5 of the Coupled Model Intercomparison Project (CMIP5)—a major international effort under the auspices of the World Climate Research Programme (WCRP) to evaluate and improve climate models and provide critical input to national and international scientific assessments.[4] Extensive analysis of these simulations by members of the international climate community has provided an important scientific basis for the IPCC's AR5 (IPCC 2013).

The Project for Climate Model Diagnosis and Intercomparison (PCMDI) at the Lawrence Livermore National Laboratory is playing a leadership role worldwide in managing CMIP5

---

[4] See http://cmip-pcmdi.llnl.gov/cmip5/.

data archival and access, including responsibility for leading the Earth System Grid Federation, which stores and distributes terascale data sets from multiple coupled ocean-atmosphere global climate model simulations and allows users to download model output from multiple locations without needing to know where the data sets physically reside—giving them faster, easier access to climate data.

Recently, NASA's Jet Propulsion Laboratory and PCMDI have worked jointly on "Obs4MIPS," an effort to identify and provide a number of appropriate satellite data sets in a format specifically tailored to facilitate model evaluation, with the initial target being CMIP5. In addition, the scenarios and emission profiles used to drive the CMIP5 models were developed as a result of international and interagency cooperation. DOE and the U.S. Environmental Protection Agency (EPA) supported the U.S. contribution to this effort, which projected socioeconomic trends, energy pathways, land use, and biogeochemical emissions and their implications for GHG concentrations at appropriate spatial scales.

Increased Representation of Processes—The scope of processes represented in such models, particularly in the area of biogeochemistry, has increased as a direct result of U.S. and international investments in basic research. A first generation of ESMs now captures representations of carbon and nitrogen cycles and dynamic vegetation, thereby allowing for feedbacks involving these processes. In addition, the simulation of cloud and aerosol processes has become more sophisticated, enabling improved modeling of aerosol effects on clouds and climate, as well as associated feedbacks. Also, until recently, ESMs have not included dynamic models for the large Greenland and Antarctic ice sheets, and have thus been unable to provide projections of future sea level rise. However, ice sheet model components have recently been added to some ESMs to provide a fully interactive and dynamic model of ice sheet melting and its contribution to sea level rise.

Decadal and Regional Climate Prediction—To advance these and related areas, NSF, DOE, and the U.S. Department of Agriculture (USDA) have developed a joint funding competition, Decadal and Regional Climate Prediction using Earth System Models (known as the EaSM program). The EaSM projects address challenges associated with the development of next-generation ESMs that include coupled and interactive representations of ecosystems, agricultural lands and forests, urban environments, Earth's biogeochemistry, atmospheric chemistry, ocean and atmospheric currents, the water cycle, land and sea ice, and human activities. These projects are expected to generate results that will lead to improved understanding of impacts at regional levels, as well as facilitate development of effective adaptation strategies on decadal time scales. Both the regional spatial scale and the earlier time frame are direct responses to the needs of decision makers, who have repeatedly requested information at the scale at which management decisions are made. Through two rounds, these three agencies have jointly supported 61 projects for a total investment of more than $90 million.

Seasonal Climate Prediction

In addition, NOAA, in partnership with NASA, DOE, NSF, and other research institutions, has initiated a research effort to improve seasonal climate prediction skill based on multiple U.S. climate models. Such a research effort follows the U.S. National Academy of Sciences' 2010 recommendation for experimentation with multi-model ensembles as a way to improve upon current predictive capabilities. The current initiative, named the National Multi-Model Ensemble (NMME), in its initial phase, is producing real-time multi-model seasonal climate predictions based on readily available models and a basic experimental design. Future NMME plans, spearheaded by NOAA, include a more comprehensive research investigation regarding the optimal design and added value of this multi-model predictive system.

High-Performance Computing Capabilities

Finally, the U.S. government has made major new investments in high-performance capabilities to support the global change modeling enterprise. For example, ARRA support for the Evergreen project and the Pacific Northwest National Laboratory Joint Global Change Research Institute enabled the creation of an advanced computing infrastructure, installed at the Research Data Center at the University of Maryland, to execute millions of simulations, conduct post-processing calculations, store input and output data, and visualize results.

USGCRP International Research Programs and Partnerships

At its core, global change is an issue that requires an international, coordinated response. Effectively advancing the understanding of global change, establishing and sustaining observations, and preparing for global environmental change require concerted international cooperation. Since its mandate includes both basic research coordination and supporting decision making about responding to global change, USGCRP finds it necessary and desirable to engage other nations and international organizations.

Congress recognized the importance of international cooperation and collaboration and codified it in the GCRA of 1990, where USGCRP is mandated to (1) coordinate U.S. activities with other nations and international organizations on global change research projects and activities, (2) promote international cooperation and access to scientific data and information, and (3) participate in international global change research by developing nations. Through this engagement, USGCRP and its member agencies leverage existing and future scientific capabilities and more effectively use resources to accomplish strategic priorities.

USGCRP engages with, and provides significant financial support for, a variety of international programs, such as the WCRP, the International Geosphere-Biosphere Program, the International Human Dimensions Program, the Earth Systems Science Partnership, DIVERSITAS, the SysTem for Analysis, Research and Training, and the Global Research Alliance on Agricultural Greenhouse Gases. U.S. agencies were among the largest sponsors of WCRP's 2011 Open Science Conference, with more than 1,900 participants from around the world. In addition, the USDA Foreign Agricultural Service sponsors the Global Research Alliance Fellowships, which to date have provided funding for 17 scientists from developing countries to come to the United States and work directly with U.S. researchers on research priorities and goals of the Alliance.

In addition, USGCRP-supported researchers continue to play critical and wide-ranging roles in the development of several major international assessments, including the IPCC AR5 (IPCC 2013). They serve as working group co-chairs, coordinating lead authors, lead authors, contributing authors, review editors, and reviewers, and they provide technical support and scientific expertise as reviewers to IPCC assessments and other international efforts. USGCRP coordinates author nominations, as well as government and expert reviews for AR5. It also provides direct financial support for the operations of the IPCC Working Group II Technical Support Unit, which is responsible for coordinating the production of the Working Group II volume, U.S. participation in the production of the Working Group I and III reports, and U.S. participation in the ongoing Scientific Assessment of Ozone Depletion,[5] the *Special Report on Renewable Energy Sources and Climate Change Mitigation* (IPCC 2011), and the *Special Report on Managing the Risks of Extreme Events and Disasters to Advance Climate Change Adaptation* (IPCC 2012).

USGCRP also supports regional activities through the Inter-American Institute for Global Change Research and the Asia-Pacific Network for Global Change Research, and is working with international partners to foster global change research cooperation in Africa. Individual USGCRP agencies provide additional support to other programs and projects that advance collaborative multidisciplinary research relevant to global environmental change and its impacts on society. These types of global partnerships maximize international scientific exchange and best practices, support complementary research efforts, and allow decision makers to make more informed science-based decisions domestically and globally. Support of these programs provides opportunities for U.S. investigators to work with their counterparts from other countries in a coordinated fashion. These activities enrich national activities on the same subjects, build capacity to conduct research and make observations of environmental change in less-developed countries, and foster advances in understanding of global environmental change in ways the investments of any single nation could not accomplish.

---

[5] *See* http://www.esrl.noaa.gov/csd/assessments/ozone/.

The mission of the USGCRP under its new decadal strategic plan aligns with efforts being undertaken recently by the international community, in which the traditional physical and biological research focus on global change is being restructured to respond to the growing demand for information and products by both the public and the private sectors. The ICSU, with the ISSC and other partners, including the International Group of Funding Agencies for Global Change Research (IGFA) and its Council of Principals, the Belmont Forum, is shaping the future of international global change research coordination.

One such initiative is Future Earth, which follows on years of planning that began with the review of a suite of ICSU-sponsored global change research programs, in particular the Earth System Science Partnership. Future Earth will merge the International Geosphere-Biosphere Program, the International Human Dimensions Program, and the DIVERSITAS program. USGCRP played a role in the Alliance Transition Team, which led an 18-month process to design a 10-year Future Earth Initiative that is the result of the visioning process led by ICSU and ISSC. USGCRP also contributes to a variety of other activities of the Belmont Forum and IGFA, including redesigning and hosting the group's Web sites and hosting the U.S. portion of the secretariat.

Another example of efforts to advance cooperation among international global environmental change communities can be found in the outcomes of the World Climate Conference-3, with a decision to establish a Global Framework for Climate Services (GFCS) to strengthen the application of science-based climate prediction and services around the world. Such a framework has the potential to offer significant economic, public health and safety, and security benefits for participating countries, and the physical, biological, and social science research and infrastructure funded by USGCRP agencies are highly relevant to the GFCS. USGCRP is already working with WCRP to develop the modeling and understanding components of the GFCS that will emphasize linkages to adaptation and observations. USGCRP can further contribute to, and benefit from, this emerging framework through increased coordination with the international community to provide global change information.

SYSTEMATIC OBSERVATIONS

Continuous, high-quality, scientific observations of the global environment are critical for defining the current state of Earth's integrated environmental system—in particular, the constantly changing conditions of the atmosphere, hydrosphere, and biosphere. A historical continuum of high-confidence data is essential to initialize forecast models, reconstruct historical variances and interrelationships, and document changes in Earth's systems. Building this knowledge base requires systematizing historical data and paleoclimatic reconstructions to modern scientific standards, as well as quantifying the ever-shifting present. The fidelity of predictions of the future is directly related to such a knowledge base being in place, accurate, and sustained over a long time period.

The term "climate observations" encompasses a broad range of environmental observations, including (1) routine weather observations, which are collected consistently over a long period; (2) observations collected as part of research investigations to elucidate processes that contribute to maintaining climate patterns or their variability; (3) highly precise, continuous observations of climate system variables collected for the express purpose of documenting long-term (decadal to centennial) changes; (4) observations to document the changing state of the oceans and atmosphere; and (5) observations of climate proxies, collected to extend the instrumental climate record to remote regions and back in time.

A critical challenge is to maintain measurements provided by current observing capabilities. To detect climate change, understand and attribute change to specific climate processes, and anticipate climate impacts on the Earth system requires a long-term (many decades), consistent, comprehensive observing system with multiple complementary components. Many climate trends are small and can only be distinguished from short-term variability through careful analysis of long time series of sufficient length, consistency, continuity, and accuracy to determine climate variability and change (e.g., climate data records). Short data records or long gaps in the records can make such detection and analysis much more uncertain and costly. To confidently detect small climate shifts requires instrument stability better than generally required for other uses.

In addition, the sustained global observing systems that are essential to global change research require international partnerships. *In situ* and satellite-based observations of the environment are of fundamental importance to understanding the Earth system. Because these observations are of great value globally, require significant investments of resources, and need to be collected outside of the United States, international partnerships are crucial to leverage investments, expand system coverage, and increase usable science. The global scientific community has recognized the value of intelligently connected and consistent observing systems that incorporate both longer-term (sustained) and shorter-term (intensive) observations. As discussed in detail in the following section, the United States plays a leadership role in a number of international observing systems.

Documentation of U.S. Climate Observations

U.S. government investments in climate observing systems provide the backbone of much of the international climate data information infrastructure. Since the *U.S. Climate Action Report 2010* (2010 CAR) (U.S. DOS 2010), the United States has maintained and improved its domestic and international investments in both satellite and nonsatellite observing systems.

The United States supports a large number of remote-sensing satellite platforms, as well as a broad network of Earth-based global atmospheric, oceanic, and terrestrial observation systems that are essential to climate monitoring. These systems are a baseline Earth-observing system and include NASA, NOAA, USGS, and DMSP Earth-observing satellites and extensive nonsatellite observational capabilities across multiple federal agencies that participate in USGCRP.

Working through the U.S. Group on Earth Observations (USGEO), the United States is a founding member of and vital contributor to the intergovernmental Group on Earth Observations (GEO). As such, it contributes to the development and operation of a number of global observing systems, both research and operational, that collectively provide a comprehensive measure of climate system variability and climate change processes. In particular, through USGEO, and through the international Committee on Earth Observation Satellites (CEOS), of which NASA, NOAA, and USGS are active members, the United States further supports cooperative, international efforts to build the Global Earth Observation System of Systems (GEOSS). GEOSS is being developed through the GEO, a partnership of 80 countries, the European Commission, and nearly 60 international organizations.

Global Climate Observing System

USGCRP also supports surface-based measurement activities that provide the data used in studies of the various climate processes necessary for better understanding of climate change. U.S. observational and monitoring activities contribute significantly to several international observing systems, including the Global Climate Observing System (GCOS), principally sponsored by the World Meteorological Organization (WMO); the Global Ocean Observing System (GOOS), sponsored by the United Nations Educational, Scientific and Cultural Organization's Intergovernmental Oceanographic Commission (IOC); and the Global Terrestrial Observing System (GTOS), sponsored by the Food and Agriculture Organization of the United Nations (FAO). The latter two have climate-related elements being developed jointly with GCOS.

Based at NOAA's National Climatic Data Center, the U.S. GCOS Program[6] has two primary areas of focus: the development and sustenance of reference-level climate observing efforts, and the contribution to a sustained climate science, observing, and associated data management program in the Pacific Islands region. U.S. support for a strong GCOS regional program in the Pacific is of critical importance for climate observation, given that the Pacific is the source of such phenomena as El Niño, coupled with the general sparseness of data from this critical climate region. The U.S. GCOS Program, via NOAA's Pacific Climate Information System (PaCIS), has partnered with the New Zealand MetService and National Institutes of Water and Atmosphere, as well as the Australian Bureau of Meteorology, in a series of bilateral efforts to help carry out a number of activities toward strengthening climate science, observation, and related data management efforts across the region.

[6] *See* http://gosic.org/gcos/USGCOS.html.

Nonsatellite Atmospheric Observations

The United States supports 114 stations in the GCOS Surface Network, 4 stations in the GCOS Reference Upper Air Network (GRUAN), and 4 stations in the Global Atmospheric Watch (GAW). These stations are distributed geographically, as prescribed in the GCOS and GAW network designs. The data (metadata and observations) from these stations are shared according to GCOS and GAW protocols.

The U.S. GCOS program's primary mission is support of nonsatellite reference observational efforts, including developing the GRUAN (Box 8-1).[7] GRUAN enhances the quality of upper-tropospheric and lower-stratospheric water vapor measurements at a subset of 30–40 global stations. Led by the GRUAN Lead Centre in Lindenberg, Germany, GRUAN began operation on January 1, 2009, and is a critical contributing network to GCOS. GRUAN contributes to the GEOSS goal of "understanding, assessing, predicting, mitigating, and adapting to climate variability and change." GRUAN is also a key element supporting the Global Space-Based Inter-Calibration System (GSICS) effort.[8] GSICS is an international collaborative effort initiated in 2005 by WMO and the Coordination Group for Meteorological Satellites to monitor, improve, and harmonize the quality of observations from operational weather and environmental satellites of the Global Observing System. Long-term surface-based reference climate sites are essential for creating a continuous and homogeneous climate data record, such as those used by the IPCC and the United Nations Framework Convention on Climate Change in global climate assessments.

This type of climate data may also be essential for use by least-developed nations for local and regional planning related to protecting and monitoring water resources; for understanding the effects of climate change on human health; and for understanding, assessing, predicting, mitigating, and adapting to climate variability and change. Additionally, this kind of data record is a key element in reducing uncertainties in global temperature and precipitation variances, providing reference ground-truth data to aid in the evaluation of climate model simulations and in the provision of quality data for the calibration and validation of satellite data.

U.S. Climate Reference Network

The United States has continued to field and commission the U.S. Climate Reference Network (USCRN). Since USCRN's beginning in 2002, 114 stations have been commissioned in the continental United States, as well as 13 in Alaska and 2 in Hawaii. The USCRN concept is also being applied toward expanding reference surface observing on an international basis as resources allow. An effort is now underway to install a USCRN station at the Russian Arctic observing station in Tiksi as part of a U.S.–Russia bilateral effort.

Cooperative Observer Program

The Cooperative Observer Program (COOP) is the nation's largest and oldest weather network, with nearly 10,250 observations taken daily, mostly by volunteers, over the course of the past 121 years. The COOP is the primary source for monitoring U.S. climate variability, including measuring weekly-to-interannual time frames on national, regional, and local scales. These data are also the primary basis for assessments of decadal and centennial climate change. The network is in stable locations of urban, suburban, and rural settings in flat, mountainous, and coastal areas. Because of the density of this observation network, the information collected can clarify how the U.S. climate has changed in the past century or more.

USCRN installed the final station in 2008, and uses historic data from the COOP network to develop pseudo-normals. Each year, these data help to inform decisions related to Federal Emergency Management Agency Disaster Declarations based on weather, insurance industry claims, water resource management, drought declarations, transportation issues, legal issues, computing model guidance to daily weather forecasts, normals and extremes, and energy consumption.

U.S. Observing Campaigns and Systems

While the large number of U.S. observing campaigns and systems makes it impractical to list all of them, the following should be noted for their global significance.

[7] See http://www.wmo.int/pages/prog/gcos/index.php?name=GRUAN.

[8] See http://www.wmo.int/pages/prog/sat/GSICS.

The Atmospheric Radiation Measurement (ARM) Climate Research Facility (ACRF) and Mobile Facilities (AMFs)[9]—The ACRF and AMFs are scientific user capabilities for obtaining continuous, long-term measurements of radiative fluxes, cloud and aerosol properties, and related atmospheric characteristics in focused clusters of instruments in diverse climate regimes for critical process-oriented studies. Operating for more than 20 years, the ARM program paradigm of long-term, continuous measurements is essential to the evaluation and enhancement of climate models that must simulate the evolution of atmospheric properties for long continuous periods, from decades to centuries. The two AMFs, which include aerial measurements that complement the ground-based measurements, expand the geographic coverage of the ACRF through deployments in major field campaigns, such as the Ganges Valley Aerosol Experiment, the ARM Madden-Julian Oscillation Experiment, the Arctic Observing eXperiment, and GOAMAZON 2014.

The AErosol RObotic NETwork[10]—AERONET is a federation of ground-based remote-sensing aerosol networks established in part by NASA and France's Centre National de la Recherche Scientifique. AERONET provides a long-term, continuous, and readily accessible public domain database of aerosol optical properties for research and characterization of aerosols; validates satellite retrievals; and provides synergy with other databases.

Advanced Global Atmospheric Gases Experiment (AGAGE)[11] *and NOAA's Carbon Cycle Greenhouse Gases Group Cooperative Global Air Sampling Network*[12]—The collaborative effort between NASA's AGAGE program and NOAA's flask monitoring network has been instrumental in measuring the composition of the global atmosphere continuously since 1978. AGAGE is distinguished by its capability to measure globally and at high frequency most of the important gases in the Montreal Protocol and almost all of the significant non-CO_2 gases in the Kyoto Protocol.

Micro-Pulse Lidar Network[13]—The NASA MPLNET is a federated network of micro-pulse light-detection and ranging (MPL lidar) systems designed to measure aerosol and cloud vertical structure continuously, day and night, over the long time periods required to contribute to climate change studies and provide ground validation for models and satellite sensors in the NASA Earth Observing System. At present, there are 18 active sites worldwide. Numerous temporary sites have also been deployed in support of field campaigns. Most MPLNET sites are co-located with AERONET sites to provide both column and vertically resolved aerosol and cloud data.

Surface Radiation Budget Network[14]—SURFRAD was established in 1993 through NOAA to support climate research with accurate, continuous, long-term measurements of the surface radiation budget. Currently, seven SURFRAD stations are operating in climatologically diverse regions across the United States. These sites provide primary measurements of upwelling and downwelling solar and infrared, along with ancillary observations of direct and diffuse solar, photosynthetically active radiation, ultraviolet B radiation, spectral solar, and meteorological parameters. SURFRAD is an important contribution to the worldwide GCOS Baseline Surface Radiation Network.

Interagency Coordinating Committee for Airborne Geosciences Research and Applications[15]—ICCAGRA is a collaboration of U.S. government agencies (NASA, NOAA, NSF, DOE, the U.S. Department of Defense [DoD], and USGS). Its primary purpose is to increase the effective utilization of the federal airborne fleet in support of airborne geoscience research and applications programs conducted by the individual agencies. ICCAGRA improves cooperation, fosters awareness, and facilitates communication among U.S. government agencies having or using aircraft and instruments for airborne research and applications, and serves as a resource to senior-level managers on airborne geoscience issues. ICCAGRA members operate and manage more than 25 aircraft across the country, including unmanned aircraft systems.

Nonsatellite Ocean Observation

Global Ocean Observing System

The United States currently provides satellite, buoy, glider, and ship coverage of the global oceans for sea-surface temperatures, surface elevation, ocean-surface vector winds, sea ice,

[9] See http://www.arm.gov/.

[10] See http://gcmd.nasa.gov/records/GCMD_AERONET_NASA.html.

[11] See http://agage.eas.gatech.edu/.

[12] See http://www.esrl.noaa.gov/gmd/ccgg/flask.html.

[13] See http://www.ndsc.ncep.noaa.gov/coop/mplnet/.

[14] See http://www.esrl.noaa.gov/gmd/grad/surfrad/.

[15] See http://www.nsf.gov/geo/ags/ulafos/laof/iccagra.jsp.

ocean color, and other climate variables (Box 8-2). These observations provide foundational support for the Global Ocean Observing System (GOOS) and other international efforts. The climate requirements of GOOS are the same as those for GCOS; like GCOS, GOOS is based on a number of nonsatellite and space-based observing components.

Completed in September 2005, the first element of the climate portion of GOOS is the global drifting buoy array, which is a network of 1,250 drifting buoys measuring sea-surface temperature and other variables as they flow in the ocean currents. At present, the United States is the world leader in implementing the nonsatellite elements of GOOS for climate, and sponsors the majority of the U.S. Integrated Ocean Observing System (IOOS) global component, which is the U.S. contribution to the international GOOS program and the GEOSS ocean baseline. Specifically, the United States sponsors nearly half of the platforms currently deployed in the global ocean (3,860 of 7,723), with 72 other countries providing the remainder.

Integrated Ocean Observing System

IOOS[16] is the U.S. coastal observing component of GOOS, envisioned as a coordinated national and international network of observations, data management, and analyses that systematically acquires and disseminates data and information on past, present, and future states of the oceans. A coordinated IOOS effort is being established by NOAA via a national IOOS Program Office. The IOOS observing subsystem employs both remote and nonsatellite sensing, including satellite-, aircraft- and land-based sensors; ships; buoys; and gliders. The United States supports IOOS's surface and marine observations through a variety of components, including fixed and surface-drifting buoys, subsurface floats, and volunteer observing ships. Expanding in coverage, currently 60 percent of the initial GOOS design is complete.

U.S.-Funded Ocean Observing Systems

While the large number of U.S.-funded ocean observing systems makes it impractical to list all of them, the following systems have global significance.

Argo[17]—In 1998, an international consortium presented plans for Argo, a global array of 3,000 autonomous instruments that would revolutionize the collection of critical, climate-relevant information from the upper 2 kilometers (1.2 miles) of the world's oceans. These instruments drift at depth, periodically rising to the sea surface, collecting data along the way, and report their observations in real time via satellite communications.

The initial deployment objective of 3,000 instruments distributed homogeneously throughout the world's oceans has been attained, and Argo now provides more than 100,000 high-quality temperature and salinity profiles annually, along with global-scale velocity data, all without a seasonal bias. The Argo array has been deployed through the collaboration of more than 40 countries plus the European Union. Argo data are openly and immediately available to anyone wishing to use them.

Argo data, coupled with global-scale satellite measurements from radar altimeters, have made possible significant advances in the representation of the oceans in coupled ocean-atmosphere models for climate forecasts and the routine analysis and forecasting of the state of the subsurface ocean. Going forward, the United States has committed to maintaining half of the array, and other contributing nations are striving to continue the array's strong international nature.

Ocean Observatories Initiative[18]—Construction is now underway on the OOI, a significant new effort funded by NSF. The OOI is planned as a networked infrastructure of sensor systems to measure the physical, chemical, geological, and biological variables in the ocean and seafloor, with the goal of improving detection and forecasting of environmental change and its effects on biodiversity, coastal ecosystems, and climate. Ultimately, the OOI will be one fully integrated system, collecting data on coastal, regional, and global scales employing advanced ocean research and sensor tools, including buoys and remotely operated and autonomous vehicles—all linked via telecommunications cables and satellites directly to laboratories. With these advances, the OOI will improve the rate and scale of ocean data collection, and its networked observatories will focus on global, regional, and coastal science questions, and provide platforms to support new types of instruments and autonomous vehicles.

Box 8-2 **Major Categories of U.S. Contribution to Nonsatellite Ocean Observations**

- Moored and floating buoy networks
- Argo floats and gliders
- Research and volunteer ships
- Tide gauge networks

---

[16] See http://www.ioos.noaa.gov/.

[17] See http://www.argo.net/.

[18] See http://www.oceanobservatories.org/.

Global Sea Level Observing System[19]—Continued upgrading of the GLOSS tidal gauge network from 43 to 170 stations is planned for the period 2006–2010. Ocean carbon inventory surveys in 10-year repeat survey cycles help determine the anthropogenic intake of carbon into the oceans.

Tropical-Atmosphere-Ocean Network[20]—The TAO network of ocean buoys includes an expansion of the network into the Indian Ocean. (The Pacific Ocean has a current array of 70 TAO buoys.) From 2005 to 2007, 8 new TAO buoys were installed in the Indian Ocean in collaboration with partners from India, Indonesia, and France. Plans call for a total of 38 TAO buoys in the Indian Ocean by 2013.

Research Moored Array for African-Asian-Australian Monsoon Analysis and Prediction[21]—The RAMA network is a multinationally supported element of the Indian Ocean Observing System, a combination of complementary satellite and nonsatellite measurement platforms for climate research and forecasting purposes. NASA is currently investing in the development of new prototype geodetic instruments for deployment later this decade to support the creation of a next-generation geodetic network for the improvement of the terrestrial reference frame.

Voluntary Observing Ship Climate Program[22]—Voluntary ship observations have been the backbone of the ocean observing system for centuries. Volunteer crew members around the world observe the weather at their location, encode each observation in a standard format, and transmit the data to national meteorological services that have responsibility for marine weather forecasts. In addition, these data are archived for future use by climatologists and other scientists. The U.S. VOS Program within the overall WMO VOS framework services about one-quarter of the world's VOS fleet, providing ships' crews with weather observer training, handbooks and forms, observation encoding software, barometer calibration, the Mariners Weather Log, and weather-observing tools. A subprogram within VOS is VOSClim, an ongoing, NOAA-supported program within the WMO Joint Technical Commission for Oceanography and Marine Meteorology's Voluntary Observing Ships' Scheme. It aims to provide a high-quality subset of marine meteorological data, with extensive associated metadata, to be available in both real-time and delayed modes to support global climate studies.

University-National Oceanographic Laboratory System[23]—UNOLS is an organization of 62 academic institutions and national laboratories involved in oceanographic research and joined for the purpose of coordinating oceanographic ships' schedules and research facilities. A major aim of UNOLS is to ensure the efficient scheduling of scientific cruises aboard the 21 research vessels located at 16 U.S. operating institutions (and numerous partner institutions) in the UNOLS organization.

Nonsatellite Terrestrial and Cryospheric Observations

Many of the most critical variables for long-term monitoring and process-level understanding of the rate and magnitude of climate change and its impacts involve *in situ* observations of terrestrial and cryospheric variables, such as soil moisture, streamflow, permafrost, glaciers, and terrestrial ecosystems (Box 8-3). Following are some major U.S. terrestrial observation programs.

National Streamgage and Groundwater Networks
Streamflow is one of the most important variables for both long-term monitoring of the impacts of climate change and real-time decision making about water availability and quality. USGS has been measuring flow in U.S. rivers and streams since 1889. In partnership with more than 850 other federal, state, and local agencies, USGS maintains a comprehensive U.S. streamgage network of consistent measurements, obtained using standard techniques and technology subject to the same quality assurance and quality control. In addition, USGS annually monitors groundwater levels in thousands of U.S. wells, and collects and stores the data either as discrete field-water-level measurements or as continuous time-series data from automated recorders. The overall USGS groundwater database consists of more than 850,000 records of wells, springs, test holes, tunnels, drains, and excavations in the United States.

SCAN (Soil Climate Analysis Network)[24]
The SCAN monitoring network provides automated comprehensive soil moisture and related climate information designed to support natural resource assessments. SCAN consists of

[19] See http://www.gloss-sealevel.org/.

[20] See http://www.pmel.noaa.gov/tao/.

[21] See http://www.pmel.noaa.gov/tao/rama/.

[22] See http://www.vos.noaa.gov/.

[23] See http://www.unols.org/.

[24] See http://www.wcc.nrcs.usda.gov/scan/.

more than 120 sites that collect and disseminate continuous, standardized soil moisture and other climate data in publicly available databases and climate reports. Uses for these data include providing inputs to global circulation models, verifying and ground-truthing satellite data, monitoring drought development, forecasting water supply, and predicting sustainability for cropping systems.

SNOTEL (SNOpack TELemetry) [25]

The SNOTEL monitoring network provides automated comprehensive snowpack and related climate information designed to support natural resource assessments. SNOTEL operates more than 660 remote sites in mountain snowpack zones of the western United States. This network collects and disseminates continuous, standardized data in publicly available databases and climate reports. Uses for these data include inputs to global circulation models, and verifying and ground-truthing satellite data.

USGS Glacier Monitoring

USGS operates a long-term benchmark glacier program to intensively monitor climate, glacier motion, glacier mass balance, glacier geometry, and stream runoff at a few select sites. The data collected are used to understand glacier-related hydrologic processes and improve the quantitative prediction of water resources, glacier-related hazards, and the consequences of climate change. Long-term mass-balance monitoring programs have been established at three widely spaced U.S. glacier basins to clearly sample different climate-glacier-runoff regimes: the South Cascade Glacier in Washington State and the Gulkana and Wolverine glaciers in Alaska. Mass-balance data are available beginning in 1959 for the South Cascade Glacier, and beginning in 1966 for the Gulkana and Wolverine glaciers.

Real-Time Permafrost and Climate Monitoring Network

For terrestrial observations, GCOS and GTOS have identified permafrost thermal state and permafrost active layer as key variables for monitoring the state of the cryosphere. The USGS Real-Time Permafrost and Climate Monitoring Network in Arctic Alaska is a collaborative effort with the Bureau of Land Management, U.S. Fish and Wildlife Service, private organizations, and universities and is a subset of a larger USGS permafrost and climate monitoring research network. Many of the stations are co-located with deep boreholes, thus forming the basis for comprehensive permafrost monitoring observatories. Data from this network contribute to several international networks as well, primarily the Global Terrestrial Network for Permafrost, part of GCOS.

IceBridge [26]

This NASA airborne mission maps the polar ice sheets to understand their contributions to sea level rise and connections to the global climate system. IceBridge uses aircraft carrying lidar, radar, and other geophysical instruments to determine changes in ice elevation, map the underlying bed, and measure other characteristics of the ice sheets. IceBridge surveys the land ice of Greenland and Antarctica, and the major glacial systems of Alaska and Canada, as well as the sea ice of the Arctic and Southern oceans.

By continuing a critical subset of the global ice elevation measurements obtained by the ICESat satellite from 2003 to 2009, the IceBridge mission also helps bridge the gap in measurements to ICESat 2, to be launched in 2016. IceBridge involved interagency partnerships with NSF, NOAA, the Office of Naval Research, and the U.S. Army Corp of Engineers.

Land Cover Characterization Program

This program was begun in 1995 to develop land cover and other land characterization databases to address national and international requirements that were becoming increasingly sophisticated and diverse. To meet these requirements, USGS develops multiscale land cover characteristics databases used by scientists, resource managers, planners, and educators, and contributes to the understanding of the patterns, characteristics, and dynamics of land cover across the United States and the globe. The program also conducts research to improve the utility and efficiency of large-area land cover characterization and land cover characteristics databases.

Box 8-3 **Major Categories of U.S. Contribution to Nonsatellite Terrestrial and Cryospheric Observations**

- Glacier, permafrost
- Snow monitoring networks
- Streamgaging
- Soil moisture networks
- Groundwater wells
- Terrestrial ecosystem and biodiversity monitoring networks

---

[25] See http://www.wcc.nrcs.usda.gov/snow/.

[26] See http://www.nasa.gov/mission_pages/icebridge/index.html#.UgVZcGRAT_4.

AmeriFLUX Network[27]

This network endeavors to establish an infrastructure for guiding, collecting, synthesizing, and disseminating long-term measurements of CO_2, water, and energy exchange from a variety of ecosystems. Its objectives are to collect critical new information to help define the current global CO_2 budget, to enable improved projections of future concentrations of atmospheric CO_2, and to enhance the understanding of carbon fluxes, net ecosystem production, and carbon sequestration in the terrestrial biosphere.

North American Carbon Program[28]

A major focus of USGCRP and the U.S. Carbon Cycle Science Program, NACP is a multidisciplinary research program established to obtain scientific understanding of North America's carbon sources and sinks, and changes in carbon stocks. NACP is supported by a number of federal agencies through a variety of intramural and extramural funding mechanisms and award instruments.

NACP relies upon a rich and diverse array of existing observational networks, monitoring sites, and experimental field studies in North America and its adjacent oceans to determine the emissions and uptake of CO_2, methane (CH_4), and carbon monoxide (CO); the changes in carbon stocks; and the factors regulating these processes for North America and adjacent ocean basins. NACP also aims to develop the scientific basis to implement full carbon accounting on regional and continental scales. This is the knowledge base needed to design monitoring programs for natural and managed CO_2 sinks and emissions of CH_4; to support long-term quantitative measurements of fluxes, sources, and sinks of atmospheric CO_2 and CH_4; and to develop forecasts for future trends.

USGS LandCarbon Project[29]

USGS has initiated the LandCarbon project, a national assessment of ecosystem carbon sequestration and GHG fluxes. This assessment focuses on carbon stored in the U.S. land surface, by region, with model-based projections of future carbon storage in the U.S. land surface by region and by land cover type. Assessments for the western and central United States have been published, the eastern U.S. assessment will be published in late 2013, and assessments for Alaska and Hawaii are under development.

National Ecological Observatory Network[30]

NEON is a planned continental-scale research platform for discovering and understanding the impacts of climate change, land-use change, and invasive species on ecology, natural resources, and biodiversity. NEON is expected to serve as a U.S. terrestrial contribution to GEOSS. Data are planned to be collected from 106 sites (60 terrestrial, 36 aquatic, and 10 aquatic experimental) across the United States (including Alaska, Hawaii, and Puerto Rico), using instrument measurements and field sampling. The sites have been strategically selected to represent different regions of vegetation, landforms, climate, and ecosystem performance. NEON will combine site-based data with remotely sensed data and other large-scale data sets to provide a range of data products that can be used to describe changes in the nation's ecosystems through space and time, linked by advanced cyber infrastructure to record and archive ecological data for at least 30 years.

NEON has successfully completed the planning and design phases, and has entered the construction and deployment phase. Constructing the entire network will take approximately five years, so NEON expects to be in full operation by approximately 2017.

Long-Term Ecological Research Program[31]

NSF has supported the LTER program for three decades, with 26 projects currently existing, including two urban sites in Phoenix, Arizona, and Baltimore, Maryland. Over this time, the U.S. Forest Service (USFS) has collaborated in supporting seven of the LTER sites, including the Baltimore Ecosystem Study site.

Recent strategic planning by the LTER community has highlighted the need for greater integration of the social and ecological sciences across the LTER network, as evidenced in its decadal plan and the strategic research initiative titled *Integrative Science for Society and the*

[27] *See* http://ameriflux.lbl.gov/ SitePages/Home.aspx.

[28] *See* http://www.nacarbon.org/.

[29] *See* http://www.usgs.gov/climate_ landuse/land_carbon/.

[30] *See* http://www.neoninc.org/.

[31] *See* http://www.lternet.edu/.

Environment (ISSE 2010). LTER planning efforts, the success of the urban LTER programs, and the success of the Dynamics of Coupled Natural and Human Systems Program (also co-funded and coordinated by NSF and USFS) have led NSF and USFS leaders to jointly explore possibilities for development of a network of large-scale Urban Long-Term Research Area (ULTRA) projects, including the funding of a series of ULTRA exploratory awards.

Long-Term Agro-Ecosystem Research Network[32]

The USDA Agricultural Research Service is coordinating a number of its well-established research watersheds and rangelands as the LTAR Network to provide a sophisticated platform for research on the sustainability of U.S. agricultural systems. Over time, the network will develop research questions that are shared and coordinated across sites; provide the capacity to address large-scale questions across sites through shared research protocols; collect compatible data sets across sites; and provide the capacity and infrastructure for cross-site data analysis.

Space-Based Observations

Satellite observations are a primary source of scientific understanding of Earth's changing environment and, thereby, form a critical component of the scientific foundation for subsequent actions by society. Space-based, remote-sensing observations of the atmosphere–ocean–land system have evolved substantially since the early 1970s, when the first operational weather satellite systems and the first land-imaging research satellites were launched (Box 8-4).

Over the last decade, satellites have proven their observational capability to accurately monitor nearly all aspects of the total Earth system on a global basis. Currently, satellite systems monitor the evolution and impacts of El Niño and La Niña weather phenomena, natural hazards, and vegetation cycles; the ozone (O_3) hole and global O_3 distribution; solar activity; snow cover, sea ice and ice sheets, ocean surface temperatures, and biological activity; coastal zones and algal blooms; deforestation and forest fires; carbon storage in tropical forests; urban development; volcanic activity; tectonic plate motions; aerosol and three-dimensional (3D) cloud distributions; water distribution; and other climate-related information.

NASA currently contributes to the operation and data analysis of 16 major satellite missions that provide high-spatial-resolution, high-accuracy, well-calibrated, sustained observation of the land surface, oceans, atmosphere, ice sheets, and biosphere. Many of these satellites involve international partnerships, illustrating the value of cooperation in the peaceful use of space. Additionally, NASA is developing 11 Earth-observing research missions for launch between 2014 and 2020, and several of these missions involve international partnerships (Box 8-5).

The next launch will be the Global Precipitation Mission in February 2014, which is a major partnership between NASA and the Japanese Aerospace Exploration Agency. The mission represents both continuity with the long-running Tropical Rainfall Measuring Mission (TRMM), launched in 1997, and a major expansion in capability through incorporation of new technology, coverage at higher latitudes resulting from the use of a higher-inclination orbit, and incorporation of other nations' satellites in a constellation of passive microwave sensors to provide better diurnal sampling of precipitation.

NOAA currently operates four geostationary satellites and six polar-orbiting satellites. NOAA recently took over operation of Suomi-NPP, which will continue weather and climate measurements and reduce risks for the next-generation polar-orbiting satellite. NOAA's partnership with the European Organization for the Exploitation of Meteorological Satellites (EUMETSAT) provides essential global coverage as well. Additionally, NOAA operates the Jason-2 ocean surface topography spacecraft, developed by NASA and France's Centre National d'Études Spatiales in collaboration with EUMETSAT. In 2012, NOAA delivered five new Climate Data Records that provide societal benefits, such as improvements in precipitation forecasts for agriculture, pollutant forecasts for health, temperature trend estimates, and fisheries impacts analyses—all essential in an era of increased climate uncertainty.

Through a partnership between NASA and USGS, the United States develops, launches, and operates the Landsat satellite series for monitoring land surfaces at a scale where natural and human-induced changes can be detected, characterized, and monitored over time. Since

---

Box 8-4 **Major Categories of U.S. Contribution to Space-Based Observations**

- Met-class infrared, vis, and multispectral imagers
- Medium-resolution imagers
- High-resolution imagers and aerial surveys
- Infrared profilers/sounders
- Microwave profilers/sounders
- Broadband/multispectral radiometers
- Doppler radar and synthetic aperture radar, radar scatterometers, other wind instruments
- Cloud/aerosol profilers
- Precipitation instruments
- Altimetry
- Global Navigation Satellite System radio occultation
- Microwave ranging systems
- Spectrometers and occultation (for atmospheric chemistry)

---

[32] *See* http://www.ars.usda.gov/research/programs/programs.htm?np_code=211&docid=22480.

1972, Landsat satellites have consistently captured moderate-resolution (e.g., 30-meter [98-foot]) data of the Earth. This archive of data has become vital for agriculture and water management, disaster response, forest carbon monitoring, and monitoring incremental effects of climate change. A cost-free and open-data policy, combined with consolidation of the Landsat Global Archive, provides current, repeatable, and historical access to more than 40 years of terrestrial land cover change.

With the successful launch of the Landsat Data Continuity Mission, which was renamed Landsat 8 once it became operational at the end of May 2013, scientists throughout the world can now make direct comparisons with the past, while taking advantage of significant advancements incorporated in the mission, including additional bands to improve atmospheric corrections to the data and higher quantization of the entire data stream to enable detection of more subtle changes.

U.S. satellite observing activities contribute significantly to several international observing systems, principally sponsored by elements of the United Nations, such as WMO, IOC, and FAO. In particular, the United States continues to work with GCOS, whose goal is to provide a comprehensive view of the total climate system. GCOS partners include NOAA and NASA, as well as three international groups strongly supported and led by the United States: GEO, CEOS, and the Coordination Group for Meteorological Satellites. GCOS constitutes the climate-observing component of GEOSS. A number of U.S. satellite operational and research missions form the basis of a robust national remote-sensing program that seeks to fully support the requirements of GCOS.

The United States continues to demonstrate the immense value of satellites for observing the changing global climate and for developing new fundamental knowledge of the global integrated Earth system. Satellite observations and the increased scientific understanding they enable can improve international security, enhance economic prosperity, mitigate impacts of short-term and climate-related hazards, and strengthen global stewardship of the environment. The U.S. policy is to maximize timely, full, and open access to data from its civil satellites and to disseminate tools and knowledge to use this information.

Suomi National Polar-orbiting Partnership[33]

In October 2011, NASA and NOAA launched the Suomi NPP satellite, with a mission to acquire a wide range of land, ocean, and atmospheric measurements. The 2,100-kilogram (4,600-pound) spacecraft carries five key instruments: the Advanced Technology Microwave Sounder, the Cross-track Infrared Sounder (CrIS), the Ozone Mapping and Profiler Suite, the Visible Infrared Imaging Radiometer Suite, and Clouds and the Earth's Radiant Energy System. The NPP mission is a bridge between NASA's Earth Observing System (EOS) satellites and the forthcoming series of Joint Polar Satellite System satellites, and will provide a wide range of data, including atmospheric and sea surface temperatures, land and ocean biological productivity measurements, cloud and aerosol property information, ozone measurements, and information about fluxes in Earth's radiation budget.

Landsat Data Continuity Mission/Landsat-8[34]

NASA launched its LDCM successfully on February 11, 2013. Following on-orbit testing, NASA turned its satellite operations over to USGS on May 30, 2013, when the mission officially became Landsat 8. Landsat data offer the longest continuous record of satellite observations of Earth's land surface at scales for detecting, characterizing, and monitoring natural and human-induced changes on the landscape. The Landsat satellite series has provided imagery of Earth's surface for more than 40 years, providing the most consistent, reliable documentation of global land surface change ever assembled.

Thousands of Landsat images are downloaded every day from the USGS archive. Government, commercial, industrial, civilian, military, and educational communities throughout the United States and worldwide rely on Landsat for a wide range of applications in such areas as global change research, agriculture, forestry, geology, resource management, geography, mapping, water quality, and oceanography. The full USGS archive holds more than four million Landsat scenes obtained continuously from July 1972 to the present day. Since December 2008, when

[33] See http://npp.gsfc.nasa.gov/.

[34] See http://landsat.usgs.gov/.

the images became available free of charge over the Internet, more than 12 million scenes have been downloaded by users in 186 countries and territories.

Jason Altimeter Series

Observation of global sea level rise through satellite altimetry, in particular the systematic collection of sea level observations gathered first by TOPEX/Poseidon and now by the ongoing Jason series of satellite missions, is a critical data stream for understanding global change. These observations suggest that sea level rise is accelerating. In particular, the value of approximately 3.1 millimeters (mm) (0.12 inches [in]) per year from altimeters over the past 15 years is almost twice the estimate of approximately 1.7 mm (0.07 in) per year from tide gauges over the past century.

The Jason series is being transitioned as a research endeavor from NASA and the Centre National d'Études Spatiales to NOAA and EUMETSAT, for joint implementation as a sustained operational capability. NOAA and EUMETSAT have already assumed responsibility for the ground system and operation of the Jason-2 satellite launched in June 2008. Jason-3, scheduled to launch in 2015, will extend this critical time series of ocean surface topography measurements.

Aquarius[35]

NASA's Aquarius mission was launched in 2011 in partnership with the Space Agency of Argentina (Comisión Nacional de Actividades Espaciales). Aquarius is the first satellite mission specifically focused on producing global observations of sea surface salinity. It delivers monthly salinity maps with an estimated accuracy of 0.2 practical salinity units, equivalent to detecting a single "pinch" of salt (about half a milliliter, or 1/8th of a teaspoon) in nearly 4 liters (1 gallon) of water.

In the fall of 2012, Aquarius measurements were complemented by the SPURS field campaign to closely monitor the saltiest region of Earth's oceans—the subtropical North Atlantic gyre—to provide a 3D view of processes that drive changes in salinity distribution. NASA, NSF, NOAA, and European partner agencies have been deploying instruments on floats, ships, moored buoys, underwater gliders, and an autonomous underwater vehicle to capture this detailed view of ocean processes.

The Gravity Recovery and Climate Experiment[36]

The twin GRACE satellites celebrated their eleventh anniversary in orbit in March 2013. This milestone exceeded by six years a successful primary mission that demonstrated a new paradigm—the spaceborne measurement of high-resolution gravity fields with sufficient accuracy to resolve the transport of mass within the Earth system.

In conjunction with other data and models, the GRACE mission provides the first global and regional measurements of monthly to interannual changes in terrestrial water storage, polar ice cap and glacial ice masses, earthquake-induced crustal deformation, and variations in ocean mass and circulation. The GRACE mission also carries a NASA global positioning system (GPS) occultation receiver to measure atmospheric and ionospheric dynamics for weather and climate studies. The mission is a collaboration with the German space agency DLR (Deutsches Zentrum für Luft- und Raumfahrt) and numerous partnering international scientific institutions.

Constellation Observing System for Meteorology, Ionosphere, and Climate[37]

COSMIC, a system of six microsatellites launched jointly by the United States and Taiwan in 2006, uses GPS radio receivers to measure the bending of GPS signals (GPS occultation) by Earth's atmosphere. The atmospheric refractivity measurements are used to estimate atmospheric temperature and humidity with unprecedented accuracy for both weather forecasting and climate studies.

GPS occultation data have improved the accuracy of long-range weather forecasts, and have become an important data source for the operational weather services. The GPS data and a sounder instrument are also used to measure ionospheric structure for communications and space weather studies.

[35] See http://www.nasa.gov/mission_pages/aquarius/main/index.html#. UgVarGRAT_4.

[36] See http://www.nasa.gov/mission_pages/Grace/index.html#. UgVa1WRAT_4.

[37] See http://www.cosmic.ucar.edu/.

The COSMIC GPS occultation receivers were designed by NASA, and the ionospheric sounder was designed by the U.S. Naval Research Laboratory. Operations and analysis of the COSMIC data are a partnership between the Taiwanese Space Agency and the University Corporation for Atmospheric Research. Domestic funding for COSMIC is coordinated by NSF, and the program also receives co-funding from NOAA, NASA, the U.S. Air Force, and the U.S. Navy.

Polar Operational Environmental Satellite System[38]

Since 1979, the NOAA POES system has provided the nation with the longest time series of essential climate variables (ECVs), including atmospheric temperature, water vapor, clouds, ozone, vegetation, and sea and land surface temperature.

Geostationary Operational Environmental Satellite System[39]

Since the 1980s, GOES has provided essential information on the diurnal cycle of clouds, and has been used as a key data set for the International Satellite Cloud Climatology Project. GOES has also been used to study the diurnal cycle of sea surface temperature.

Aqua[40]

Aqua is part of the "Afternoon Train" (A-Train) constellation, a key Sun-synchronous satellite formation that studies the atmosphere and consists of five satellites flying in close proximity to each other—Aqua, Aura, CALIPSO,[41] CloudSat, and now the Japanese GCOM-W1. The French mission PARASOL[42] exited the A-Train after five years of concurrent operations in the constellation.

Aqua is designed to acquire precise measurements that provide a greater understanding of the Earth's atmosphere and oceans. Operational agencies around the world are also using Aqua data to improve weather prediction. The six Aqua instruments were carefully selected to make measurements for the improved characterization and understanding of atmospheric temperature and humidity profiles, clouds, global precipitation, and Earth's thermal radiation balance; terrestrial snow and sea ice; sea surface temperature and ocean productivity; and soil moisture.

Global thermal sounder retrievals from the Atmospheric InfraRed Sounder (AIRS) instrument on Aqua help to increase understanding of the distribution and transport mechanisms of CO, CH_4, and CO_2 in the middle troposphere. NOAA has incorporated the lessons learned from AIRS into operational carbon products from the EUMETSAT Infrared Atmospheric Sounding Interferometer (IASI), which launched aboard the MetOp-A satellite in 2006.

NOAA is planning to continue these products with CrIS aboard Suomi-NPP. The IASI and CrIS missions will allow the creation of a 20-year record of satellite thermal sounder-derived carbon trace gases, along with self-consistent ozone, temperature, moisture, and cloud information. The Moderate Resolution Imaging Spectroradiometer (MODIS) instrument provides regional-to-global land cover, sea surface temperature, ocean color, clouds, and aerosols. Data from the A-Train instruments help answer important questions related to aerosols, clouds, and atmospheric processes.

Aura[43]

The NASA Aura satellite, also part of the A-Train, was launched with four instruments to extensively monitor the composition of the atmosphere. The Microwave Limb Sounder obtains highly resolved altitude profiles of the stratosphere and upper troposphere for understanding photochemical and dynamical processes in these altitude ranges. The Tropospheric Emission Spectrometer obtains column and partial altitude profiles for ozone and tropospheric trace gases, while the Ozone Monitoring Instrument obtains nearly daily global ozone column maps, as well as columns for other important air quality parameters.

CALIPSO[44] and CloudSat[45]

NASA's highly complementary CALIPSO and CloudSat satellites provide unprecedented information on the vertical profile of clouds, cloud liquid water, and aerosol particles over the globe, leading to improved 3D perspectives of how clouds and aerosols form, evolve, and affect weather and climate. Both satellites have been flying in formation as part of the NASA A-Train

[38] See http://poes.gsfc.nasa.gov/.

[39] See http://www.goes.noaa.gov/.

[40] See http://aqua.nasa.gov/.

[41] Cloud-Aerosol Lidar and Infrared Pathfinder Satellite Observation.

[42] Polarization and Anisotropy of Reflectances for Atmospheric Science coupled with Observations from a Lidar.

[43] See http://aura.gsfc.nasa.gov/.

[44] See http://www-calipso.larc.nasa.gov/.

[45] See http://www.nasa.gov/mission_pages/cloudsat/main/index.html.

constellation since their launch in 2006, providing the benefits of near simultaneity, and thus the opportunity for synergistic measurements made with complementary techniques.

Solar Radiation and Climate Experiment[46]

Launched in 2003, SORCE is equipped with four instruments that measure variations in solar radiation much more accurately than previous measurements and observe some of the spectral properties of solar radiation for the first time. These measurements have been a critical part of the long-term record of total solar irradiance observations, which also include those from, for example, the ACRIMSat mission, launched in 1999.

Terra[47]

Launched in 1999, Terra flies in the morning constellation with the Landsat missions, to complement the A-Train constellation. Like Aqua, Terra carries the multidisciplinary MODIS sensor. Terra emphasizes observations of terrestrial surface features and carries four additional sensors, all of which continue to operate successfully to provide decade-plus data sets of terrestrial and oceanic properties, clouds, water vapor, aerosols, and the radiation budget.

Tropical Rainfall Measuring Mission[48]

Launched in 1997, TRMM carries the innovative Precipitation Radar, contributed by Japan and designed to provide 3D maps of storm structure. The ongoing 15-year data set provides information on the intensity and distribution of rain.

Quick Scatterometer[49]

Launched in June 1999, QuikSCAT, remained fully operational until November 2009, when the primary instrument (SeaWinds) antenna stopped rotating due to a mechanical failure of the antenna spin mechanism. During its nominal mission, QuikSCAT was a primary data source for science applications and studies involving climate models; interactions between the atmosphere and ocean; and weather/climate phenomena, such as hurricanes and El Niño.

Although SeaWinds' radar performance was not affected by the spin mechanism failure, QuikSCAT now tracks an operational data path swath significantly reduced from its original capability. Nevertheless, these data are continuing to provide an accurate and reliable transfer standard for cross-calibration of other ocean vector wind sensors, and for establishing the measurement stability needed for continuity with future scatterometer missions.

Earth Observing-1[50]

Launched in 1999, EO-1 validated technologies contributing to future land-imaging missions. The hyperspectral instrument Hyperion is the first of its kind to provide images of land surface in more than 220 spectral colors. In the future, an operational version of the Hyperion will allow complex land ecosystems to be imaged and accurately classified.

Data Management and Information Systems

Data management is an important aspect of any systematic observing effort. While U.S. agencies have unique mandates for climate-focused and -related systematic observations, and for the attendant data processing, archiving, and use of the important information from these observing systems, it is clear that the climate observations portfolio must be handled in an integrated way. A robust strategy for management of the climate observations portfolio must capture the critical interaction between climate system components, as well as sustain this observations strategy over time.

On May 9, 2013, President Obama signed the Executive Order (E.O.) "Making Open and Machine Readable the New Default for Government Information."[51] The E.O. directed federal agencies to make government-held data more accessible to the public and to entrepreneurs and others, as fuel for innovation and economic growth. Under the terms of the E.O. and a new Open Data Policy released by the White House Office of Management and Budget and Office of Science and Technology Policy,[52] the new default for newly generated U.S. government data is that data will be open and machine readable to enhance accessibility and usefulness where possible and will be consistent with the law, while continuing to ensure privacy, confidentiality, and national security. As part of this initiative, USGEO is leading an effort to transform federal holdings of environmental observation data to machine-readable formats.

[46] See http://lasp.colorado.edu/home/sorce/.

[47] See http://www.nasa.gov/mission_pages/terra/index.html#.UgVb_GRAT_4.

[48] See http://trmm.gsfc.nasa.gov/.

[49] See http://science1.nasa.gov/missions/quikscat/.

[50] See http://eo1.gsfc.nasa.gov/.

[51] See http://www.whitehouse.gov/the-press-office/2013/05/09/executive-order-making-open-and-machine-readable-new-default-government-.

[52] See http://www.whitehouse.gov/open.

These efforts complement, and interact with, the Earth systems data aspects of the administration's "Big Data" initiative, launched in 2012.[53]

In addition, U.S. government agencies partner with nongovernmental organizations and the private sector on issues related to information management and systems through the Federation of Earth Science Information Partners (ESIP). Over the past 14 years, this open-networked community has brought together science, Earth system data, and information technology practitioners into an intellectual commons.

ESIP is a broad-based consortium of Earth scientists, representing the entire research spectrum from data collection, to research, to applications development. ESIP includes distributors of satellite- and ground-based data sets; providers of data and information products, technology, or services aimed primarily at the Earth science and research communities; commercial and noncommercial organizations engaged in developing tools for Earth science; and strategic funding partners.

Earth Observing System Data and Information System[54]

NASA's EOSDIS provides convenient mechanisms for discovering and accessing Earth science data products, almost all of which are available online at no cost to the user. EOSDIS has an operational search-and-order client, called Reverb, which provides access to all data holdings at all the Distributed Active Archive Centers (DAACs). A middleware layer called the EOS ClearingHOuse (ECHO) provides interfaces that allow other user communities to build their own search-and order-clients for EOSDIS data tailored to their needs.

EOSDIS data abide by a NASA Earth Science Data Policy[55] that promotes the full and open sharing of all data with the research and applications communities, private industry, academia, and the general public. Ten geographically distributed NASA DAACs, representing a wide range of Earth science disciplines, have the responsibility for archiving and distributing data products. The Science Investigator-led Processing Systems are responsible for processing certain standard science data products from instrument data, and the DAACs are responsible for their archiving and distribution. The DAACs also provide a full range of user support tailored for the discipline-oriented user communities they serve.

Almost 7,000 distinct data products are archived at and distributed from the DAACs, an archive volume of 7.4 petabytes in aggregate. These institutions are stewards of Earth science mission data until the data are moved to long-term archives. They ensure that data will be easily accessible to users.

Socioeconomic Data and Applications Center[56]

The recent priority given by USGCRP under its new strategic plan to integration of knowledge and models about both the natural and the human components of the Earth system underscores the need for access to and integration of relevant natural and social science data. Key in this effort is SEDAC, established more than a decade ago as part of EOSDIS.

SEDAC provides interdisciplinary data resources about human systems and their interactions with the environment, including data on population, urbanization, agriculture, natural hazards, public health, income distribution, infrastructure, climate change effects, natural resource management, and environmental governance. Data products and services are designed to complement remote-sensing data (e.g., by identifying population distribution relative to measures of land cover, air quality, or ice extent). SEDAC also provides spatial data sets, maps, and online mapping tools to promote data access, visualization, and analysis, as well as policy-relevant indicator data sets, including the Natural Resource Management Index, one of the indicators used by the Millennium Challenge Corporation in determining aid allocations. SEDAC is promoting interoperable access to its data products and services through GEOSS.

National Integrated Drought Information System[57]

With the likelihood of drier, warmer seasons and the possibility of increased frequency, duration, and intensity of droughts in some parts of the country in the future as a result of climate change, society is faced with the challenge of continuing to supply adequate amounts of fresh,

[53] See http://www.whitehouse.gov/sites/default/files/microsites/ostp/big_data_press_release_final_2.pdf.

[54] See https://earthdata.nasa.gov/.

[55] See http://science1.nasa.gov/earth-science/earth-science-data/data-information-policy/.

[56] See http://sedac.ciesin.columbia.edu/.

[57] See http://www.drought.gov/drought/.

clean water to growing populations. This is a particular concern in the arid U.S. Southwest, where the population has nearly doubled over the past 30 years.

Eight USGCRP member agencies are part of a federal consortium that supports NIDIS by providing scientific underpinnings, including new observing and modeling capabilities and products. NIDIS provides the best available information to enable users to determine risks associated with drought and provides supporting data and tools to inform drought mitigation. Programs such as NIDIS are crucial input to decision makers who manage scarce natural resources, particularly in the face of the large uncertainties about the pace and magnitude of future climate change. NIDIS continues to be a major contributor to GEOSS.

Global Change Information System

USGCRP is developing a new, systematic approach to global change information provision. This new approach is in response to the challenge that there is no single point of access for authoritative information on interrelated, multidisciplinary global change issues, such as the coastal impacts of sea level rise, the health costs associated with temperature extremes, and other topics with large user communities. GCIS uses linked data approaches to facilitate the needed aggregation and synthesis. As a first step, GCIS will provide data related to the forthcoming National Climate Assessment, which is scheduled for release in 2014.

Metadata Access Tool for Climate and Health[58]

Addressing the effects of climate change on human health is especially challenging, because both the surrounding environment and the decisions that people make influence health. In 2012, USGCRP began development of MATCH, an interactive clearinghouse of data sets and tools related to the human health impacts of global climate change. The MATCH project is a pilot data-integration effort that will inform development of the broader GCIS described above. It presents a publicly accessible user search interface for federal data sets, and allows for automated deposition of metadata into Data.gov and other existing federal portals.

Integrated Data and Environmental Applications Center[59]

NOAA's IDEA Center helps meet critical regional needs for ocean, climate, and ecosystem information to protect lives and property, support economic development, and enhance the resilience of Pacific Island communities in the face of changing environmental conditions. This activity integrates regional observations, research, assessment, and services, and provides a prototype for a next-generation NOAA data center, as well as strengthens the delivery of ocean, climate, and ecosystem data products and information services to the diverse Pacific Island user community. The IDEA Center supports the emergence of regional ocean- and climate-observing systems and information services that are responsive to the needs of communities, governments, and businesses via the evolving PaCIS program, and continues U.S. leadership in the emergence of a thematic and multi-purpose observing system (e.g., GCOS, GOOS, IOOS, and GEOSS).

State of the Climate Report[60]

Produced in partnership with WMO and numerous national and international partners, the annual "State of the Climate Report–Using Earth Observations to Monitor the Global Climate" combines historical data with current observations to place today's climate in historical context and provide perspectives on the extent to which the climate continues to vary and change.[61] More than 150 scientists from over 30 countries are now part of an annual process of turning raw observations collected from the global array of observing systems into information that enhances the ability of decision makers to understand the state of Earth's climate and its variation and change during the past year.

The report is published in the *Bulletin of the American Meteorological Society* each year and is translated into other languages and distributed to all 187 WMO member nations. The report provides details on as many of the ECVs as possible, as identified in the GCOS Second Adequacy Report (WMO 2003). Since this report began monitoring ECVs in 2001, and in line with the recently published 2008 edition, the number of reported ECVs has more than doubled to nearly 25.

[58] *See* http://match.globalchange.gov/geoportal/catalog/main/home.page.

[59] *See* http://www.ideademo.org/.

[60] *See* http://www.ncdc.noaa.gov/bams-state-of-the-climate/2012.php.

[61] *See* www.ncdc.noaa.gov/bams-state-of-the-climate/2011.php.

EPA is working with many other organizations to better understand the causes and effects of climate change. With help from these partners, EPA has compiled a set of 26 indicators tracking signs of climate change.[62] Most of these indicators focus on the United States, but some include global trends to provide context or a basis for comparison. These indicators represent a selected set of key climate change measurements related to GHGs, weather and climate, oceans, snow and ice, and society and ecosystems. These indicators are based on peer-reviewed data from various U.S. government agencies, academic institutions, and other organizations. EPA selected these indicators based on the quality of the data and other criteria.

TECHNOLOGY FOR GLOBAL CHANGE

The United States is committed not only to improving the science to better understand global climate change, but also to promoting the accelerated development and deployment of clean energy technologies to reduce GHG emissions. These efforts are targeted at increasing energy end-use efficiency and supplying energy with greatly reduced GHG emissions to meet the nation's goals of reducing GHG emissions and stabilizing GHG atmospheric concentrations at a level that avoids dangerous human interference with the climate system.

The 2011 DOE Quadrennial Technology Review (QTR) articulated six strategies for energy technology innovation for the nation: increasing vehicle efficiency, electrifying the fleet, deploying alternative hydrocarbon fuels, increasing building and industrial efficiency, modernizing the grid, and deploying clean electricity (Figure 8-1). The QTR affirms that DOE will only support technologies that emit less carbon than incumbents, in keeping with these national goals. The QTR also stresses the importance of investing in innovation as a means to this end.

To address this challenge, the Obama administration and Congress are working to spur a revolution in clean energy technologies. The technology research and innovation activities in this arena, which spans multiple federal agencies, can be organized into four areas for reducing emissions: using alternative fuels, decarbonizing the U.S. electricity supply, implementing end-use efficiency measures, and bolstering the contributions of basic science.

Figure 8-1 **Six Strategies for Address National Energy Challenges**

DOE's 2011 *Report on the First Quadrennial Technology Review* articulates six national strategies for U.S. energy technology innovation. In keeping with these national goals, DOE will support only technologies that emit less carbon than incumbents.

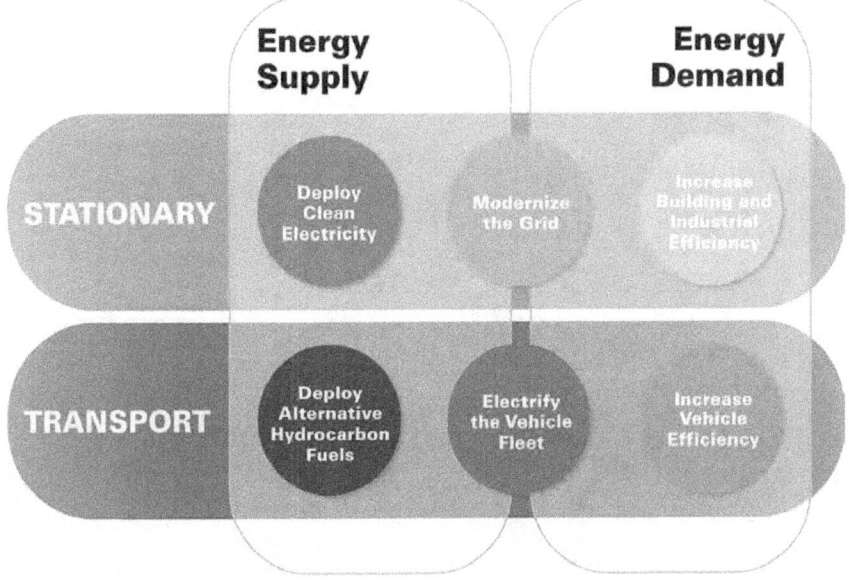

Source: U.S. DOE 2011.

[62] *See* http://www.epa.gov/climatechange/science/indicators/.

ARRA has provided more than $25 billion in additional funding for R&D activities across a broad portfolio of GHG mitigation options, including high-performance buildings; efficient manufacturing; advanced vehicles; clean biofuels; wind, solar, geothermal, hydropower, and nuclear energy; carbon capture and sequestration; advanced energy storage; a more intelligent electric grid; and techniques for reducing emissions and/or increasing uptake of CO_2 in agriculture and forestry. ARRA has also provided $400 million for establishing ARPA-E within DOE to overcome the long-term, high-risk technological barriers to the development of clean energy technologies.[63]

Alternative Fuels

The United States invests in several key pathways to reduce GHG emissions from the combustion of petroleum-derived fuels, taking a life-cycle perspective that considers both direct and indirect environmental and economic impacts. Alternative fuel options include bioenergy and hydrogen, as well as electrification of the light-duty vehicle fleet.

Bioenergy

Bioenergy R&D focuses on technologies and practices to sustainably produce biomass feedstocks and convert them to biofuels and value-added products with lower carbon intensity than petroleum-based fuels and products.[64] USDA's Biomass Research and Development Initiative and DOE's Bioenergy Technologies Office address feedstock development, biofuels, and bio-based product development, and multiple types of biomass conversion technologies that can provide drop-in replacements for gasoline, diesel, jet fuel, and other petroleum-based products.

Hydrogen and Fuel Cells

DOE's hydrogen and fuel cell R&D focuses on enabling the production of low-cost hydrogen fuel from diverse renewable pathways, addressing key challenges to hydrogen delivery and storage, and lowering the cost and improving the reliability of fuel cell technologies. Together, these efforts work to enable hydrogen-fueled vehicles to be comparable with conventional vehicles in terms of cost, convenience, and reliability (Figure 8-2).

Figure 8-2 **Hydrogen-Fueled Vehicles**

DOE's hydrogen and fuel cell R&D focuses on enabling the production of low-cost hydrogen fuel from diverse renewable pathways, such as this fuel cell vehicle powered by hydrogen fuel from renewable sources of energy.

Photo: Chris Ainscough, National Renewable Energy Laboratory.

[63] See http://arpa-e.energy.gov/.

[64] See http://www1.eere.energy.gov/biomass/.

In addition, the U.S. Department of Transportation's (DOT's) Federal Transit Administration supports research activities to improve the performance of public transportation through development, testing, and deployment of innovative technologies, such as low-emission and no-emission vehicles.[65] DOE's Vehicle Technologies Program also supports R&D to make vehicles more efficient and capable of operating on nonpetroleum fuels.[66] Other DOT programs include efforts to improve travel activity, reduce vehicle miles traveled, and enhance vehicle and system operations.

Vehicle Electrification

Vehicle electrification offers near-term efficiency gains through hybrid systems and long-term benefits as a low-emission petroleum alternative when deployed in conjunction with clean electricity generation. R&D of electric vehicles seeks to make them as affordable and convenient as today's gasoline-powered vehicles.

U.S. Electricity Supply

Global and domestic electricity generation sources are dominated by fossil fuels that emit CO_2 when burned. The transition to a low-carbon energy future will require cost-competitive, low- or zero-carbon electricity supply technologies. DOE supports R&D across a wide range of innovative low-carbon technologies in advanced fossil fuel and renewable energy, and modernization of the electric grid.

Advanced Fossil Energy, Including Carbon Capture and Storage

DOE is focused on lowering the impact of traditional fossil fuel energy production and use. The United States is actively funding applied R&D on advanced coal utilization technologies that improve efficiency and capture and store CO_2 emissions. These activities are conducted through a combination of research and demonstration programs that are primarily cost-shared partnerships between the federal government and the private sector.[67]

Carbon capture and storage (CCS) captures CO_2 emissions from stationary sources, such as power plants and factories, and permanently stores the CO_2 in the soil's subsurface. DOE classifies CCS technology as either first or second generation, or transformational. First-generation technologies are being pursued in the United States and elsewhere to demonstrate that CCS can be integrated at commercial scale while maintaining reliable, predictable, and safe plant operations. DOE currently has 16 large-scale demonstrations in this category featuring both fully integrated CCS projects and stand-alone CO_2 injections. Seven of these projects are either under construction or operating.

To reduce the cost of CCS, DOE's carbon capture research is also pursuing a new generation of solvents, solid sorbents, and membranes to greatly reduce the energy needed to separate CO_2, both for post-combustion CO_2 capture as well as for pre-combustion capture associated with coal gasification technology. In addition, enhanced oil recovery (EOR), or the process of pumping CO_2 into the ground to drive out petroleum, is being used to help enhance the economics of CCS and accelerate development of a CCS industry once a significant market incentive materializes for reducing CO_2 emissions. For DOE's first-generation demonstration projects, 12 of the 16 projects involve EOR.

Carbon storage research seeks to improve the predictability of CO_2 storage (e.g., migration and trapping of CO_2) and reduce the risk of unanticipated events (e.g., inadequate storage capacity, CO_2 leakage, induced seismicity) that could be expensive to remediate. DOE's program includes a core R&D component, as well as the Regional Carbon Sequestration Partnerships Initiative, which involves 7 partnerships, 43 states, 4 Canadian provinces, and more than 400 independent organizations. The program is entering its final, demonstration-oriented phase.

In the longer term, CCS is expected to rely on vast domestic saline and other geologic formations for CO_2 storage. When transformational CCS technologies emerge, a relatively modest "price" for CO_2 is expected to be adequate for CCS to be cost-effective without CO_2 utilization. While DOE's CO_2 capture research, development, and demonstration (RD&D)

[65] See http://www.nrel.gov/hydrogen/proj_fc_bus_eval.html.

[66] See http://www1.eere.energy.gov/vehiclesandfuels/.

[67] See http://www.netl.doe.gov.

historically has focused primarily on coal power plants, most of the innovations under investigation (and everything related to CO_2 storage) are equally applicable to large stationary facilities that use natural gas. Advanced concepts under study may be particularly effective for natural gas.

USGS has been playing a major role in the national assessment of geologic CO_2 storage resources. Several USGS assessment products have been completed since 2009, including an assessment methodology for hydrocarbon recovery potential using CO_2 and associated carbon sequestration, CO_2 fluid-flow modeling and injectivity calculations, and implementation of the methodology for the entire United States.

The Carbon Sequestration Leadership Forum (CSLF) is a multilateral U.S. initiative that provides a framework for international collaboration on sequestration technologies. The CSLF's main focus is promoting the development of improved cost-effective technologies for the separation and capture of CO_2 for its transport, utilization, and long-term safe storage. The CSLF seeks to make these technologies available internationally, and identify and address broader issues relating to carbon capture, utilization, and storage.

Nuclear Energy

A key mission of DOE's nuclear energy R&D program is to plan and conduct applied research in advanced reactor and fuel and waste management technologies. The aim of these efforts is to enable nuclear energy to be used as a safe, advanced, cost-effective source of reliable energy that will help address climate change by reducing GHG emissions. Small modular reactor designs offer attractive safety, manufacturing, and operational innovations that can be available in the next decade. DOE is investigating the next-generation reactor and fuel-cycle systems, which could represent a significant leap in economic performance, safety, and proliferation resistance.

Renewable Energy

The United States has abundant renewable energy resources. In recent years, enabling policies at the state and federal levels have driven rapid deployment of renewable electricity generation capacity. The combined impacts of private-sector investments and publicly funded R&D are continuing to push down the cost of renewable electricity technologies and improve their performance.

The federal government invests in a broad portfolio of renewable electricity technologies, including solar, wind, geothermal, and water power, with the goal of making cost-competitive renewable electricity options available in every region of the country (Figure 8-3). Some examples of these activities follow.

Solar—The DOE SunShot Initiative[68] is a national collaborative to make solar energy cost-competitive with other forms of electricity by 2020, reducing solar energy systems by 50–75 percent from 2010 baseline costs. With rapid photovoltaic module cost declines experienced in recent years, a key challenge is reducing nonhardware costs, such as permitting and installation, which can now account for more than 50 percent of a system. For concentrating solar power, R&D targets advanced thermal storage technologies to enable solar energy to provide electricity that can be dispatched when needed.

Wind—Wind power R&D by DOE's Wind Technologies Office works on advances in new wind energy system designs and technologies to increase energy capture, reliability, and survivability for reduced life-cycle costs for land-based and offshore wind turbines. Next-generation advanced rotors can enable higher wind turbine blade tip speeds with lower acoustic emissions. System-level research can lead to substantial efficiency gains, for instance, by understanding complex wind plant aerodynamics to improve overall wind plant capacity factors. DOE's Wind and Water Power Technologies Office and the U.S. Department of the Interior's Bureau of Ocean Energy Management are working to advance a coordinated strategy for offshore wind research and development.[69]

---

[68] *See* http://www1.eere.energy.gov/ solar/sunshot/.

[69] *See* eere.energy.gov/analysis/pdfs/ winds_energy_r_and_d_linkages.pdf; wind.energy.gov/; and http://www1. eere.energy.gov/windandhydro/.

Geothermal—DOE develops innovative technologies to locate, access, and develop the nation's substantial geothermal resources by advancing (1) hydrothermal power production, where fluid flow and hot rock occur naturally, and (2) enhanced geothermal systems technologies, where fluid is injected into deep, hot rock formations to create a geothermal reservoir. Development risks and costs are key barriers, and DOE's Geothermal Technologies Office supports innovative technologies for resource development and demonstrations that enable field testing and validation, committed to demonstrating ways to achieve sustained, enhanced geothermal reservoirs.

Water—Water power investments by DOE's Wind and Water Power Technologies Office enable the development of innovative technologies and improve the reliability and technology readiness of marine and hydrokinetic systems using ocean wave, current, and tidal resources. Collaborations with industry and federal agencies are working to accelerate the development and deployment of sustainable hydropower technologies utilizing domestic river, stream, and water conveyance system resources for clean generation.

Grid Modernization

Grid modernization is a key component in the transition to a cleaner supply of electricity. Improving the infrastructure of the electricity transmission and distribution grid can reduce GHG emissions by making power delivery more efficient and by enabling higher penetrations of low-emission electricity from renewable energy. Key research activities include DOE's nationwide plan to modernize the electric grid, enhance the security of the U.S. energy infrastructure, and ensure reliable electricity delivery to meet growing demand. The emphasis is on developing advanced transmission technologies, including advanced sensors and monitors, thereby strengthening the reliability of the electric grid by enabling wide-area situational awareness through real-time measurement of the system, advancing real-time visualization and operational support tools, and developing a "smart grid" system with enhanced intelligence and connectivity. In addition, DOE is investing in advanced technology research, including smart grid devices and infrastructure and more efficient grid storage and microgrids to improve resiliency. These improvements will reduce GHG emissions and increase U.S. energy independence and economic growth.

U.S. Energy End Use

Major U.S. sources of GHGs are closely tied to the use of energy in transportation, residential and commercial buildings, and industrial processes. Improving energy efficiency and reducing the intensity of GHG emissions in these sectors can significantly reduce overall GHG emissions. DOE invests in R&D for technologies that enable high-performance buildings, advance clean and efficient industrial technologies and processes, and create more efficient transportation options. These investments will significantly reduce both U.S. energy consumption and domestic and global GHG emissions.

Residential and Commercial Buildings

The Emerging Technologies Program within DOE's Building Technologies Office partners with national laboratories, industry, and universities to advance research, development, and commercialization of energy-efficient, cost-effective building technologies that could be

Figure 8-3 **U.S. Renewable Energy**

The United States is capitalizing on its abundant renewable energy resources, including this photovoltaic array and wind turbines at the National Wind Technology Center near Boulder, Colorado.

Photo: Dennis Schroeder, National Renewable Energy Laboratory.

market-ready in less than five years. Areas of research include commercial and residential building appliances; building envelope, windows, skylights, and doors; space heating and cooling; solid-state lighting; building sensors and controls; and building energy modeling (Figure 8-4).

Industry

DOE's Advanced Manufacturing Office (AMO) works with diverse partners to develop and deploy next-generation manufacturing technologies and processes that will help U.S. manufacturers succeed in global markets. The goal of AMO is to reduce the life-cycle energy consumption of manufactured goods by 50 percent over 10 years for supported technologies, compared with conventional manufacturing processes, and encourage a culture of continuous improvement in manufacturing energy efficiency.

AMO is working toward this goal through several initiatives, including the R&D of advanced manufacturing process and materials technologies. DOE is also supporting innovation through the establishment of Clean Energy Manufacturing Innovation Institutes; the Critical Materials Hub; and Manufacturing Demonstration Facilities, which provide American small and medium-sized enterprises, in addition to large businesses, timely and affordable access to cutting-edge physical and virtual advanced tools. At the same time, DOE works to increase American competitiveness in clean energy manufacturing, by strategically investing in technologies that leverage American competitive advantages and overcome competitive disadvantages.

Transportation

Transportation R&D by DOE's Vehicle Technologies Office focuses on reducing the cost and improving the performance of a mix of near- and long-term vehicle technologies, including advanced batteries, power electronics and electric motors, light-weight and propulsion materials, advanced combustion engines, advanced fuels and lubricants, and other enabling technologies. Research partnerships with industry leverage technical expertise, prevent duplication, ensure public funding remains focused on the most critical barriers to technology commercialization, and accelerate progress.

The DOE SuperTruck Initiative aims to develop technologies to improve the fuel economy (freight-hauling efficiency) of heavy-duty, class 8 vehicles by 50 percent by 2015, compared with a comparable 2009 vehicle. SuperTruck project teams are using a variety of approaches to meet this goal, and have made significant progress in the areas of engine efficiency and emission control, advanced transmissions and hybridization, aerodynamic drag of the tractor and trailer, tire rolling resistance, light-weight materials, and auxiliary power units to reduce engine idling (Figure 8-5).

Aviation activity is another source of GHG emissions. To identify opportunities for GHG emission reductions in the aviation sector, DOT's Federal Aviation Administration (FAA) launched the Aviation Climate Change Research Initiative (ACCRI). ACCRI research helps to assess emission-reducing improvements in aircraft and engine technology, operational procedures, and the airspace management system by measuring and tracking fuel efficiency from aircraft operations. FAA's Commercial Aviation Alternative Fuels Initiative is a government–private-sector coalition that works to bring commercially viable, environmentally friendly alternative aviation fuels to market.[70] With support from NASA, FAA launched the Continuous Lower Energy Emissions and Noise Program to advance maturing engine and aircraft technologies for quick inclusion into the U.S. aviation fleet, to increase fuel efficiency (which is directly related to CO_2 emissions), and to reduce nitrogen oxide emissions (which affect distributions of ozone and methane.)

These strategies to improve the transportation system can reduce GHG emissions, lead to environmental benefits, reduce oil use, improve America's energy security, and benefit the U.S. economy. Other DOT programs include efforts to improve travel activity, reduce vehicle miles traveled, and enhance vehicle and system operations.

---

[70] See http://www.caafi.org/.

Basic Science

Basic scientific research is a fundamental element of DOE's efforts, supported by President Obama's commitment to increased federal investment in this area. Tackling the dual challenges of addressing climate change and meeting growing world energy demand is likely to require discoveries and innovations that can shape the future in often unexpected ways. DOE's approach aims to strengthen the basic research enterprise through strategic research that supports ongoing or future activities and exploratory research involving innovative concepts.

DOE supports three multidisciplinary BRCs that conduct fundamental research underpinning the development of advanced sustainable biofuel production strategies: improvements in plant feedstocks, plant deconstruction, and fuel synthesis. DOE core research in genomic sciences also includes biosystem design tools and biodesign technologies for bioenergy research, and advances a predictive understanding of the design, function, and regulation of plants, microbes, and biological communities contributing to the cost-effective production of next-generation biofuels as a major secure national energy resource.

Figure 8-4   **Energy-Efficient, Cost-Effective Building Technologies**

Daylighting, natural ventilation design, solar water heating, and rainwater reuse systems are among the technologies employed at this commercial building in Annapolis, Maryland.

Photo: Robb Williamson, National Renewable Energy Laboratory.

Figure 8-5   **Supercomputing Simulations**

Supercomputing simulations have enabled engineers to develop a system that dramatically reduces drag and increases fuel mileage in trucks.

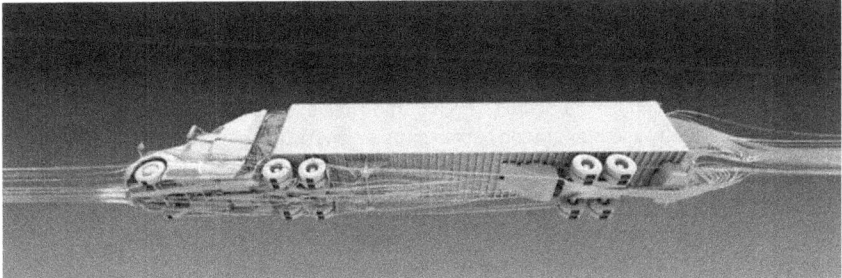

Photo: Michael Matheson, Oak Ridge National Laboratory.

In addition, DOE will continue to support a number of EFRCs that are addressing current fundamental scientific roadblocks to clean energy and energy security.[71] These centers address a range of energy research challenges in renewable and low-carbon energy, energy efficiency, energy storage, and cross-cutting science. The EFRCs are taking advantage of new capabilities in nanotechnology, light sources and neutron scattering sources, supercomputers, and other advanced instrumentation.

DOE's multidisciplinary Energy Innovation Hubs are also addressing basic science, technology, and economic and policy issues. The hubs support cross-disciplinary R&D focused on the barriers to transforming energy technologies into commercially deployable materials, devices, and systems. Current hubs focus on fuels from sunlight, energy-efficient buildings, modeling for nuclear reactors, critical materials, and batteries and electrical energy storage. These hubs are advancing promising areas of energy science and technology from their early stages of research to the point where the risk of investing in them will be low enough for industry to deploy them into the marketplace.

Established by DOE in 2009, ARPA-E is modeled after the Defense Advanced Research Projects Agency, created during the Eisenhower administration in response to the Russian Sputnik program, which launched the world's first artificial satellite. The purpose of ARPA-E is to advance high-risk energy research projects that can yield revolutionary changes in how energy is produced, distributed, and used.[72] ARRA has provided $400 million for ARPA-E, and the program received funding for 2010–2013 that greatly expanded the number of projects it supports.

Multilateral Research and Collaboration

The United States believes that well-designed multilateral collaborations focused on achieving practical results can accelerate development and commercialization of new technologies. Thus, the United States has initiated or joined a number of multilateral technology collaborations in hydrogen energy, carbon sequestration, nuclear energy, and fusion that address many energy-related concerns (e.g., energy security, climate change, and environmental protection). The following initiatives are examples of U.S. multinational collaboration.

Carbon Sequestration Leadership Forum

The CSLF is a multilateral U.S. initiative that provides a framework for international collaboration on sequestration technologies.[73] Established at a June 2003 ministerial meeting held in Washington, D.C., the CSLF consists of 23 members, including 22 national governments representing both developed and developing countries, as well as the European Commission. The CSLF's main focus is assisting the development of technologies to separate, capture, transport, and store CO_2 safely over the long term; making carbon sequestration technologies broadly available internationally; and addressing broader issues, such as regulation and policy. To date, the CSLF has endorsed 20 international research projects, five of which involve the United States.

ITER

In 2003, the United States joined the negotiations for agreeing on the construction and operation of ITER, an international experiment to design and build a fusion reactor.[74] The goal of this collaborative project is to demonstrate the scientific and technological feasibility of fusion as an energy source. If successful, ITER will advance progress toward producing clean, abundant, commercially available fusion energy by the end of the 21st century.

Toward this goal, the seven ITER partners signed an agreement in November 2006 to build the project; site preparation began in Saint-Paul-lez-Durance, France, in January 2007; and civil construction began in 2009 and continues today. The ITER Agreement established the ITER Organization, a public international organization managed by a Director General, as the ITER design authority and operator on behalf of the seven partner members. ITER has approximately 500 full-time staff.

The ITER Organization has secured nuclear regulatory approval for full facility construction. Fabrication of in-kind components by ITER members is accelerating. U.S. long-lead, early-delivery items are currently being fabricated, and. some U.S. in-kind components have been delivered.

[71] See http://.science.energydoe.gov/bes/efrc/.

[72] See http://arpa-e.energy.gov/.

[73] See http://www.cslforum.org/.

[74] See http://www.iter.org/.

Major Economies Forum Global Partnership

At the 2009 Group of Eight (G8) meeting in L'Aquila, Italy, the Major Economies Forum countries (G8 plus China, India, South Africa, Brazil, Mexico, and Indonesia) announced a global partnership for transformational low-carbon, climate-friendly technologies. The partners committed to significantly increase and coordinate public-sector investments in RD&D of these technologies. The partnership's ultimate goal is to double these investments by 2015, while recognizing the importance of private investment, public–private partnerships, and international cooperation, including regional innovation centers.

The United States will lead on "efficiency," which includes commercial and residential buildings and the industrial sector. Technology action plans and roadmaps will be developed, along with recommendations for further progress. Drawing on global best practice policies, the Global Partnership will strive to remove barriers to, establish incentives for, and enhance capacity building of U.S. climate-friendly technologies, and implement appropriate measures to aggressively accelerate deployment and transfer of key existing and new low-carbon technologies, in accordance with national circumstances.

Education, Training, and Outreach

"Understand this is not just a job for politicians. So I'm going to need all of you to educate your classmates, your colleagues, your parents, your friends. Tell them what's at stake. Speak up at town halls, church groups, PTA meetings. Push back on misinformation. Speak up for the facts. Broaden the circle of those who are willing to stand up for our future."[1]

—President Barack Obama

In 2012, the U.S. Global Change Research Program (USGCRP) expanded its mission statement to include education as a critical component of the nation's response to global change. This new mission articulates USGCRP's role in addressing the mandated scope of the Global Change Research Act of 1990 over the next decade: "To build a knowledge base that informs human responses to climate and global change through coordinated and integrated Federal programs of research, education, communication, and decision support." The resulting USGCRP strategic plan emphasizes better integration of social, ecological, and physical sciences to understand changing conditions, increased utilization of scientific information and knowledge, and better communication and education (USGCRP 2012b).

The increased strategic focus of the federal government and its partners on climate change communication and education programs in the United States seeks to promote a deeper understanding of the science of climate change, behavioral change, and stewardship, and to support informed decision making by individuals, organizations, and institutions—all of which are summarized under the term "climate literacy."[2] The ultimate goal of climate literacy is to enable individuals, businesses, and communities to address climate change, in terms of stabilizing and reducing emissions of greenhouse gases (GHGs), and also increasing capacity to adapt to and prepare for the consequences of climate change.

U.S. educational efforts focus on three distinct, but related, areas: the science of climate change, the human–climate interaction, and using climate education to promote behavioral change. Each of these approaches is represented in the *Atlas of Science Literacy* (AAAS and NSTA 2007) and in the conceptual framework for science education developed at the National Research Council (NRC) in 2011 (Quinn et al. 2013). These approaches also informed the development of the *Next Generation Science Standards for Today's Students and Tomorrow's Workforce*—an innovative way to address climate change education within the decentralized U.S. education system (Figure 9-1) (NAS et al. 2013).

Climate change communication faces many challenges. Federal agencies, civil society, and individuals have invested in numerous initiatives to develop a climate-literate citizenry and skilled workforce. The authors of *America's Climate Choices* found that although "climate change is difficult to communicate by its very nature, ... education and communication are among the most powerful tools the nation has to bring hidden hazards to public attention, understanding, and action" (NRC 2011).

[1] President Barack Obama's speech at Georgetown University announcing his new climate change policy, June 25, 2013. See http://www.georgetown.edu/landing/1242711958096.html.

[2] See http://www.climate.gov/teaching/teaching-climate-literacy-and-energy-awareness.

Figure 9-1 **University of Maryland Awarded Grant for Renewable Energy Systems**

UMD was selected as a Maryland Energy Administration Project Sunburst Initiative Partner and awarded a grant aimed at promoting the installation of renewable energy systems on public buildings in Maryland. This photo shows a part of the Severn Solar Array, which was installed in 2011 with more than 2,600 solar panels.

Photo: Frances Avendano.

Numerous federal agencies, nongovernmental organizations (NGOs), and individuals have supported sustained and robust educational and communication initiatives to harness these tools. When citizens have knowledge of the causes, likelihood, and severity of climate impacts, as well as of the range, cost, and efficacy of options to adapt to impacts, they are more prepared to effectively address the risks and opportunities of climate change. Furthermore, since 2010, more Americans than ever before have experienced the impacts of climate change first-hand in the form of extreme events, such as Superstorm Sandy and prolonged drought, resulting in increased public interest in and an opportunity for engagement on climate literacy issues.

UPDATES SINCE THE 2010 *U.S. CLIMATE ACTION REPORT*

Climate change education, training, and outreach efforts have matured significantly since the *U.S. Climate Action Report 2010* (2010 CAR) (U.S. DOS 2010), even in the recently constrained budgetary environment. Since the 2010 CAR, federal programs that support formal educational initiatives on climate change have begun to develop a coordinated national network of regionally or thematically based partnerships devoted to increasing the adoption of effective, high-quality educational programs and resources related to the science of climate change and its impacts. These programs involve kindergarten through grade 12 (K–12) and undergraduate curricula and postgraduate professional development programs, as well as informal education programs conducted in museums, parks, nature centers, zoos, and aquariums across the country.

Federal Program Coordination

Federal agencies coordinate climate change educational efforts through USGCRP and other cross-cutting initiatives. USGCRP, which coordinates and integrates climate research across 13 government agencies, included education in its 10-year strategic plan (USGCRP 2012b). USGCRP has committed its focus over the next decade not only to encouraging greater public understanding of the science through the dissemination of relevant, timely, and credible global change information, but also to gaining further understanding of the public's science and information needs through engagement and dialogue. This two-pronged approach will help decision makers at all levels to make informed decisions. This strategy is being implemented through the integration of communication, education, and engagement into core USGCRP activities.

As the leading federal authority on global change science, USGCRP, together with its member agencies, is uniquely positioned to serve as the gateway to global change information for the nation, and has taken a leadership role in the development of the scientific workforce of the future. Many other federal agencies, such as the U.S. Environmental Protection Agency (EPA), National Park Service (NPS), National Oceanic and Atmospheric Administration (NOAA), and National Institute of Food and Agriculture (NIFA), also have the capacity to communicate with citizens on specific aspects of global change related to their respective missions. Many of these agencies have supported educational institutions in developing a pipeline of the scientific workforce relevant to global change.

While individual agency actions are important and their contributions in the aggregate are significant, one of the greatest strengths of USGCRP is its ability to develop synergies across federal agencies to coordinate efforts in communication and education. The USGCRP strategy for communication, education, and engagement efforts over the next decade will build on the strengths of the participating agencies. USGCRP will coordinate the development of multi-agency products and programs, grow and expand the reach of information beyond single agencies, and ensure that feedback from public engagement is shared broadly within the federal global change science community.

The coordination in climate change communication and education across the federal departments and programs contained in the 2010 CAR has continued through the USGCRP Communication and Education Interagency Working Group. This group develops a national climate change education communication strategy that includes all USGCRP members, and coordinates climate education, communication, and engagement activities and priorities across the USGCRP members.

For example, the National Science Foundation (NSF), the National Aeronautics and Space Administration (NASA), and NOAA coordinated in the Tri-Agency Climate Change Education grant effort.[3] In another example, discussions among NSF, NOAA, NIFA, EPA, and NASA in 2009 led to the development of the NSF Climate Change Education Partnership (CCEP) Program, which develops transdisciplinary collaborations among climate scientists, learning scientists, and education practitioners working in formal and informal learning environments, discussed in more detail below.

Sample Partnerships in Climate Change Education

In fiscal year (FY) 2010, NSF launched an innovative science education program focused on educating students, teachers, and the public about global climate change and its impacts. Structured as a two-phase competition, the CCEP Program established new transdisciplinary collaborations among climate scientists, learning scientists, and education practitioners working in formal and informal learning environments. Numerous federal agencies partner with these NSF-funded projects, including NASA, NOAA, and the U.S. Department of the Interior (DOI). The following initiatives are examples of federal partnerships in climate change education.

Climate Literacy Zoo Education Network

The NSF-funded Climate Literacy Zoo Education Network (CLiZEN) highlights some of the important results of CCEP. The overarching purpose of CLiZEN was to develop and evaluate a new approach to climate change education that connects zoo visitors to polar animals currently endangered by climate change, leveraging the associative and affective pathways known to dominate the general public's decision making. CLiZEN built on interagency principal investigator meetings, and the NOAA-funded research on American attitudes about the ocean and climate change (Boyle and Mott 2009).

[3] See http://gcce.larc.nasa.gov/trace/trace_catalog.php and https://nice.larc.nasa.gov/tri_pi/.

Utilizing a polar theme, the network brings together a strong multidisciplinary team led by the Chicago Zoological Society, with a geographically distributed consortium of nine partners: Columbus Zoo & Aquarium, Ohio; Como Zoo & Conservatory, Minnesota; Indianapolis Zoo, Indiana; Louisville Zoological Garden, Kentucky; Oregon Zoo, Oregon; Pittsburgh Zoo & PPG Aquarium, Pennsylvania; Roger Williams Park Zoo, Rhode Island; Toledo Zoological Gardens, Ohio; and the Polar Bears International.

The project's long-term vision focuses on the development of a network of U.S. zoos, in partnership with climate change domain scientists, learning scientists, conservation psychologists, and other stakeholders, that fosters changes in public attitudes, understanding, and behavior surrounding climate change. This vision was captured in the e-book *Climate Change Education: A Primer for Zoos and Aquariums* (Grajal and Goldman 2012). Much of this work has been continued by the NSF-funded National Network for Ocean and Climate Change Interpretation.[4]

NSF, NOAA, and NASA Grant Collaboration

Since FY 2009, NSF has also participated in a multi-agency effort to coordinate U.S. government investments in climate change education through a collaboration with NOAA and NASA, which also have grant programs related to climate and environmental education. The three agencies now jointly convene annual meetings of the awardees of their respective grant programs—representing more than 120 projects—to share insights, resources, tools, and strategies. This event has provided a crucial mechanism for coordination, and has enhanced learning among practitioners of climate change education at a range of levels.

Climate Change Education Roundtable

To support and strengthen these education initiatives, and in response to a 2009 congressional mandate connected to NSF's funding for a climate change education program, NRC's Board on Science Education, in collaboration with the Committee on Human Dimensions of Global Change and the Division on Earth and Life Studies, created the Climate Change Education Roundtable.[5] The roundtable provides a forum for dialogue among practitioners and experts in multiple disciplines relevant to climate change education. It facilitates collaboration among federal agencies and private organizations, helping to promote unique contributions and align overall education strategies. Two NRC Roundtable reports provide significant input for this chapter:

- *Climate Change Education: Goals, Audiences, and Strategies: A Workshop Summary* (Forest and Feder 2013) and

- *Climate Change Education: Formal Settings, K–14: A Workshop Summary* (Beatty et al. 2013).

Table 9-1 at the end of this chapter presents an extensive listing of federal agencies' online, climate-relevant education resources.

Climate Literacy and Energy Awareness Network[6]

CLEAN is an important community-based informal network of scientists, educators, policymakers, community leaders, students, and citizens who are engaged in fostering climate and energy literacy in the United States and abroad. CLEAN provides a forum for organizations, agencies, and individuals to collaborate for climate education. Members share ideas, coordinate efforts, promote policy reform, develop learning resources, and support integration of climate literacy into formal and informal education venues. Initiatives of CLEAN feature accurate scientific information, engaging learning experiences, and multiple formal and informal pathways to reach broad and diverse audiences.

National Efforts to Engage Americans on Climate Change

Since the publication of the 2010 CAR, NGOs and federal, state, and local governments have conducted major communications campaigns to raise awareness and educate the nation about a variety of climate issues. As noted above, this chapter focuses on federal efforts, and is therefore not an exhaustive compilation of all of these actions.

---

[4] See http://support.neaq.org/site/PageNavigator/prof_devel_study_circle.html.

[5] See http://sites.nationalacademies.org/DBASSE/BOSE/CurrentProjects/DBASSE_072014#.UgTobfmR-So.

[6] See http://cleanet.org/clean/community/cln/index.html.

Connecting the Dots between Climate Change and Extreme Weather Events

The extreme weather events in the United States over the last four years have presented perhaps the most effective educational opportunities.

In 2012, the nation was struck by 11 individual weather and climate disasters with impacts of at least $1 billion. Cumulatively, these 11 events resulted in more than $110 billion in damages and 377 deaths, and directly affected major population centers and key industries and economic sectors.

The impact of these events on Americans' perceptions of climate change is described in the April 2013 report *Extreme Weather and Climate Change in the American Mind* (Leiserowitz et al. 2013). This report notes that 85 percent of Americans stated that they experienced one or more types of extreme weather in the past year. Additionally, 6 in 10 Americans (58 percent) believe global warming is affecting U.S. weather.

Superstorm Sandy provides insights into how extreme events have increased Americans' eagerness to learn more about climate change and how the U.S. government has leveraged this interest. On October 25, 2012, extratropical Hurricane Sandy struck the Mid-Atlantic states of New Jersey, New York, Connecticut, and Rhode Island. As a result, the national conversation regarding climate changed dramatically.

The nation's educators and communicators have been working with federal Web portals—e.g., NOAA's Climate.gov,[7] NASA's Climate Portal,[8] EPA's Climate Change Portal,[9] and the Climate Change Indicators in the United States site,[10] and NGOs like Climate Nexus,[11] Climate Access,[12] and Climate Central[13]—to help citizens connect the dots between climate change and extreme weather events in scientifically correct and meaningful ways. As extreme events continue to increase, these sorts of combined efforts will be needed to better serve the public's need for timely and trusted scientifically based information about how such extreme events may change in frequency or intensity in the future, and what people can to prepare for and become more resilient to their impacts.

Capitalizing on Public Survey Research

During the past four years, numerous organizations and federal programs have used public survey research on beliefs and attitudes from Yale University,[14] George Mason University,[15] and elsewhere to differentiate their climate and global change education and communication projects. As a result, these programs realize that people actively interpret information and construct their own mental models based on what they personally know, value, and feel. Using this research, the U.S. climate and global change communication and education community can be much more strategic in designing and implementing programs with limited resources.

Developing Data-Driven, User-Friendly Web Sites

To support growing public requests for meaningful and timely scientific information regarding climate and extreme weather, NOAA developed Climate.gov to provide climate data and information to help build a climate-smart nation. This user-friendly, online source of timely and authoritative scientific data and information about climate is designed to serve four segments of the public: the science-interested public, scientists and specialists, formal and informal educators, and planners and policy leaders.

Since the site's prototype launch in 2010, the Climate.gov team has engaged in direct dialogue with data users and site visitors in the public and private sectors. The Web analytics from Climate.gov show significant visit spikes after each high-impact extreme event, similar to other climate change Web sites.

In May 2013, Climate.gov was redesigned based on user feedback for each of the four main audiences. New data browse and access tools, such as the Global Climate Dashboard and the Integrated Map Application, make it easier for visitors to find and use climate data. The site's scope of contents has also expanded to serve hundreds of educational resources, decision-support tools, articles, and videos.

[7] See http://www.climate.gov/.

[8] See http://www.climate.nasa.gov/.

[9] See http://www.epa.gov/climatechange/.

[10] See http://www.epa.gov/climatechange/science/indicators/.

[11] See http://www.climatenexus.org/.

[12] See http://www.climateaccess.org.

[13] See http://www.climateccentral.org.

[14] See http://environment.yale.edu/climate-communication/.

[15] See http://www.climatechangecommunication.org/.

Figure 9-2 **2000-2013 U.S. Newspaper Coverage of Climate Change or Global Warming**

The Center for Science and Technology Policy Research at the University of Colorado has tracked media coverage of climate change since 2000. Researchers there saw a worldwide uptick across all media in 2012 in Europe, Asia, Africa, and South America and the five largest U.S. daily newspapers.

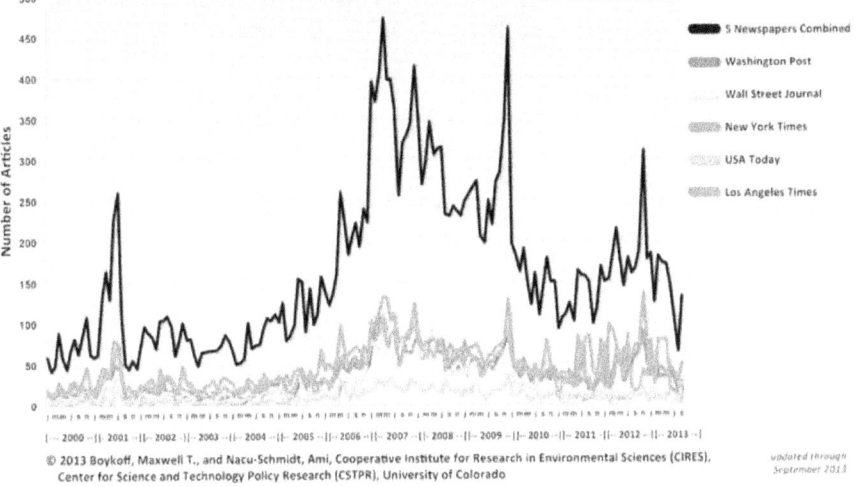

© 2013 Boykoff, Maxwell T., and Nacu-Schmidt, Ami, Cooperative Institute for Research in Environmental Sciences (CIRES), Center for Science and Technology Policy Research (CSTPR), University of Colorado

updated through September 2013

Increasing Media Coverage

In 2009, news media coverage of climate change increased substantially (Figure 9-2). Recent studies on the role of mass media in communicating climate science, mitigation, and adaptation have been mixed or more positive. The Center for Science and Technology Policy Research[16] at the University of Colorado has tracked media coverage of climate change since 2000. Researchers there saw a worldwide uptick across all media in 2012 in Europe, Asia, Africa, and South America and the five largest U.S. daily newspapers (Fisher 2013).

Developing Next-Generation Science Standards

One of the most significant advances in K–12 climate change educational efforts is the Next-Generation Science Standards (NGSS) for teaching science in the United States (NAS et al. 2013).[17] Developed in collaboration with 26 states and several scientific organizations, these transformative guidelines for the first time recommend climate change as a core concept for U.S. science curricula, including an emphasis on anthropogenic or "human-caused" effects in both middle and high school science standards.

In the next four years, significant work in educator professional development and curricular design is planned to support this critical advancement in the nation's climate education. States across the nation have begun to adopt NGSS, which will improve overall climate literacy among all Americans, and build in the next generation a firm foundation of knowledge and discourse as the nation faces decisions on how to best deal with a changing climate.

Preparing for the Challenges and Opportunities of Climate Change

Higher education also has a key role to play in developing graduates with the skills, background, and knowledge to meet the challenges of climate change. A 2010 report by the Association for the Advancement of Sustainability in Higher Education (AASHE) calls for "ensuring that all students in higher education have access to education for sustainability and opportunities to learn how to participate in and to lead the sustainability transformation" (AASHE 2010). Over the last 20 years, scholars, activists, and others have noted that through the research they conduct, their engagement with the broader community, and the operations they oversee, colleges and universities can serve as test sites and models for sustainable practices and societies. Where colleges and universities may have the largest impact, however, is with the students they educate.

[16] *See* http://sciencepolicy.colorado.edu/media_coverage/us/index.html.

[17] *See* http://www.nextgenscience.org/.

Through the leadership of AASHE, ecoAmerica, Second Nature, and the American College & University Presidents' Climate Commitment's (ACUPCC's) 665 signatory institutions, higher education is beginning to provide college and university graduates with the skills, background, knowledge, and habits of mind that will prepare them to meet the challenges presented by climate change. ACUPCC signatories continue their ongoing efforts to publicly report on progress made to eliminate operational GHG emissions and to provide the education, research, and community engagement to enable the rest of society to do the same.

The ACUPCC Reporting System allows signatories to track, assess, and communicate progress to their campus community and beyond, demonstrating to prospective students, foundations, and potential private-sector partners that their institution is serious about its commitment to climate change and sustainability. Since the last data summary in June 2012, the number of Progress Reports on Climate Action Plans has increased from 240 to 306, providing significantly more data to draw from and demonstrating continued growth in climate and sustainability action.

To date, 68 percent of the 306 institutions that submitted a Progress Report have affirmed that their Climate Action Plan has helped them realize significant financial savings, including $119 million in savings from implemented projects. Another 137 signatories reported that they have secured funding from outside resources totaling more than $305 million to implement climate and sustainability efforts. ACUPCC signatories are building institutional capacity to foster career preparedness for their students through curriculum development, securing funding for and from climate and sustainability efforts and advancing innovation through institutional research (Figure 9-1).[18]

Audience Segmentation Strategies

The United States is using audience segmentation to prioritize strategies for communication and education about climate change, as demonstrated in the report *Climate Change Education: Goals, Audiences, and Strategies: A Workshop Summary* (Forest and Feder 2013). One of the key steps in ensuring the effective use of communication and education practices is to know the target audiences—who they are, what they already know, how they learn, and their preferred methods of communication and education.

Studies have found that different audiences have different information gaps and misconceptions and want to know different things. To this end, U.S. federal, state, and NGO programs have identified high-priority audiences, like formal educators, informal educators (e.g., weather forecasters), and decision makers. This outreach helps convey clear and concise information through appropriate communication and education channels. Following are some examples of programs using audience segmentation.

NOAA Climate.gov portal

The NOAA Climate.gov portal used an audience-focused approach to refine its design, enhance its functionality, and expand its scope of contents in response to user feedback. NOAA defines the "public" as any nongovernmental segment of society that can be characterized by its specific need for climate information and services, and its information-seeking behaviors. NOAA's Climate Literacy Objective targets six priority publics: (1) decision makers and policy leaders, (2) scientists and applications-oriented data users, (3) educators, (4) students and lifelong learners, (5) journalists and TV meteorologists, and (6) the climate-interested public.

National Wildlife Federation

The National Wildlife Federation (NWF) engages leaders in influential communities as voices for both personal and civic actions on climate and broader policy reforms (Coyle 2010). From 2007 to 2010, NWF trained 5,000 leaders in climate education from selected constituent groups. The training programs reflected lessons learned from a previous effort focused on hunters and anglers. Based on this success, NWF staff used survey research to identify and develop training aligned with the cultural sensitivities, conceptual frames, and informational needs of several other constituencies. Training was targeted to the unique interests and concerns of environmental and civic activists, master gardeners, conservative faith-based organizations, watershed conservationists, land trust leaders, birders, university groups, coastal wetland conservation organizations, and business leaders.

[18] See more at http://www.secondnature.org/blog/2013-04-04/second-nature-applauds-unprecedented-progress-made-signatories#sthash.SiBUHXe2.dpuf.

Interfaith Power and Light

Interfaith Power and Light (IPL), the largest faith-based climate change organization in the United States, works with more than 10,000 congregations in 38 states. The community of faith-based organizations is growing to include the National Religious Partnership for the Environment, the National Council of Churches Eco-Justice Programs, the Evangelical Environmental Network, and the Coalition on the Environment in Jewish Life. IPL has identified several key barriers to the acceptance of climate change information in faith-based audiences. State directors of IPL also reported success across audiences using messages framed in terms of certain values, including stewardship and eco-justice.

Center for Climate and Energy Solutions

The Center for Climate and Energy Solutions' (C2ES') primary mission is to engage the business community on climate change issues, by providing credible information and workable solutions and framing appropriate messages. C2ES programs have found that although climate change remains a polarizing issue in the United States, there are ways to communicate effectively about the challenges and engage government, business, and individuals in finding solutions. C2ES has found that peer-to-peer learning is very effective for climate change education.

Effective education and communication efforts directed toward the public and decision makers are interactive and ongoing. Effective programs allow for feedback of shared knowledge, provide a forum for sustained discussions of climate change impacts, and build trust between the public and policymakers. Decision makers reflect community values, needs, and interests. Recent U.S. climate education, communication, training, and engagement allow the public and policymakers to engage in a dialogue in which all viewpoints are understood and considered.

FEDERAL AGENCY EDUCATION, TRAINING, AND OUTREACH PROGRAM OVERVIEWS

A significant number of federal agencies provide state and local governments, industry, NGOs, and the public with information about national and global climate change research and risk assessments studies, U.S. mitigation activities, and policy developments. They work both independently and in partnership with other agencies, NGOs, and industry toward the common goal of increasing awareness and understanding about the potential environmental and societal challenges posed by climate change and opportunities for solutions. As President Obama said in the June 25, 2013, release of his *Climate Action Plan*: "We've got to look after our children; we have to look after our future; and we have to grow the economy and create jobs. We can do all of that as long as we don't fear the future; instead we seize it [EOP 2013a]."

U.S. Global Change Research Program

USGCRP is responsible for communicating with a variety of stakeholders nationally and globally on issues related to climate variability and climate change science, and for coordinating the federal agencies' climate change communications and education programs. The Communications and Education Interagency Working Group leads efforts to coordinate interagency education and communications activities.

U.S. Department of Commerce

National Oceanic and Atmospheric Administration

NOAA is committed to developing a society that is environmentally responsible and uses effective, science-based problem-solving skills. NOAA recognizes that improvements in societal stewardship of natural resources extend directly from effective stakeholder engagement, training, extension, and formal and informal education systems.

NOAA's climate education programs support the development of strong and comprehensive educational materials about climate and oceanic and atmospheric sciences. NOAA works to facilitate a formal education system that produces climate-literate citizens by engaging participation from policymakers, academic institutions, professional associations, teachers, and students.

In addition, informal education plays a critical role in developing climate-literate citizens. To help equip informal education institutions with modern instructional resources and interdisciplinary methods for teaching Earth system science, NOAA partners with aquariums, zoos, national parks, national marine sanctuaries, national estuarine research reserves, and National Sea Grant colleges (Figure 9-3). NOAA also works with other informal science education centers addressing climate change through the Climate Interpreter Network, which is funded by the Institute of Museum and Library Services, NOAA, and NSF. NOAA is engaged in improving both formal and informal education systems because these venues are important to the development of literate citizens and to the long-term maintenance of their skills, knowledge, and attitudes. Partnerships and collaboration are integral to sustaining and scaling up NOAA's ability to promote public climate literacy.

NOAA's Regional Integrated Science and Assessments (RISA) program and the National Integrated Drought Information System (NIDIS) support research teams that help expand and build the nation's capacity to prepare for and adapt to climate variability and change. Central to the RISA and NIDIS approaches are commitments to process, partnership, and trust building, with the goal of translating science into actionable knowledge and increasing capacity for making decisions in a rapidly changing environment. As societal awareness of climate risk grows, climate information is being infused into public spheres in richer ways, placing more emphasis on innovation of different methods for providing actionable knowledge. The dialogue between scientists and stakeholders also provides the perfect setting for social scientists and outreach experts to evaluate how well science is informing societal outcomes. RISA and NIDIS work closely with applied scientists who provide predictions and projections of weather and climate, with cooperative extension and outreach professionals, and with communications experts.

NOAA addresses growing societal challenges and the need for enhanced information products and services through integrated research, monitoring, and services development, including regional climate assessments, early-warning information systems, and training and education activities.

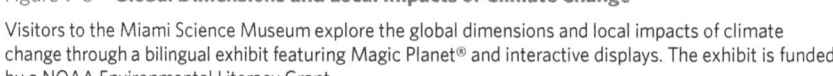

Figure 9-3 **Global Dimensions and Local Impacts of Climate Change**

Visitors to the Miami Science Museum explore the global dimensions and local impacts of climate change through a bilingual exhibit featuring Magic Planet® and interactive displays. The exhibit is funded by a NOAA Environmental Literacy Grant.

Photo: Juan Manuel Garcia Studio.

U.S. Department of Energy

Global Change Education Program

The U.S. Department of Energy's (DOE's) Office of Energy Efficiency and Renewable Energy (EERE) funds partners to develop curricula and implement standardized, high-quality training programs. These projects are aimed at creating a pipeline beginning at the K–12 level and extending through the postgraduate level to ensure the ongoing development of a workforce to invent and scale up clean energy and energy efficiency technologies and processes over the long term. Education and workforce training are critical parts of EERE's mission, which is to create an energy-literate generation of skilled workers, leaders, and innovators who will produce affordable, abundant, and clean energy, thus accelerating the transition to a low-carbon economy and ensuring U.S. global competitiveness.

U.S. Department of Health and Human Services

Centers for Disease Control and Prevention and National Institute of Environmental Health Sciences

Climate change information, education, and outreach from the U.S. Department of Health and Human Services (HHS) center around impacts of climate change on human health, with a particular focus on vulnerable populations. These activities are primarily coordinated through programs at the Centers for Disease Control and Prevention (CDC) and the National Institute of Environmental Health Sciences (NIEHS). CDC's program is aimed at state and local public health departments, and NIEHS serves the education and research communities.

Both institutions, along with The National Library of Medicine, also maintain Web sites with climate change and health information links designed for use by the general public. NIEHS also serves as the HHS principal to the USGCRP and, along with CDC and NOAA, co-leads USGCRP's Climate Change and Human Health Working Group, through which many inter-agency communications and outreach activities are planned and implemented.

U.S. Department of the Interior

DOI has an integrated climate change research and adaptation strategy for itself and its agencies. DOI agencies include its research arm, the U.S. Geological Survey USGS), and land- and resource-managing agencies, such as the National Park Service (NPS), U.S. Fish and Wildlife Service (FWS), Bureau of Land Management (BLM), Bureau of Reclamation (BOR), Bureau of Indian Affairs (BIA), Bureau of Ocean Energy Management (BOEM), and related agencies and offices.

DOI's climate strategy has a hub and spokes, with the hub located in the USGS National Climate Change and Wildlife Science Center, and the spokes located in the eight regional DOI climate science centers (CSCs) set up since 2010. The CSCs are operated in conjunction with universities in each region of the United States. In addition to advising land managers about research related to their regions, the CSCs coordinate with 22 landscape conservation cooperatives (LCCs) composed of landowners near U.S. parks, refuges, and other lands; government officials at the federal, state, and local levels; tribal leaders; and nonprofit and citizens' groups.

DOI, its bureaus, and CSCs and LCCs maintain climate Web pages and sites, as well as create social media, press releases, and publications related to climate change. In addition, parks and refuges have public education programs.

National Park Service

The NPS manages 3.4 million hectares (ha) (84 million acres [ac]) of land, including more than 400 national parks and other units; almost a million historic structures and archeological sites; thousands of kilometers of rivers; and 69,463 kilometers (43,162 miles) of shoreline. Because some of America's greatest wildlands, wildlife, and cultural treasures are especially vulnerable to climate change, the NPS considers it one of the agency's greatest challenges.

The NPS *Climate Change Action Plan: 2012–2014* builds on a strategy released in 2010, stating that by articulating "a set of high-priority, no-regrets actions the NPS is currently undertaking or committed to undertake, in the next one to two years" to help park managers and staff effectively plan for and respond to climate change (U.S. DOI/NPS 2010). Near-term priorities include enhancing workforce climate literacy; engaging youths and their families in climate change research, education, and hands-on projects; providing climate change science in parks; implementing a *Green Parks Plan* (U.S. DOI/NPS 2012); applying appropriate adaptation tools and options; and strengthening communication with the public within the "natural classrooms" in the parks and through a wide variety of interpretive and educational media.

Between 2007 and 2012, NPS held 17 workshops to train park managers on scenario planning. In addition, NPS's Climate Change Response Program has provided climate change-related training to NPS staff since 2007. Over the longer term, NPS planning is flexible to adapt to ongoing and emerging developments, such as climate change research, new advances in media and technology, and extreme events and disasters.

U.S. Fish and Wildlife Service

FWS administers the U.S. wildlife conservation laws, monitors and manages migratory birds, restores nationally significant fisheries, conserves wetlands, and regulates international wildlife trade. FWS also manages the 4-million-ha (96-million-ac) National Wildlife Refuge System. All of these responsibilities require preparation for climate change and adaptation, contained in the FWS climate strategy.

In addition, FWS has taken the lead in setting up the interagency LCCs that work in conjunction with the CSCs. These cooperatives address the challenges that are too great for any single national wildlife refuge, national park, or other community to manage alone—such as drought, climate change, and large-scale habitat fragmentation. The 22 LCCs work together on mutual conservation goals, benefitting from scientific and technical expertise beyond the reach of any one group.

FWS also co-led the development of the March 2013 *National Fish, Wildlife and Plants Climate Adaptation Strategy* (U.S. DOI/FWS 2012). This is the first nationwide strategy to help public and private decision makers address the impacts that climate change is having on wildlife and other natural resources and the people and economies that depend on them. The strategy's development was guided by an innovative partnership of federal, state, and tribal fish and wildlife conservation agencies in response to a 2010 call by the U.S. Congress for a national, government-wide climate adaptation strategy to assist fish, wildlife, and plants, and related ecological processes in becoming more resilient to, adapting to, and surviving the impacts of climate change.

The partnership was co-led by FWS, NOAA, and the New York State Department of Environmental Conservation (representing state fish and wildlife agencies). An intergovernmental steering committee that included representatives from 15 federal agencies, five state fish and wildlife agencies, and two inter-tribal commissions oversaw development of the strategy, with extensive public input and support from the Association of Fish and Wildlife Agencies.

U.S. Geological Survey

USGS is a multidisciplinary science arm of the U.S. government that undertakes scientific research, monitoring, remote sensing, modeling, synthesis, and forecasting to address the effects of climate and land-use change on the nation's resources. The resulting research and products are provided as the scientific foundation upon which policymakers, natural resource managers, and the public make informed decisions.

USGS runs the National Climate Change and Wildlife Science Center, which provides scientific and technical support to other agencies on the impacts of climate change. USGS also helped DOI establish the CSCs.

The USGS Land Remote Sensing Program operates the Landsat satellites (which are built and launched by NASA) and provides the nation's portal to the largest archive of remotely sensed land data in the world. These images serve many purposes, including tracking climate change. In addition, the Earth Resources Observation and Science Center contributes to USGS's climate and land-use programs with basic and applied research, data acquisition, systems engineering, and information access and management. USGS also conducts research to assess the potential capacities and limitations of various forms of carbon sequestration.

Bureau of Land Management

BLM manages more than 9.9 million ha (245 million ac) of public land for a wide variety of uses, including conservation, energy development, and recreation. Most of this land is found in the West, where average temperatures are rising, droughts are increasing, snowpack is declining, water supplies are diminishing in key areas, and wildfires have become larger and more frequent. BLM is undertaking two connected initiatives to understand, anticipate, and respond to the effects of climate change on the public lands: Rapid Ecoregional Assessments, which are currently being prepared, and a landscape approach for managing public lands.

Bureau of Reclamation

BOR conducts research on the effects of climate change on water supplies that is useful to water managers and decision makers. The WaterSMART program provides grants and other resources to help communities improve climate analysis tools and stretch water supplies through various conservation and water recycling projects. The WaterSMART Clearinghouse provides water resource planners and managers with tools related to water conservation and sustainability, arranged by term, topic, state, river basin, or tribal area.

U.S. Department of Transportation

DOT has developed many programs to educate the public, government employees, state and local agencies, and other transportation stakeholders about climate change.

Federal Transit Administration

FTA has several programs that provide information about the benefits of public transit and how to reduce the environmental impacts of transportation. The Environmental Management Systems Training, in particular, offers training for public transit agencies to assess and reduce the environmental impacts of their operations, including their carbon footprint.

FTA organizes, sponsors, and participates in numerous conferences as part of its outreach efforts, including conferences and sessions geared toward education on environmental and climate change issues. During 2013, FTA sponsored and participated in climate change panels at the annual Transportation Research Board conference, the Rail-volution conference, the American Public Transportation Association sustainability workshop, and the New Partners for Smart Growth Conference.

Funded by FTA, the National Transit Institute (NTI) at Rutgers, The State University of New Jersey, was established under the Intermodal Surface Transportation Efficiency Act of 1991 to provide training, education, and clearinghouse services in support of U.S. public transportation and quality of life. NTI courses on transportation planning, environmental review, transit-oriented development, and transportation and land use are particularly relevant to climate change issues.

FTA's climate change adaptation initiative Web page provides the public and transit agencies with information on FTA efforts with regard to climate change adaptation; published reports, policy statements, and letters; past events and workshops focusing on transit adaptation to climate change; and current activities taking place (including information on the seven FTA climate adaptation pilot projects).

Federal Highway Administration

FHWA targets metropolitan planning organizations (MPOs) and local transportation agencies to provide information on their climate science and mitigation strategies. Recently, FHWA unveiled the Energy and Emissions Reduction Policy Analysis Tool (EERPAT). FHWA developed EERPAT for use by state departments of transportation (DOTs) to model a large number of inputs and policy scenarios to support strategic transportation and visioning, including GHG emission reduction alternatives. EERPAT can be used to assist state DOTs in analyzing GHG reduction scenarios and alternatives for use in the transportation planning process, climate action plans, scenario planning exercises, and meeting state GHG reduction targets and goals. FHWA has also developed a mitigation reference sourcebook to accompany the tool, which is currently being revised to highlight the GHG reduction strategies that can be analyzed by the tool (Kalra et al. 2012).

More recently, in September 2013, FHWA hosted two peer exchanges for information sharing among 19 climate resilience pilots at state DOTs and MPOs. Previously, between June 2011 and April 2012, FHWA convened three peer exchanges for transportation agencies to share information related to climate change mitigation activities. These efforts are in addition to a DOT-wide effort to educate federal and state employees about a variety of transportation and climate change issues. For example, the Transportation and Climate Change Clearinghouse Web site, a one-stop source of information on transportation and climate change issues, includes information on GHG inventories, analytic methods and tools, GHG reduction strategies, potential impacts of climate change on transportation infrastructure, and approaches for integrating climate change considerations into transportation decision making.

National Aeronautics and Space Administration

NASA supports extensive education, training, and public awareness on climate change that take advantage of NASA's capabilities of observing the Earth system from space. In addition to programs targeted at training at the graduate and early-career levels, NASA is committed to building partnerships in communication and education to effectively reach various segments of the public.

The Global Learning and Observations to Benefit the Environment (GLOBE) program, jointly sponsored by NASA and NSF, continues to support teachers and students to conduct hands-on research projects about their local environment across 109 countries worldwide.[19] The NASA Innovations in Climate Education project offers opportunities to educational institutions in climate education.[20] Through Earth to Sky, NASA also works with interpretation experts at NPS, FWS, BLM, and other agencies to connect the wonder of science with the power of place by providing relevant and integrative information about climate change to the public.[21] Finally, NASA participates in public events and engages the public online, to promote broader understanding of climate change and its impacts on society.[22]

National Science Foundation

Consistent with its mission to support research and education across a broad range of science and engineering disciplines, NSF funds research in numerous areas related to global climate change. NSF's Directorates for Geosciences; Biological Sciences; Social, Behavioral, and Economic Sciences; Education and Human Resources; Mathematics and Physical Sciences; Computer and Information Science and Engineering; and the former Office of Polar Programs (recently merged with Geosciences) participate in the USGCRP and provide access to climate-related results from principal investigators.

NSF is the principal federal agency charged with promoting science, technology, engineering, and math education. To this end, NSF supports the development of a diverse and well-prepared scientific and technical workforce, and a scientifically literate citizenry.

Smithsonian Institution

The Smithsonian is addressing the global challenge of climate change with special exhibitions and ongoing research. Smithsonian collections related to the evidence about the impacts of and responses to climate change provide a unique and accessible resource for public education. Smithsonian scientists and curators regularly engage the museums' visitors with evidence about climate change issues, from the perspectives of science, history, and art.

U.S. Agency for International Development

As a the foreign assistance arm of the U.S. government, USAID plays a leadership role in delivering climate change-related international assistance to more than 40 developing and transition countries. With headquarters in Washington, D.C., USAID has field offices in many regions of the world—namely, sub-Saharan Africa, Asia, the Middle East, Latin America and the Caribbean, and Europe and Eurasia. USAID works in close partnership with private voluntary organizations, indigenous groups, universities, American businesses, international organizations, other governments, trade and professional associations, faith-based organizations, and other U.S. government agencies.

---

[19] *See* http://www.globe.gov.

[20] *See* http://gcce.larc.nasa.gov/.

[21] *See* http://earthtosky.org/.

[22] *See, for example,* http://climate.nasa.gov, http://earthobservatory.nasa.gov, social media.

USAID's foreign assistance work incorporates climate change considerations into development projects, supporting on-the-ground programs to achieve climate change results and strengthen economic growth. Climate change education, training, and outreach are a cornerstone of USAID's activities, providing the foundation for sustainable actions (Figure 9-4). Capacity building for improved decision making through applied science and access to information is increasingly important. (This work is highlighted in Chapter 7.) Building on clean energy, sustainable landscapes, and adaptation strategies, USAID will continue to integrate education, outreach, and training into its development mission to contribute to reducing the threat of climate change around the world.

U.S. Department of Agriculture

Agricultural Research Service

As USDA's chief intramural scientific research body, ARS is responsible for research on the impacts of agricultural practices on potential climate change or disruptions and vice versa. Although ARS has no formal educational mechanism to disseminate research information to the general public, it employs a number of less formal means to communicate and make use of research advances. All USDA scientific research publications are submitted with an Interpretive Summary that is used for timely news releases. In addition, through collaboration with university scientists, climate change research information is provided to state and county cooperative extension agencies for release to identified producers. Also, all USDA field locations publish informative brochures and technical reports that describe their work and the impacts of the research findings on stakeholders' interests.

National Institute of Food and Agriculture

Established by the Food, Conservation, and Energy Act of 2008, NIFA replaced the former Cooperative State Research, Education, and Extension Service, which had been in existence since 1994. NIFA is the primary USDA agency that supports extramural research, extension, and education activities by providing competitive and capacity funds in such areas as agriculture and natural resources science for climate variability and change. The NIFA Coordinated

Figure 9-4 **The Role of Mangrove Ecosystems in Building Resilience to the Effects of Climate Change**

During a field trip and workshop for Utwe elementary school teachers in Kosrae in the Federated States of Micronesia, participants learned about the connections between water salinity and the resilience of mangrove ecosystems to climate change, and role of mangroves in protecting island coasts from sea level rise and storm surges.

Photo: Julian Sachs.

Agricultural Project awards support projects to deliver the best tools available to accurately measure and respond to the effects of climate, and better understand how to work with and educate farmers, landowners, and foresters about regional climate change issues. Through federal funding and leadership for research, education, and extension programs, NIFA focuses on investing in science and solving critical issues affecting people's daily lives and the nation's future.

Climate Change Research Centers

Similar to DOI's and NOAA's regional climate centers strategy, the new USDA climate change research centers have a stated mission to educate the public about regional climate change issues.

U.S. Forest Service

USFS national efforts in climate change education, training, and public awareness are based on the scientific expertise and findings of the agency's more than 500 scientists. The USFS Research and Development program conducts research investigating how climate change is and may be affecting terrestrial and freshwater natural resources and ecosystems. These results are made available to professional resource managers and the public through a variety of Web sites and publications.

USFS also provides climate change education resources to educators and students through a variety of programs. One of these is *The Natural Inquirer*, a science education journal based on published USFS science, targeted for U.S. and international middle school students. Climate change editions of *The Natural Inquirer* have focused on contemporary research findings regarding climate change and wildfires and the impact of a changing climate on wildlife and stream temperatures.

In its most recent project, USFS has partnered with 18 other agencies and organizations to offer ClimateChangeLIVE,[23] a distance learning adventure. This project brings climate learning through a series of science-based, televised webcasts, webinars, and online climate education resources. In addition, EUGENE (Ecological Understanding as a Guideline for Evaluation of Nonformal Education)[24]—a broadly applicable, user-friendly Web-based environmental education evaluation instrument that assesses student knowledge on limits, regulation, and adaptation related to climate change—will assist educators in evaluating and improving their climate change programs and will increase accountability in climate change education.

U.S. Environmental Protection Agency

Climate change information, education, and outreach are an important part of EPA's work. EPA maintains a Climate Change Web site and a Student's Guide to Climate Change Web site, and has produced educational and informational materials that reach a wide range of audiences. In addition, EPA provides outreach programs that educate decision makers and the public about opportunities to reduce GHG emissions and adapt to the impacts of climate change that humans and nature are already facing.

EPA also runs a grant program that distributes more than $3 million a year to formal and informal education programs across the country that educate learners of all ages about the causes of and solutions to environmental problems. For the last several years, a significant percentage of those funds went specifically to climate change education programs.

---

[23] *See* http://www.climatechangelive.org/.

[24] *See* https://projecteugene.org/cgi-bin/eugene.

Table 9-1 **Federal Climate Change Programs Grouped by Primary Audience**

Program Name	Description	Audiences	Learning Setting	Web Site
K–12 Students and Teachers				
National Aeronautics and Space Administration (NASA)				
Global Climate Change Education/Earth Science Education Alliance (ESSEA)	Implemented by the Institute for Global Environmental Strategies to improve the quality of geoscience instruction for pre-service and in-service K–12 teachers, ESSEA is based on a series of online courses for teachers offered by several participating universities. The inquiry-based courses provide teachers with the content knowledge and tools they need to incorporate Earth system science into their curricula. ESSEA modules are also available online as teacher resources. Many of the course modules use NASA data and content. Some examples of ESSEA course modules include black carbon, Brazilian deforestation, coral reefs, Hurricane Katrina, stratospheric ozone, and sea ice. Partners: NSF	K–12 teachers, pre-service teachers	Formal	http://esseacourses.strategies.org/
Global Learning and Observations to Benefit the Environment (GLOBE)	GLOBE is a worldwide hands-on, primary and secondary school-based science and education program. GLOBE observations and measurements include atmosphere and climate, hydrology, land cover and phenology, and soils. GLOBE students, teachers, and scientists collaborate on inquiry-based investigations of the environment and the Earth system, working in close partnership with NASA and NSF Earth System Science Projects, on research topics related to the carbon cycle, watersheds, seasons, and biomes and extreme environments. Understanding Earth as an interconnected system is at the core of the GLOBE program. Partners: NASA, NSF	K–12 students, K–12 teachers	Formal/informal	http://www.globe.gov/
Students' Cloud Observations On-Line (S'COOL)	S'COOL is a component of NASA's CERES (Clouds and the Earth's Radiant Energy System). The CERES instrument measures the amount of energy reflected and emitted by the Earth system, focusing on understanding how clouds affect these energy transfers. Participating students make basic weather observations and record the types and features of clouds in the sky at the time the satellite passes over their location, and submit the data to NASA for entry into an online database. Students can access their results as well as those from other participating schools via the S'COOL Web site, which is available in seven languages. Satellite observations for matching times are also posted, so that students can compare their observations with those of the satellite, and scientists can evaluate CERES' performance. Participants receive instructional materials and information necessary for reporting results.	K–12 students, K–12 teachers	Formal/informal	http://science-edu.larc.nasa.gov/SCOOL/index.php
National Oceanic and Atmospheric Administration (NOAA)				
American Meteorological Society (AMS) Education Program	This program promotes the teaching of atmospheric, oceanographic, and hydrologic sciences through pre-college teacher training and instructional resource material development. It also promotes instructional innovation at the introductory college course level; hence, the K–13 designation for the program. All programs promote activity directed toward greater human resource diversity in the sciences AMS represents. To date, more than 100,000 teachers have received training and instructional resources, which have benefited millions of students. Partners: NSF, NASA	K–12 teachers	Formal	http://www.ametsoc.org/amsedu/

Table 9-1 (Continued)   **Federal Climate Change Programs Grouped by Primary Audience**

Program Name	Description	Audiences	Learning Setting	Web Site
Climate Stewards Education Project	This project increases understanding of essential climate concepts, providing educators with ready access to reliable scientific information through an array of professional development opportunities. Through direct interaction with scientists and education specialists, participants receive instruction in the use of data resources, digital tools, and other innovative technologies. Educators benefit from an active online learning community that offers collaborative space, Web seminars, conference symposia, workshops, and virtual conferences. Armed with this knowledge, NOAA Climate Stewards design and implement environmentally friendly action plans to reduce their communities' carbon footprint.	K–12 teachers, informal educators	Formal/ informal	http://oceanservice. noaa.gov/education /climate-stewards/
Communications and Education Program	This program takes an audience-focused approach to promoting climate science literacy among priority publics, including educators. It communicates the challenges, processes, and results of NOAA-supported climate science through stories and data visualizations on the Web and in popular media, and provides information to a range of audiences to enhance society's ability to plan for and respond to climate variability and change. Partner: USGCRP	K–12 teachers, undergraduate students, graduate students, public, professionals	Formal	http://climate.noaa. gov/education/
National Science Teachers Association's (NSTA's) The Learning Center, E-professional development portal	NSTA collaborates with NOAA, NASA, and NSF on the Learning Center to provide a variety of climate-focused online learning experiences to fit any teacher's learning style and content need. Teachers can access the center 24/7. NSTA is committed to providing the very best online professional development to science teachers. Partners: NSF, NASA, USDA/USFS, EPA	K–12 teachers	Formal/ informal	http:// learningcenter.nsta. org/
National Science Foundation (NSF)				
Discovery Research K–12 (DR K–12)	DR K–12 seeks to enable significant advances in pre-K–12 student and teacher learning of the STEM disciplines through projects that study the development, testing, deployment, effectiveness, and/or scale-up of innovative resources, models, and technologies for use by students, teachers, and policymakers.	K–12 students, K–12 teachers	Formal (K–12)	http://www.nsf. gov/funding/pgm_ summ.jsp?pims_ id=500047
Smithsonian Institution				
Climate Change Distance Learning	The National Zoological Park (NZP) is collaborating with the U.S. Forest Service (USFS), Prince William Network, NOAA, and many other groups that are working on climate change distance-learning experiences during 2014. The objectives of this effort are to (1) provide credible, science-based climate change education resources; (2) educate students about climate change; (3) share success stories about what students, schools, and communities are doing to help protect and conserve natural resources and to encourage viewers to take action; and (4) share information about what partner agencies are doing to address climate change. The plans include (1) distance-learning broadcasts and webcasts, (2) resource Web sites, (3) webinars for teachers, and (4) educational resources. The culminating student-driven event of ClimateChangeLIVE™ is planned for early March 2014. Partners: USFS, NOAA, USFWS, SI, EPA, NIST, NSF, DOE	K–12 students, K–12 teachers	Formal	http://www. climatechangelive. org/

Table 9-1 (Continued) **Federal Climate Change Programs Grouped by Primary Audience**

Program Name	Description	Audiences	Learning Setting	Web Site
David H. Koch Hall of Human Origins	Based on decades of cutting-edge research by Smithsonian scientists, the David H. Koch Hall of Human Origins exhibition is the result of an international collaboration with more than 60 research and educational organizations and more than 100 researchers from around the world. Visitors are taken on an immersive, interactive journey through six million years of scientific evidence of human origins and the stories of survival and extinction in humanity's family tree during times of dramatic climate instability. Visitors can explore actual archaeological field sites at interactive snapshots in time, examine more than 75 cast reproductions of real skulls from around the world, engage with an interactive family tree of evolutionary evidence, and address pressing questions and issues surrounding climate change and humans' impact on Earth in the "One Species Living Worldwide" theatre and the "Changing the World" gallery. Educational resources, public programs, and an immersive online experience accompany the exhibit.	K-12 students, K-12 teachers, public	Informal	http://humanorigins.si.edu/exhibit
Demography of Songbird Populations in a Rapidly Changing World: The Importance of Long-Term Studies	The Migratory Bird Center staff published a paper in *American Biology Teacher* in 2011 describing a Web-based teaching module based on a long-term study of a migratory songbird, the black-throated blue warbler. The module describes this species and the ecological factors that affect its population growth and provides exercises developed to span a range of student levels. It discusses the results of the study in the context of climate change, and prompts students to consider the impact of climate change on the study population.	K-12 students, K-12 teachers	Formal	http://www.jstor.org/discover/10.1525/abt.2011.73.5.8?uid=3739256&uid=2&uid=4&sid=21102869689333
Ecosystems on the Edge	This is a 16-part mini-video series on threats to coastal ecosystems, many of which deal with climate change. The Smithsonian Environmental Research Center is building a Web site to host the video series, which will include a section specifically devoted to climate change and how it will affect the plants, animals, and people living on the coast. A curriculum to facilitate the use of the videos by high school teachers in the classroom is also under development.	K-12 students, K-12 teachers, public	Formal, informal	http://ecosystems.serc.si.edu/
Forces of Change Program	Nearly every scientific and social issue today involves change: climate change, ecological change, cultural change, etc. What forces drive these changes? What are the tempo and mode of these changes? Are these changes natural or the result of human tampering? Are they to be feared or welcomed? How do we—and all life on this planet—adapt to these changes? This program seeks to address these questions through a variety of resources, including online exhibits and educational products.	K-12 students, K-12 teachers, public	Informal	http://forces.si.edu/index.html
Gabon Biodiversity Program	As part of ongoing conservation education efforts in Gabon, Africa, the Smithsonian Conservation Biology Institute (SCBI) conducted a climate change program in 2011 for high school students in Gamba. This topic will continue to be an important component of the education and outreach programs offered through the SCBI program, based in Gabon.	K-12 students	Formal	http://nationalzoo.si.edu/SCBI/Collaborative-Research-Initiatives/Gabon-Biodiversity-Program.cfm

Table 9-1 (Continued)   **Federal Climate Change Programs Grouped by Primary Audience**

Program Name	Description	Audiences	Learning Setting	Web Site
The Habitable Planet: A Systems Approach to Environmental Science	This multimedia course for high school teachers and adult learners interested in studying environmental science was developed by the Smithsonian Center for Astrophysics (and is currently hosted by Annenberg Learner). It includes a unit on Earth's changing climate, which examines the science behind global climate change and explores its potential impacts on natural ecosystems and human societies.	K-12 teachers, public	Formal	http://www.learner.org/courses/envsci/
Looking at Earth Exhibition Gallery	Understanding Earth's environment and how climate conditions are changing with time requires the collection of weather data from all over the globe. The National Air and Space Museum (NASM) is home to a variety of historical and modern examples of the satellites, cameras, and other hardware used to examine Earth from above. NASM's Looking at Earth exhibition gallery showcases the use of the tools and tactics that have been developed over time for scrutinizing the surface of Earth from the highest of "high-ground" perspectives then attainable.	K-12 students, K-12 teachers, public	Informal	http://airandspace.si.edu/exhibitions/gal110/index.cfm
Marine Environmental Education Program	The Smithsonian Topical Research Institute (STRI) has environmental education programs that address issues of climate change. These take place at the Culebra Point Marine Exhibition Center, the Bocas del Toro Marine Laboratory, and the Galeta Point Marine Laboratory, whose marine environmental education program links STRI's research to Panama's classrooms. This program has reached 95,000 students from Panama and abroad.	Graduate students	Informal	http://www.stri.si.edu/english/education_fellowships/field_courses/index.php
Ocean Portal	The Ocean Portal is part of the Smithsonian's Ocean Initiative. Together with the National Museum of Natural History's (NMNH's) Sant Ocean Hall and the Sant Marine Science Chair, the Ocean Portal supports the Smithsonian's mission to increase the public's understanding and stewardship of the ocean. This portal includes a variety of resources related to climate change. Partner: NOAA	K-12 students, K-12 teachers, public	Informal	http://ocean.si.edu/
Punta Culebra Nature Center (PCNC)	This nonprofit initiative of STRI offers visitors an open-air museum focusing mainly on marine science and education and on conservation and interpretation of marine coastal environments. More than 700,000 students and visitors have come to PCNC since it opened in 1996, and hundreds of schools have taken part in its educational programs. PCNC addresses climate change through teacher workshops, lesson plans for students who visit the center, and summer camps. Specific topics addressed include ocean acidification, studies on carbon dioxide (CO_2) storage in tropical rainforests, changes in ocean level, and the greenhouse effect. An educational activity titled "The CO_2 Eaters," which has been included in several educational programs conducted by PCNC, was also published in the educational teacher package "Native Trees of Panama and Neotrópico" in collaboration with BioMuseo, Aprendo, and *La Prensa*.	K-12 students, K-12 teachers, public	Informal	http://www.stri.si.edu/english/visit_us/culebra/
Sant Ocean Hall	The Sant Ocean Hall is NMNH's largest exhibit, providing visitors with a unique and breathtaking introduction to the majesty of the ocean. The hall's combination of over 675 marine specimens and models, high-definition video, and the newest technology allows visitors to explore the ocean's past, present, and future. The exhibit addresses climate change through graphics and interactive features. Partner: NOAA	K-12 students, K-12 teachers, public	Informal	http://www.mnh.si.edu/exhibits/ocean_hall/

Table 9-1 (Continued)   **Federal Climate Change Programs Grouped by Primary Audience**

Program Name	Description	Audiences	Learning Setting	Web Site
Science on a Sphere® (SOS)	NMNH and NZP participate in SOS, which is a room-sized, global display system that uses computers and video projectors to display planetary data onto a 6-foot-diameter sphere, analogous to a giant animated globe. Researchers at NOAA developed SOS as an educational tool to help illustrate Earth system science to people of all ages. Animated images of atmospheric storms, climate change, and ocean temperature can be shown on the sphere, which is used to explain what are sometimes complex environmental processes, in a way that is simultaneously intuitive and captivating. Partner: NOAA	K–12 students, K–12 teachers, public	Informal	http://sos.noaa.gov
Smithsonian Science Education Academy for Teachers on Earth's History & Global Change	This week-long summer academy is for Earth science school teachers from grades 6 through 12 and interested educators from museums and science centers. It examines global climate change from the perspective of the history of Earth from its formation through the origin of life. Topics include planetary processes, volcanism and plate tectonics, and the oceans and atmosphere. Each day participants engage Earth scientists at the Smithsonian and elsewhere in hands-on content sessions that take them behind the scenes and explore current research on Earth's past environments. Participants learn about resources available for teachers at the Smithsonian's museums and facilities and federal science agencies. Participants also have the opportunity to earn graduate credit through Virginia Commonwealth University. Partners: NASA, NOAA	K–12 teachers	Formal/ informal	http://www. scienceteachers academies.si.edu/
Smithsonian Treebanding Project	This project recruited schools from across the world to measure how fast local trees were growing, partly to track how trees respond to climate change. Students and teachers received a kit and instructions in the mail. Later they added their data to an online database and could compare what they found with what classrooms in other countries found. At its peak, the program had 490 schools in 38 countries participating.	K–12 students, K–12 teachers	Formal	https://treebanding. si.edu/
Teacher Training Program	STRI conducts a teacher training program in coordination with Panama's Ministry of Education (MEDUCA) and with funding from the International Community Foundation. In 2013, Galeta Point Laboratory conducted its VI Teacher Training Course on Tropical Marine and Coastal Ecosystems. Altogether, 42 docents from all provinces and comarcas participated, from public and private schools. The MEDUCA participants include the national director for elementary education, the national director for science education, and the science supervisors from seven of Panama's nine provinces. During this two-week intensive course, docents receive the latest scientific material in Spanish and presentations by STRI researchers, complemented by hands-on field trips. A theme of several of those presentations was climate change and its implications for the nations of Central America and the Caribbean. Partners: Panama's Ministry of Education, International Community Foundation	K–12 teachers	Formal	http://stri.si.edu/ english/about_stri/ headline_news/ news/article. php?id=1660

Table 9-1 (Continued)   **Federal Climate Change Programs Grouped by Primary Audience**

Program Name	Description	Audiences	Learning Setting	Web Site
Understanding Weather and Climate	This instructional unit for middle school students explores the atmospheric events and oceanic processes that affect Earth and its inhabitants. Students experiment with factors that determine storms and daily weather, explore the impact of the oceans on Earth, and examine the influences that produce climate zones and changes. Throughout the unit, students make predictions, collect data to test hypotheses, and draw conclusions based on evidence. This unit is part of the Smithsonian Science Education Center's Science and Technology Concepts Program, a research-based science curriculum for grades K–8.	K–12 students, K–12 teachers	Formal	http://www.carolinacurriculum.com/stc/Secondary/Weather+Climate/index.asp
U.S. Department of Agriculture (USDA)				
Chugach National Forest Children's Forest	In the summer of 2008, the Chugach National Forest was designated as a Children's Forest. One of the key programs for the Children's Forest will be a climate change research program in which students will shadow researchers from the U.S. Forest Service (USFS) and the University of Alaska. The researchers are conducting quantifiable, inquiry-based research to monitor the impacts of climate change on Alaska's forest and wetland ecosystems.	K–12 students	Formal/ informal	http://www.alaskageographic.org/static/1040/programs
Forest Service Climate Change Educator Web Site	USFS has several inter-related programs to help forests, grasslands, and humans mitigate and adapt to global climate change. This site contains a variety of resources for researchers, managers, educators, and the public on climate change issues and science, and provides links to cost-free climate change education resources.	K–12 teachers, K–12 students, public	Formal/ informal	http://www.fs.fed.us/climatechange/
GreenSchools! Initiative	In partnership with the American Forest Foundation, USFS selected five schools in Washington, D.C., to pilot the GreenSchools! Initiative. This program provides training and funding for diverse and underserved pre-K-12 public schools. Students and teachers investigate environmental issues at their schools and engage with their communities in ongoing service-learning projects that create green and healthy learning environments. Partner: American Forest Foundation—Project Learning Tree	Pre-K-12 teachers, K-12 students, public	Formal/ informal	http://www.plt.org/cms/pages/21_23_242.html
The Investigator	Based on published USFS science, this science education journal is intended for upper elementary students in the U.S. and abroad. All resources are correlated to National Science Education Standards. Teachers can use The Investigator to introduce students to the concept that rising levels of ozone will affect tree growth.	K–12 students	Formal	http://www.scienceinvestigator.org
The Mayor's Green Summer Job Corps Program	USFS is joining Anacostia Urban Tree House partners to train the on-the-ground supervisors of this Washington, D.C., program, which introduces local youths to green-collar career paths. The program uses a combination of substantive work projects and traditional educational sessions to increase job readiness, connect youths to the environment within their communities, and improve the District's environment overall. Broadly, this program complements the District's efforts in combating climate change, restoring its waterways, and increasing its green infrastructure. Partner: Anacostia Urban Tree House	K–12 teachers, K–12 students, public	Formal/ informal	http://green.dc.gov/green/cwpview,a,1233,q,461478.asp
The Natural Inquirer	This science education journal, based on published USFS science, is for middle school students both in the U.S. and abroad. Resources are correlated to national science education standards and are available in English and Spanish.	K–12 students	Formal	http://www.naturalinquirer.org

Table 9-1 (Continued)　　**Federal Climate Change Programs Grouped by Primary Audience**

Program Name	Description	Audiences	Learning Setting	Web Site
U.S. Department of Energy (DOE)				
America's Home Energy Education Challenge	DOE and the National Science Teachers Association host a competition for grades 3–8 through their schools or informal education to learn about energy and apply energy-efficient behavior changes at home.	K–12 teachers and students	Online Web site for competition, including energy curricula	www.Aheec.org
Climate Literacy and Energy Awareness Network (CLEAN)	The Teaching Climate section of climate.gov partnered with CLEAN to use the *Climate Literacy* guide to identify and integrate effective resources across different educational levels. The CLEAN framework for vetting, reviewing, and ensuring the scientific quality of climate and global change education materials on climate, energy, and related topics will be very useful to teachers and educational systems across the nation. Partners: NSF, DOE	K–12 teachers, informal educators	Formal/ informal	http://www.climate.gov/teaching
K–12 Clean Energy Activities and Curricula	One-stop shop for K–12 lesson plans, curricula, and activities.	K–12 teachers and students	Online Web site and app	http://www1.eere.energy.gov/education/lessonplans/default.aspx
U.S. Environmental Protection Agency (EPA)				
EPA Student's Guide to Climate Change Web Site	This popular environmental education site provides a wealth of resources for students and educators. Graphically engaging and interactive, this site includes information about climate change science, interactive "Climate Expeditions" to learn about climate change impacts around the globe, a section on what people can do to make a difference, resources for educators and administrators, and more. One feature is a GHG calculator, which instructs students about steps they can take to reduce their carbon footprint and what those reductions can mean for the environment. Revamped in 2011, the site has more than 30,000 unique visitors each month.	Students grades 6–8	Formal	www.epa.gov/climatechange/kids
Undergraduate Students				
National Oceanic and Atmospheric Administration (NOAA)				
Significant Opportunities in Atmospheric Research in Science (SOARS) (UCAR)	SOARS is an undergraduate-to-graduate bridge program designed to broaden participation in the atmospheric and related sciences. The program is equal parts research internship, learning community, and mentoring. Partner: NSF	Undergraduate students	Formal	http://www.soars.ucar.edu/
National Science Foundation (NSF)				
Applied Conservation Strategies and Ecology for Effective Conservation Practices	The Smithsonian-Mason School of Conservation holds two residential undergraduate semesters, in which climate change is a learning module in both semesters (one focuses on ecological studies evaluating ecosystem responses to climate change; the other focuses primarily on the science of climate and community engagement related to climate change).	Undergraduate students	Formal	http://smconservation.gmu.edu/programs/undergraduate/

Table 9-1 (Continued)   **Federal Climate Change Programs Grouped by Primary Audience**

Program Name	Description	Audiences	Learning Setting	Web Site
Smithsonian Institution				
Advanced Technological Education (ATE)	With an emphasis on two-year colleges, ATE focuses on the education of technicians for the high-technology fields that drive the nation's economy, through support for curriculum development, professional development of college faculty and secondary school teachers, and articulation of career pathways between high school, two-year colleges, and four-year institutions.	Undergraduate students	Formal (undergrad)	http://www.nsf.gov/funding/pgm_summ.jsp?pims_id=5464
U.S. Department of Agriculture (USDA)				
Higher Education Challenge Grants	Grant program addressing national priorities in the development of higher education programs and curricula.	Land grant colleges and universities	Formal (undergrad/grad)	http://www.csrees.usda.gov/
U.S. Department of Energy (DOE)				
Advanced Vehicle Competitions	Multi-year competitions to challenge undergraduate engineering students to reengineer existing cars with advanced vehicle technologies to reduce fuel consumption and lower emissions. The objective is to stimulate the development of advanced-propulsion and alternative-fuel technologies and provide the training ground for the next generation of automotive engineers. The current competition—EcoCAR 2: Plugging In to the Future—which began in 2011 and will conclude in 2014, includes 15 universities from across North America. Student teams are using a 2013 Chevrolet Malibu as the integration platform for their advanced vehicle design.	Undergraduate students	Hands-on learning using an actual vehicle	http://www1.eere.energy.gov/vehiclesandfuels/deployment/education/index.html
Energy 101 Course Framework	A model, interdisciplinary, general energy course for college students in two- and four-year schools to explore systematically the science and social science behind sound energy decision making. Based on the Energy Literacy Framework.	Undergraduate students	Customizable online course framework	http://www1eere.energy.gov education/energy_101.html
Graduate Students, Professionals				
National Oceanic and Atmospheric Administration (NOAA)				
Climate and Society Masters Program	This program enables understanding of climate science, decision processes, and social needs to deliver management strategies that incorporate climate. Its core courses have been developed by the International Research Institute for Climate and Society, in collaboration with renowned Columbia University faculty in climate, engineering, policy, public health, economics, political science, statistics, psychology, sociology, and anthropology.	Graduate students	Formal	http://www.columbia.edu/cu/climatesociety/
National Weather Service (NWS) Training Program in Climate Services	NWS initiated a training program in climate services in 2001 to increase the knowledge base of its field staff. It included about 25 hours of online distance learning material, a 5-day virtual course on Climate Variability and Change, and a 3-day residence course on Operational Climate Services. Because of the continuing interest in global and regional climate variability and change, as well as their local impacts on socioeconomic development, the NWS training program is expanding.	Professionals, graduate students, educators	Training	http://www.nws.noaa.gov/om/csd/pds/DistanceLearning.shtml

Table 9-1 (Continued)   **Federal Climate Change Programs Grouped by Primary Audience**

Program Name	Description	Audiences	Learning Setting	Web Site
National Science Foundation (NSF)				
Integrative Graduate Education and Research Traineeship Program (IGERT)	IGERT was developed to meet the challenges of educating U.S. Ph.D. scientists and engineers who will pursue careers in research and education, with interdisciplinary backgrounds; deep knowledge in chosen disciplines; and technical, professional, and personal skills to become, in their own careers, leaders and creative agents for change. IGERT has a strong focus on new models for graduate education that prepare students to contribute in new ways to benefit society.	Graduate students, professionals	Formal (grad)	http://www.nsf.gov/funding/pgm_summ.jsp?pims_id=12759
Transforming Undergraduate Education in STEM (TUES)	TUES seeks to improve the quality of STEM education for all undergraduate students through projects with potential to transform education by bringing about widespread adoption of classroom practices that embody understanding of how students learn most effectively, develop faculty expertise, implement educational innovations, assess learning and evaluate innovations, prepare K-12 teachers, or conduct research on STEM teaching and learning.	Undergraduate students, professionals	Formal (undergrad)	http://www.nsf.gov/funding/pgm_summ.jsp?pims_id=5741
U.S. Department of Energy (DOE)				
Graduate Automotive Technology Education (GATE)	GATE Centers of Excellence support the development of advanced multidisciplinary coursework, certificate, and degree programs in advanced vehicle technologies at competitively selected universities. Funds are also provided for research and development and laboratory experiences in critical automotive technologies to help develop next-generation expertise to overcome technology barriers preventing the development and production of cost-effective, high-efficiency U.S. vehicles. The awardees will focus on three critical automotive technology areas: hybrid propulsion, energy storage, and lightweight materials.	Graduate students	Classroom/ laboratory research	http://www1.eere.energy.gov/vehiclesandfuels/deployment/education/fcvt_gate.html
Undergraduate, Graduate, and Postgraduate Students, Professionals				
National Science Foundation (NSF)				
Antarctic Earth Sciences Program	Beneath its thick ice sheets, Antarctica is a dynamic and diverse continent with mountains, volcanoes, deserts, meteorites, dinosaur fossils, and some of Earth's most ancient crust. This program supports research to interpret this rich history and the processes that shape Antarctica today.	Undergraduate students, graduate students, professionals	Formal (undergrad, grad)	http://www.nsf.gov/funding/pgm_summ.jsp?pims_id=8173
Antarctic Glaciology Program	This program is concerned with the study of the history and dynamics of all naturally occurring forms of snow and ice, including floating ice shelves, glaciers, and continental and marine ice sheets. Program emphases include paleoenvironments from ice cores, ice dynamics, numerical modeling, glacial geology, and remote sensing of ice sheets.	Undergraduate students, graduate students, professionals	Formal (undergrad, grad)	http://www.nsf.gov/funding/pgm_summ.jsp?pims_id=12798
Antarctic Ocean and Atmospheric Sciences	Antarctic oceanic and tropospheric studies focus on the structure and processes of the ocean–atmosphere environment and their relationships with the global ocean, atmosphere, and marine biosphere. As part of the global heat engine, the Antarctic has a major role in the world's transfer of energy. Its ocean–atmosphere system is known to be both an indicator and a component of climate change.	Undergraduate students, graduate students, professionals	Formal (undergrad, grad)	http://www.nsf.gov/funding/pgm_summ.jsp?pims_id=13422

Table 9-1 (Continued) **Federal Climate Change Programs Grouped by Primary Audience**

Program Name	Description	Audiences	Learning Setting	Web Site
Arctic Natural Sciences Program	Areas of special interest include marine and terrestrial ecosystems, Arctic atmospheric and oceanic dynamics and climatology, Arctic geological and glaciological processes, and their connectivity to lower latitudes.	Undergraduate students, graduate students, professionals	Formal (undergrad, grad)	http://www.nsf.gov/funding/pgm_summ.jsp?pims_id=13424&org=NSF
Arctic Observing Network (AON)	Compared with much of the rest of Earth, the Arctic is a data-sparse region where large, rapid, and system-wide environmental change is occurring. The goal of AON is to enhance the environmental observing infrastructure required for the scientific investigation of Arctic environmental change and its global connections.	Undergraduate students, graduate students, professionals	Formal (undergrad, grad)	http://www.nsf.gov/funding/pgm_summ.jsp?pims_id=503222&org=NSF
Arctic Research Support and Logistics Program (RSL)	RSL supports the field component of research projects funded through NSF science programs. RSL accepts proposals that support long-term observations of the Arctic; support the acquisition of data sets useful to a broad segment of the Arctic research community; will lead to Cooperative Agreements to operate multi-use Arctic research facilities; or provide services that broadly support the Arctic research community, such as facilitating communication, developing research ideas in an Arctic-wide community setting, and cooperating with Arctic communities.	Undergraduate students, graduate students, professionals	Formal (undergrad, grad)	http://www.nsf.gov/funding/pgm_summ.jsp?pims_id=13437&org=NSF
Arctic-SEES	This multi-year, interdisciplinary program seeks both fundamental research that improves the ability to evaluate the sustainability of the Arctic human-environmental system, as well as integrated efforts that will provide community-relevant sustainability pathways and engineering solutions. For this competition, interdisciplinary research is focused in four thematic areas: the natural and living environment, the built environment, natural resource development, and governance.	Undergraduate students, graduate students, professionals	Formal (undergrad, grad)	http://www.nsf.gov/funding/pgm_summ.jsp?pims_id=503604
Arctic Social Sciences (ASSP)	ASSP encompasses all social sciences supported by NSF, including anthropology, archaeology, economics, geography, linguistics, political science, psychology, science and technology studies, sociology, traditional knowledge, and related subjects.	Undergraduate students, graduate students, professionals	Formal (undergrad, grad)	http://www.nsf.gov/funding/pgm_summ.jsp?pims_id=13425
Arctic System Science (ARCSS) Program	The Arctic comprises a complex, tightly coupled system of air, ice, ocean, land, and people. The system behaves in ways not fully understood, and has demonstrated the capacity for rapid and unpredictable change with global ramifications. Because the Arctic is pivotal to Earth's dynamics, ARCSS's goal is to advance understanding of this complex and interactive system.	Undergraduate students, graduate students, professionals	Formal (undergrad, grad)	http://www.nsf.gov/funding/pgm_summ.jsp?pims_id=13426
Climate and Large-Scale Dynamics (CLD)	CLD's goals are to (1) advance knowledge about the processes that force and regulate the atmosphere's synoptic and planetary circulation, weather, and climate; and (2) sustain the pool of human resources required for excellence in synoptic and global atmospheric dynamics and climate research.	Undergraduate students, graduate students, professionals	Formal (undergrad, grad)	http://www.nsf.gov/funding/pgm_summ.jsp?pims_id=11699
Decadal and Regional Climate Prediction Using Earth System Models (EaSMs)	This program supports the development and application of next-generation EaSMs that include coupled and interactive representations of such things as ocean and atmospheric currents, human activities, agricultural working lands and forests, urban environments, biogeochemistry, atmospheric chemistry, the water cycle, and land ice. The program seeks to attract scientists from the disciplines of geosciences, social sciences, agricultural and biological sciences, mathematics and statistics, physics, and chemistry. Partners: USDA, DOE	Undergraduate students, graduate students, professionals	Formal (K–12, undergrad, grad)	http://www.nsf.gov/funding/pgm_summ.jsp?pims_id=503399%5Barchived%5D

Table 9-1 (Continued)   **Federal Climate Change Programs Grouped by Primary Audience**

Program Name	Description	Audiences	Learning Setting	Web Site
Decision, Risk and Management Sciences (DRMS)	DRMS supports scientific research directed at increasing the understanding and effectiveness of decision making by individuals, groups, organizations, and society. Disciplinary and interdisciplinary research, doctoral dissertation research, and workshops are funded in the areas of judgment and decision making; decision analysis and decision aids; risk analysis, perception, and communication; societal and public policy decision making; and management science and organizational design.	Undergraduate students, graduate students, professionals	Formal (undergrad, grad)	http://www.nsf.gov/funding/pgm_summ.jsp?pims_id=5423
Dimensions of Biodiversity	This campaign's goal is to transform, by 2020, how the scope and role of life on Earth are described and understood. The campaign promotes novel, integrated approaches to identify and understand the evolutionary and ecological significance of biodiversity amidst the changing environment of the present and in the geologic past. This campaign seeks to characterize Earth's biodiversity by using integrative, innovative approaches to fill the most substantial gaps in understanding of the diversity of life on Earth. The campaign takes a broad view of biodiversity, and currently focuses on the integration of genetic, taxonomic/phylogenetic, and functional dimensions of biodiversity. Partner: NASA	Undergraduate students, graduate students, professionals	Formal (undergrad, grad)	http://www.nsf.gov/funding/pgm_summ.jsp?pims_id=503446
Emerging Topics in Biogeochemical Cycles	Proposals should be interdisciplinary and should address biogeochemical processes and dynamics within and/or across one or more of the following systems: terrestrial, aquatic, and atmospheric. NSF encourages proposals that focus on nonlinear dynamics and/or on interactions and thresholds in climate, ecological, and/or hydrological systems. Goals of this effort are to increase understanding of how biological systems respond to changing physical and chemical conditions, and how biological systems influence the physical and chemical characteristics of soils and sediments, air, or water.	Undergraduate students, graduate students, professionals	Formal (K–12, undergrad, grad)	http://www.nsf.gov/pubs/2009/nsf09030/nsf09030.jsp
Energy for Sustainability	This program supports fundamental research and education in energy production, conversion, and storage, and is focused on environmentally friendly and renewable energy sources.	Undergraduate students, graduate students, professionals	Formal (K–12, undergrad, grad)	http://www.nsf.gov/funding/pgm_summ.jsp?pims_id=501026
Environmental Engineering	This program encourages transformative research that applies scientific and engineering principles to avoid or minimize solid, liquid, and gaseous discharges resulting from human activity into land, inland and coastal waters, and air, while promoting resource and energy conservation and recovery.	Undergraduate students, graduate students, professionals	Formal (undergrad, grad)	http://www.nsf.gov/funding/pgm_summ.jsp?pims_id=501029
Environmental Sustainability	This program supports engineering research with the goal of promoting sustainable, engineered systems that enhance human well-being and are compatible with sustaining natural (environmental) systems that provide ecological services vital for human survival. The long-term viability of natural capital is critical for many areas of human endeavor. Environmental sustainability research typically considers long time horizons and may incorporate contributions from the social sciences and ethics. Research areas include industrial ecology, green engineering, ecological engineering, and Earth systems engineering.	Undergraduate students, graduate students, professionals	Formal (undergrad, grad)	http://www.nsf.gov/funding/pgm_summ.jsp?pims_id=501027

Table 9-1 (Continued)   **Federal Climate Change Programs Grouped by Primary Audience**

Program Name	Description	Audiences	Learning Setting	Web Site
Ethics Education in Science and Engineering (EESE)	EESE funds research and education projects that improve ethics education in all fields of science and engineering supported by NSF, with priority given to interdisciplinary, inter-institutional, and international contexts.	Undergraduate students, graduate students, professionals	Formal (K–12, undergrad, grad)	http://www.nsf. gov/funding/pgm_ summ.jsp?pims_ id=13338
Frontiers in Earth System Dynamics (FESD)	Earth is often characterized as "dynamic," because its systems are variable over space and time, and they can respond rapidly to multiple perturbations. FESD's goals are to (1) foster an interdisciplinary and multiscale understanding of the interplay among and within Earth's various subsystems, (2) catalyze research in areas poised for a major advance, (3) improve data resolution and modeling capabilities to more realistically simulate complex processes and forecast disruptive or threshold events, and (4) improve knowledge of the resilience of Earth and its subsystems.	Undergraduate students, graduate students, professionals	Formal (undergrad, grad)	http://www.nsf. gov/funding/pgm_ summ.jsp?pims_ id=503525
Integrated Earth Systems (IES)	IES focuses on the continental, terrestrial, and deep Earth subsystems of the whole Earth system, with the goal of supporting collaborative, multidisciplinary research into the operation, dynamics, and complexity of Earth systems and subsystems at all temporal and spatial scales.	Undergraduate students, graduate students, professionals	Formal (undergrad, grad)	http://www.nsf. gov/funding/pgm_ summ.jsp?pims_ id=504833
Multi-scale Modeling (MSM)	MSM supports projects that focus on the development and/or integration of environmental models that link local, regional, and global scales. Proposals are encouraged that have the potential to dramatically improve understanding of how small- and large-scale processes lead to nonlinearities and activation thresholds, as well as to improve predictive capabilities. Projects could address such topics as the carbon cycle, climate, population dynamics, food webs, biodiversity, biogeochemical cycles, and hydrological processes.	Undergraduate students, graduate students, professionals	Formal (undergrad, grad)	http://www.nsf. gov/pubs/2009/ nsf09032/ nsf09032.jsp
NSF Science, Engineering and Education for Sustainability Fellows	Through this program, NSF seeks to advance science, engineering, and education to inform the societal actions needed for environmental and economic sustainability and human well-being, while creating the necessary workforce to address these challenges. The program's emphasis is to facilitate investigations that cross traditional disciplinary boundaries and address issues of sustainability through a systems approach, building bridges among academic inquiry, economic growth, and societal needs.	Professionals	Formal (post-doctoral)	http://www.nsf. gov/funding/pgm_ summ.jsp?pims_ id=504673
Paleoclimate	This program supports research on the natural evolution of Earth's climate, with the goal of providing a baseline for present variability and future trends through improved understanding of the physical, chemical, and biological processes that influence climate over the long term.	Undergraduate students, graduate students, professionals	Formal (undergrad, grad)	http://www.nsf. gov/funding/pgm_ summ.jsp?pims_ id=12727
Paleo Perspectives on Climate Change	The goal of research is to utilize key geological, chemical, and biological records of climate system variability to provide insights into the mechanisms and rates of change that characterized Earth's past climate variability, the sensitivity of Earth's climate system to changes in forcing, and the response of key components of Earth's system to these changes.	Undergraduate students, graduate students, professionals	Formal (undergrad, grad)	http://www.nsf. gov/funding/pgm_ summ.jsp?pims_ id=5750

Table 9-1 (Continued) **Federal Climate Change Programs Grouped by Primary Audience**

Program Name	Description	Audiences	Learning Setting	Web Site
Research Coordination Networks (RCN)	RCN's goal is to advance a field or create new directions in research or education by supporting groups of investigators to communicate and coordinate their research, training, and educational activities across disciplinary, organizational, geographic, and international boundaries. RCN provides opportunities to foster new collaborations, including international partnerships, and address interdisciplinary topics. Innovative ideas for implementing novel networking strategies, collaborative technologies, and development of community standards for data and meta-data are especially encouraged.	Undergraduate students, graduate students, professionals	Formal (undergrad, grad)	http://www.nsf.gov/funding/pgm_summ.jsp?pims_id=11691&org=GEO&from=home
Sustainable Energy Pathways (SEP)	SEP calls for innovative, interdisciplinary basic research in science, engineering, and education by teams of researchers for developing systems approaches to sustainable energy pathways based on a comprehensive understanding of the scientific, technical, environmental, economic, and societal issues. The SEP solicitation considers scalable approaches for sustainable energy conversion to useful forms, as well as its storage, transmission, distribution, and use.	Undergraduate students, graduate students, professionals	Formal (K-12, undergrad, grad)	http://www.nsf.gov/funding/pgm_summ.jsp?pims_id=504690
Water Sustainability and Climate (WSC)	WSC's goal is to enhance the understanding of and predict the interactions between the water system and land-use changes (including agriculture, managed forests, and rangeland systems), the built environment, ecosystem function and services, and climate change/variability through place-based research and integrative models. Studies of the water system using models and/or observations at specific sites, singly or in combination, that allow for spatial and temporal extrapolation to other regions, as well as integration across the different processes in that system, are encouraged, especially to the extent that they advance the development of theoretical frameworks and predictive understanding.	Undergraduate students, graduate students, professionals	Formal (undergrad, grad)	http://www.nsf.gov/funding/pgm_summ.jsp?pims_id=503452

Smithsonian Environmental Research Center (SERC)

Internship Program	The theme of SERC's 2013 NSF-funded internship program is "Global Change Ecology at the Smithsonian Environmental Research Center." Many of the intern research projects will relate to climate change.	Undergraduate students, graduate students	Formal	http://www.serc.si.edu/pro_training/index.aspx
Professional Development	SERC research labs, including the Biogeochemistry Lab working in the Global Change Research Wetland, bring on 30-50 interns and postdoctoral students annually for professional training. The Biogeochemistry Lab is currently testing how higher carbon dioxide and sea level rise could change the research site in the year 2100.	Undergraduate students, graduate students, professionals	Formal, training	http://www.serc.si.edu/labs/biogeochem/research_wetland.aspx

U.S. Department of Energy (DOE)

EERE Postdoctoral Research Awards	The awards aim to create the next generation of scientific leaders in energy efficiency and renewable energy by attracting the best scientists and engineers to pursue breakthrough technologies.	Postgraduate students	Laboratory, conferences	http://www.windpoweringamerica.gov/schools/projects.asp
Department of Energy Solar Decathlon	The decathlon educates student participants and the public about the environmental benefits and cost-saving opportunities presented by clean-energy products; demonstrates to the public the accessibility and affordability of cost-effective homes that combine energy-efficient construction and appliances with renewable energy systems that are available today; and provides participating students with unique training that prepares them to enter the nation's clean-energy workforce.	Undergraduate and graduate students, educators	National competition	http://www.solardecathlon.gov/

Table 9-1 (Continued)　　**Federal Climate Change Programs Grouped by Primary Audience**

Program Name	Description	Audiences	Learning Setting	Web Site
Geothermal Student Competition	This competition is designed to advance the understanding of geothermal energy. Students form an interdisciplinary team to develop a business plan for creating a geothermal enterprise in their local areas.	Undergraduate and graduate students	Online	http://orise.orau.gov/science-education/capabilities/science-education-events/eere-geothermal-student-competition.aspx
Hydro Research Fellowships Program	The program is designed to stimulate student research and academic interest in research and careers in conventional or pumped storage hydropower. The research seeks to advance knowledge about hydroelectric technology, including efficiency improvements and environmental mitigation.	Postgraduate students	Research and training	http://www.hydrofoundation.org/fellowshipOverview.html

Informal Educators, Public

National Aeronautics and Space Administration (NASA)

| Earth to Sky: Climate Change Professional Development for Informal Educators | This ongoing and expanding partnership provides professional development for informal educators to access and use relevant NASA science, data, and educational products in their work. Partners: DOI/NPS, USFWS | Informal educators | Training | http://www.earthtosky.org/ |

National Oceanic and Atmospheric Administration (NOAA), Office of Education (OEd)

| Ocean Education Grants for AZA Aquariums | OEd issued a request for applications to support education projects designed to engage the public in activities that increase ocean and/or climate literacy and the adoption of a stewardship ethic. | Informal educators | Informal | http://www.oesd.noaa.gov/funding_opps.html |
| Science On a Sphere Collaborative Users Network | Science On a Sphere (SOS)® is a spherical display system, approximately 6 feet in diameter, that shows "movies" of animated Earth system dynamics (http://www.sos.noaa.gov/). NOAA's Office of Education supports the use of spherical display systems, such as SOS, in public exhibits as part of a focused effort to increase environmental literacy. The institutions that currently have NOAA's SOS, as well as other partners who are creating content and educational programming for these systems, have formed a collaborative network. Partners: NASA, DOE | Informal educators, public | Informal | http://www.oesd.noaa.gov/network/ |

National Science Foundation (NSF)

| Advancing Informal STEM Learning (AISL) | AISL invests in research and development of innovative and field-advancing out-of-school STEM learning and emerging STEM learning environments. Funding is provided for projects that advance understanding of informal STEM learning, develop and implement innovative strategies and resources for informal STEM education, and build the national professional capacity for research, development, and practice in the field. | Informal education, public | Informal | http://www.nsf.gov/funding/pgm_summ.jsp?pims_id=504793 |

Formal/Informal Educators

Smithsonian Institution

| Arctic Studies Center | Established in 1988, the center is the only U.S. government program with a special focus on northern cultural research and education. In keeping with this mandate, the center specifically studies northern people, exploring history, archaeology, social change, and human lifeways across the circumpolar world. The center conducts various outreach activities that relate to climate change issues, including exhibitions and conferences. | K–12 students, K–12 teachers, public, professionals | Informal | http://www.mnh.si.edu/arctic/index.html |

Table 9-1 (Continued)   **Federal Climate Change Programs Grouped by Primary Audience**

Program Name	Description	Audiences	Learning Setting	Web Site
Evolution of Terrestrial Ecosystems (ETE) Program	ETE investigates Earth's land biotas throughout their 400-million-year history. The program's goal is to understand how terrestrial ecosystems have been structured and have changed over geologic time. Using the fossil record, ETE scientists study the characteristics of ecological communities and the changing dynamics of ecosystems. Paleoecological analyses determine patterns through time in community structure and composition, investigate the effects of ecological change on individual lineages, and relate patterns of stasis or change to environmental and other processes that influence ecosystem formation, sustainability, and collapse. The ETE program conducts a variety of outreach activities, including hosting workshops, meetings, and conferences; teaching courses at area universities; and providing content for various museum exhibits.	K–12 students, K–12 teachers, undergraduate students, graduate students, public, professionals	Formal, informal	http://www.mnh.si.edu/ete/
U.S. Environmental Protection Agency (EPA)				
Environmental Education Grant Program	EPA's Environmental Education Division distributes more than $3 million annually to formal and informal education organizations across the nation to provide environmental education programs to learners of all ages. Many of these grants have gone to climate change education programs over the last several years, including public school districts, privately run nature centers, public and private colleges and universities, and community organizations.	Formal/informal educators	Formal/informal	www.epa.gov/enviroed/
U.S. Global Change Research Program (USGCRP)				
Climate Change, Wildlife and Wildlands Toolkit for Formal and Informal Educators	The new toolkit is an updated and expanded version of the award-winning *Climate Change, Wildlife and Wildlands Toolkit for Teachers and Interpreters*, first published in 2001 (2001 Public Relations Society of America Bronze Anvil Award for Interactive Communications and 2002 Telly Award). The toolkit is very popular, with more than 40,000 kits distributed in all 50 states and U.S. territories and over a dozen countries across the world. The toolkit profiles climate stewards in all 11 ecoregions. Here, students participate in the Baldwin County Grasses in Classes program to help grow native plants for wetland and dune restoration projects. The new kit is designed for classroom teachers and informal educators in parks, refuges, forestlands, nature centers, zoos, aquariums, science centers, etc., and is aimed at the middle school grade level. In partnership with the National Park Service, the U.S. Fish and Wildlife Service, NOAA, NASA, the U.S. Forest Service, and the Bureau of Land Management, EPA developed this toolkit to aid educators in teaching how climate change is affecting the nation's wildlife and public lands, and how everyone can become climate stewards.	Formal/informal educators	Formal/informal	http://globalchange.gov/resources/educators/toolkit/
Climate Literacy: The Essential Principles of Climate Sciences—A Guide for Individuals and Communities	This guide presents important information for individuals and communities to understand Earth's climate, impacts of climate change, and approaches for adapting to and mitigating climate change. Principles in the guide can serve as discussion starters or launching points for scientific inquiry. The guide can also serve educators who teach climate science as part of their science curricula. A guide is available to help individuals of all ages understand how climate influences them—and how they influence climate. A product of USGCRP, the guide was compiled by an interagency group led by NOAA.	Formal/informal educators	Formal/informal	http://globalchange.gov/resources/educators/climate-literacy

Table 9-1 (Continued)   **Federal Climate Change Programs Grouped by Primary Audience**

Program Name	Description	Audiences	Learning Setting	Web Site
Office of Education: Environmental Literacy Grants program	NOAA's Office of Education issued a request for applications for projects designed to build the capacity of informal educators (including interpreters and docents) and/or formal educators (pre- or in-service) to use NOAA data and data access tools to help K–12 students and/or the public understand and respond to global change. Successful projects will enhance educators' ability to use the wealth of scientific data, data visualizations, data access technologies, information products, and other assets available through NOAA (plus additional sources, if desired) to engage K–12 students and/or other members of the public.	Formal/ informal educators	Formal/ informal	http://www.oesd. noaa.gov/funding_ opps.html

Professionals

National Oceanic and Atmospheric Administration (NOAA)

Program Name	Description	Audiences	Learning Setting	Web Site
Building Capacity for Communicating about Climate	NOAA established a voluntary team to enhance the ability of NOAA personnel and partners to communicate about climate science issues. The team creates opportunities for interested staff and partners to learn about Earth's climate and how it influences our lives, and to become more conversant about NOAA's climate products, information, and services. The team engages staff through webinars, workshops, and an e-newsletter.	Professionals	Informal	https://sites.google. com/a/noaa.gov/ building-capacity- for-conversing- about-climate/
Climate Communications Workshops	In the spring of 2013, NOAA and the Cooperative Institute for Climate and Satellites organized several workshops at locations around the nation to (1) build climate communications capacity among NOAA staff and partners so that they are better able to converse about climate science issues, (2) provide communications and climate resources to staff that will help them prepare and respond to questions about climate, and (3) empower staff with the tools, techniques, and tactics to respond to questions about climate science.	Professionals	Informal	http://cicsnc.org/ events/
Coastal Resource Managers Training and Capacity Building	NOAA's National Ocean Service Coastal Services Center works with other federal agencies to impart information, services, and technology to the nation's coastal resource managers. This community includes state coastal zone management and natural resource management offices, research reserves, sanctuaries, and Sea Grant offices. Each of these organizations has the difficult task of helping coastal communities balance the often competing demands for coastal resources.	Professionals	Training	http://oceanservice. noaa.gov/topics/ coasts/training/
Coastal Training Program	The program provides up-to-date scientific information and skill-building opportunities to individuals who are responsible for making decisions that affect coastal resources. The program helps National Estuarine Research Reserves ensure that coastal decision makers have the knowledge and tools they need to address critical resource management issues of concern to local communities.	Professionals	Training	http://www8. nos.noaa.gov/ publicnerrs/ training.aspx

Table 9-1 (Continued)   **Federal Climate Change Programs Grouped by Primary Audience**

Program Name	Description	Audiences	Learning Setting	Web Site
Monthly U.S. and Global Climate Report	NOAA's National Climate Center develops monthly U.S. and global climate reports to analyze the previous month's conditions and provide additional seasonal and annual analyses. The reports present monthly statistics on surface temperature and precipitation, including ranks and patterns, as well as comparable data for the last three and six months and year to date. They also include subreports on tornadoes, wildfire, snow cover, major winter storms, and typically an update to the most recent U.S. Drought Monitor Report. A monthly call to media and stakeholders supplements these reports.	Public, professionals	Informal	http://www.ncdc.noaa.gov/sotc/
Regional Integrated Science and Assessment (RISA) Program	RISA supports research teams that conduct interdisciplinary and regionally relevant assessments to inform resource management, planning, and public policy. RISA teams help build the nation's capacity to prepare for and adapt to climate variability and change by providing cutting-edge scientific information to public and private user communities.	Public, professionals	Informal	http://cpo.noaa.gov/ClimatePrograms/ClimateSocietalInteractionsCSI/RISAProgram.aspx
Responding to Climate Change: A Workshop for Coral Reef Managers	Resources from a global series of workshops are distributed to coral reef managers to support their learning of how to predict where coral bleaching will occur, measure coral reef resilience, and assess the socioeconomic impacts of climate damage. The workshops aim is to help managers develop response strategies for coping with climate change. The workshops are hosted by NOAA, the Great Barrier Reef Marine Park Authority, and The Nature Conservancy, who partnered with the World Conservation Union in producing *A Reef Manager's Guide to Coral Bleaching*, the book that inspired these workshops.	Professionals	Training	http://coralreefwatch.noaa.gov/satellite/education/workshop/index.html
Yearly State of the Climate Report	NOAA scientists serve as the lead editors on this international, peer-reviewed annual report, which is the authoritative summary of the global climate of the previous year. In 2011, the report used 43 climate indicators to track and identify changes and overall trends in the global climate system, and was compiled by 378 scientists from 48 countries around the world.	Public, professionals	Informal	http://www.ncdc.noaa.gov/bams-state-of-the-climate/2011.php
Smithsonian Institution				
The Anthropocene: Planet Earth in the Age of Humans	The Consortia hosted a symposium on October 11, 2012, to address the tremendous scope of transformations now occurring on Earth with profound effects on plants, animals, and natural habitats. Geologists have proposed the term Anthropocene, or "Age of Man," for this new period in the history of the planet. The symposium focused on the arrival and impact of this new era through the lenses of science, history, art, culture, philosophy, and economics, and promoted discussion, debate, and deliberation on these issues of change.	Public, professionals	Informal	http://www.si.edu/consortia
Environmental Leadership Training Initiative (ELTI)	The Smithsonian Tropical Research Institute's (STRI's) Environmental Leadership Training Initiative, in partnership with Yale University, has hosted a variety of training programs related to climate change. ELTI provides policy-makers, individuals in technical positions, community representatives, indigenous leaders, and others with the knowledge, skills, and tools to conserve and restore forest ecosystems and biodiversity in tropical regions of Latin America and Asia. Partner: Yale University	Professionals	Training	http://environment.yale.edu/elti/en/

Table 9-1 (Continued)   **Federal Climate Change Programs Grouped by Primary Audience**

Program Name	Description	Audiences	Learning Setting	Web Site
[Hong Kong and Shanghai Banking Corporation] HSBC Climate Partnership	The HSBC Climate Partnership was a multi-year initiative that brought together STRI, The Climate Group, Earth-watch Institute, and the World Wildlife Fund in a partnership to address the threat of climate change. This initiative employed a participatory citizen science model in which HSBC employees worked alongside scientists from STRI and other partners to collect data from five distinct forest sites around the world to better understand how global forests respond to climate change. This citizen science model was used to educate HSBC employees about climate change and inspire them to take action to address it. Partner: HSBC	Professionals	Training	http://www.theclimategroup.org/programs/hsbc-climate-partnership/
International Outreach	Smithsonian Environmental Research Center (SERC) ecologists conduct a variety of international outreach activities related to climate change. For example, Dr. John Parker spent two weeks in India teaching Buddhist monks about climate change as part of the Science for Monks program. In addition, Dr. Pat Megonigal and his lab went to Abu Dhabi to research how well its coasts were burying carbon, and in the process conducted professional training with approximately 20 volunteers.	Public, professionals	Informal, training	http://www.serc.si.edu/index.aspx
Roger Revelle Commemorative Lecture Presented by The National Academies' Ocean Studies Board	The 2013 Roger Revelle Commemorative Lecture, "Melting Ice: What is happening to Arctic sea ice and what does it mean for us," explored the impacts of recent decreases in Arctic summer sea ice and how these decreases may already be affecting the larger climate system through a variety of physical, dynamical, and ecological processes. The featured speaker was Dr. John E. Walsh, Chief Scientist at the International Arctic Research Center. The lecture was sponsored by several organizations, including the University of Wisconsin-Madison's Space Science and Engineering Center. Partners: ONR, USGS, NSF, NASA, NOAA	Public, professionals	Informal	http://nas-sites.org/revellelecture/
Smithsonian Institution Global Earth Observatory (SI-GEO)	SI-GEO is a worldwide tree survey involving roughly 48 forest plots across the globe. SERC recruits volunteer "citizen scientists" to help survey the 33,500 trees in their SI-GEO forest plot. Through 2011, SERC partnered with Earthwatch on volunteer recruitment. Earthwatch recruited HSBC employees to participate, and volunteers spent a week at SERC learning about climate change and helping scientists in the field.	Public, professionals	Training	http://www.sigeo.si.edu/

U.S. Department of Agriculture (USDA), U.S. Forest Service (USFS)

Program Name	Description	Audiences	Learning Setting	Web Site
Climate Change Resource Center: Information and Tools for Land Managers	The center is a joint project of USFS's Pacific Northwest and Rocky Mountain Research Stations. This Web-based resource summarizes climate change research for resource managers and provides implications for management based on the scientific findings. It also contains video presentations from scientists describing their findings.	Professionals	Formal/informal	http://www.fs.fed.us/ccrc/
Eastern Forest Environmental Threat Assessment Center	The center provides regional online access to the general public and land managers.	Professionals	Formal/informal	http://www.forestthreats.org/climate-change

Table 9-1 (Continued)   **Federal Climate Change Programs Grouped by Primary Audience**

Program Name	Description	Audiences	Learning Setting	Web Site
i-Tree	i-Tree is a state-of-the-art, peer-reviewed software suite from USFS that provides urban forestry analysis and benefits assessment tools. The i-Tree Tools help professionals in communities of all sizes to strengthen their urban forest management and advocacy efforts by quantifying the structure of community trees and the environmental services that trees provide, including those that mitigate the effects of climate change.	Public/ professionals	Formal/ informal	http://www. itreetools.org/
USFS/IUFRO Task Force on Traditional Forest Knowledge	The USFS Research & Development and IUFRO (International Union of Forest Research Organizations) Task Force provide information on traditional forest knowledge and practices related to climate change.	Public/ professionals	Formal/ informal	http://www.iufro. org/science/task-forces/traditional-forest-knowledge/
Western Wildland Environmental Threat Assessment Center	The center provides regional online access to the general public and land managers.	Professionals	Formal/ informal	http://www.fs.fed. us/wwetac/threats/ climate_change. html

U.S. Department of Energy (DOE)

Program Name	Description	Audiences	Learning Setting	Web Site
Energy 101	This on-demand training course provides an introduction to federal energy management. The training is designed for new federal energy managers and others wanting an overall introduction to renewable energy, energy efficiency, and water efficiency. Attendees receive an overview of energy management, energy efficiency, renewable energy, and water efficiency; and learn about the legislative basis for federal energy management, the process for starting energy management projects, how to establish baseline energy and water measurements, developing action plans, and project financing mechanisms and options.	Federal energy managers and their support contractors	Online course/ on-demand e-training	http://apps1.eere. energy.gov/femp/ training/course_ detail_ondemand. cfm/CourseId=6
Federal Greenhouse Gas Accounting and Reporting	This no-cost, on-demand training course provides an update on greenhouse gas (GHG) regulatory requirements, as well as strategies, models, and technology tools to measure GHG emissions. Attendees learn to identify key types and sources of federal GHG emissions; understand the emerging GHG accounting and reporting framework; align and integrate diverse agency activities, processes, and resources related to GHG reductions; and adopt and implement accepted methods for gathering reliable data to measure progress, evaluate results, and improve performance.	Federal energy managers and their support contractors	Online course/ on-demand e-training	http://apps1.eere. energy.gov/femp/ training/course_ detail_ondemand. cfm/CourseId=14
Home Energy Score	This tool provides homeowners with resources to identify trusted contractors who can help them understand their home's energy use, as well as identify home improvements that increase energy performance and improve comfort.	Assessors and auditors, potential partners	Webinars	http://www1. eere.energy.gov/ buildings/ residential/hes_ past_webinars.html
Residential Building Retrofit Information	This Web site provides information about guidelines for effective training for the following residential building retrofit careers: energy auditor, retrofit installer, technician, crew leader, and quality control inspector. It also provides a link to Guidelines for Quality Work for Single-Family, Multifamily, and Manufactured Housing Energy Upgrades.	Those interested in working to upgrade/ retrofit residential buildings	Online	http://www1.eere. energy.gov/wip/ retrofit_guidelines_ overview.html

Table 9-1 (Continued)   **Federal Climate Change Programs Grouped by Primary Audience**

Program Name	Description	Audiences	Learning Setting	Web Site
Solar Instructor Training Network (SITN)	Increasing quality and access to accredited photovoltaic (PV) training, SITN partners with more than 260 community colleges in eight regions (all 50 states, 2 U.S. territories) to train instructors in PV and electrical skills to a national standard. SITN also provides free inspection training to local, county, and state code officials regarding rooftop PV inspection practices that comply with all national building codes.	Community college PV instructors, municipal building inspectors	Web site	http://www1.eere.energy.gov/solar/sunshot/instructor_training_network.html

U.S. Department of Health and Human Services (HHS)

Program Name	Description	Audiences	Learning Setting	Web Site
CDC Climate and Health Program	This Web site provides information on Centers for Disease Control and Prevention activities and funding in climate and health, including resources and links for state and local health departments.	State and local health departments, public health professionals, and general public	Web site	http://www.cdc.gov/climateandhealth/default.htm
Climate Change and Extreme Heat Events Guidebook	Provides information on extreme heat events, projected impacts from increased extreme heat events, and how the public health community can protect the nation from these impacts.	Public health community	Outreach	http://www.cdc.gov/climateandhealth/pubs/ClimateChangeandExtremeHeatEvents.pdf
Climate Change: Mastering the Public Health Role	A series of webinars developed in conjunction with the American Public Health Association and other key national organizations on climate change topics of interest to public health practitioners, featuring presentations from leading experts and public health leaders.	State and local health departments, public health professionals	Live and archived online webinar series	http://www.cdc.gov/climatechange/webinar_series.htm
Climate and Health Program	Helps communities prepare for extreme heat. For example, approximately 1,000 U.S. public health officials participated in the May 23, 2013, webinar "Beating the Heat: Preparing for Extreme Heat Events at the State and Local Level," with presentations from representatives from the New York and North Carolina health departments.	State and local health departments, public health professionals, and general public	Outreach	N/A
Climate-Ready States and Cities Initiative (CRSCI) Launch	CRSCI aims to strengthen the capabilities of state and local health agencies to deal with the challenges associated with climate change; identify and forecast the public health impacts of climate change specific to their communities and geographic areas; understand gaps in their knowledge and program capabilities to respond to the forecasted public health impacts; identify new programs or tailored program adaptations needed to counter the forecasted impacts; and collect critical information to guide resource decisions that protect their communities.	State and local health departments, public health professionals	Live and online	http://www.cdc.gov/climateandhealth/climate_ready.htm
Health Impact Assessment (HIA) Training	This day-long training course demonstrates how to undertake an HIA on a climate change-related policy, with emphasis on understanding how climate change-related policies can impact public health, and key considerations when assessing and providing recommendations based on the health impacts of a policy relevant to climate change.	State and local health departments, public health professionals	Training	N/A

U.S. Department of the Interior (DOI)

Program Name	Description	Audiences	Learning Setting	Web Site
Climate Friendly Parks Program	Through this joint partnership between EPA and the National Park Service (NPS), Climate Friendly Parks from around the country are leading the way to protect U.S. parks' natural and cultural resources and ensure their preservation for future generations. Partner: EPA	Professionals	Training	http://www.nps.gov/climatefriendlyparks/index.html

Table 9-1 (Continued)   **Federal Climate Change Programs Grouped by Primary Audience**

Program Name	Description	Audiences	Learning Setting	Web Site
Climate Leadership In Parks (CLIP)	CLIP is an Excel-based calculator designed for parks to assess their own greenhouse gas (GHG) emissions. It focuses on in-park operational activities—electricity use, transportation, waste and wastewater treatment, and "other" GHG-emitting activities inside parks. While the tool has a method to calculate forest carbon flux, it is not up to date or specific enough to adequately represent park forest carbon storage/emissions. For parks that want to include forest carbon in their reporting, NPS recommends that they use the latest forest models to calculate the flux, and then enter the numbers into the CLIP tool.	Professionals	Training	http://www.nps.gov/climatefriendlyparks/index.html
Regional Climate Change Response Centers	Eight DOI regional Climate Change Response Centers—serving Alaska, the Northeast, the Southeast, the Southwest, the Midwest, the West, Northwest, and Pacific regions—will synthesize existing climate change impact data and management strategies, help resource managers put them into action on the ground, and engage the public through education initiatives.	Professionals	Training	http://www.doi.gov/news/09_News_Releases/091409.html

U.S. Department of Transportation (DOT)

Program Name	Description	Audiences	Learning Setting	Web Site
Adaptation Peer Exchange (June 29 and November 6, 2012, and February 12 and May 21, 2013)	These webinars provide an opportunity for information exchange and peer review/input from each of the pilot projects. Each pilot presents to all others (about 30-40 people) on the webinar (usually for about 10-15 minutes) regarding the work they have completed thus far, the information they have gathered, and lessons learned/best practices.	Professionals (transit agencies, state and local governments)	Webinars	http://www.fhwa.dot.gov/environment/climate_change/adaptation/webinars/
Aviation Climate Change Research Initiative (ACCRI)	Measures and tracks fuel efficiency from aircraft operations, and provides the data for assessing improvements in aircraft and engine technology, operational procedures, and the airspace transportation system that reduce aviation's contribution to CO_2 emissions. A major ACCRI goal is to reduce key scientific uncertainties in quantifying aviation-related climate impacts and provide timely scientific input to inform policymaking decisions for the Federal Aviation Administration's (FAA's) NextGen Program.	Professionals (aviation stakeholders)	Web site	http://www.faa.gov/about/office_org/headquarters_offices/apl/research/science_integrated_modeling/accri/
Climate Change Adaptation Initiative	Web page provides information on Federal Transit Administration (FTA) efforts regarding climate change adaptation; published reports, policy statements, and letters; past events and workshops related to transit adaptation to climate change; and current activities taking place (including information on the seven FTA climate adaptation pilot projects).	Professionals (transit agencies, state and local governments, public)	Web site	http://www.fta.dot.gov/adaptation
Getting on the Right Track: Real-World Approaches to Climate Change Adaptation (Workshop March 21-22, 2012)	This workshop was held in conjunction with the American Public Transportation Association and included a discussion of the 2012-2013 FTA climate adaptation pilot projects.	Professionals (transit agencies, state and local governments)	Formal/informal	http://www.fta.dot.gov/sitemap_14257.html
Climate Change Forums	An ongoing series produced by the Center for Climate Change and Environmental Forecasting to raise the awareness of American industry, government, and nonprofit organizations. In 2011, the center hosted two sessions for all DOT employees on the need for climate adaptation in transportation and on regional climate projections and why they matter to transportation.	Professionals (government employees)	Classroom/briefing style	http://www.climate.dot.gov

Table 9-1 (Continued)   **Federal Climate Change Programs Grouped by Primary Audience**

Program Name	Description	Audiences	Learning Setting	Web Site
Environmental Management Systems (EMS) Training	Organizations use EMS to continually assess and reduce the environmental impact of their operations, including their carbon footprint. Training and technical assistance include workshops, on-site technical support visits, electronic software, and consultation. During the 18-month training period, each agency will develop an EMS suited to its needs.	Professionals (transit agencies)	Workshops, on-site technical support visits, electronic software, and consultation	http://www.fta.dot.gov/planning/environment/planning_environment_227.html
Highways and Climate Change	Provides information on Federal Highway Administration (FHWA) research, publications, and resources related to climate change science, policies, and actions. Also presents some current state and local practices for adapting to climate change and reducing greenhouse gas emissions.	Professionals (state DOTs, local transportation agencies, MPOs, public)	Web site	http://www.fhwa.dot.gov/hep/climate/index.htm
Highways and Climate Change Newsletter	Provides information on the most recent issues and activities related to transportation and climate change.	Professionals (state DOTs, local transportation agencies, MPOs, public)	Formal/informal	http://www.fhwa.dot.gov/hep/climatechange/newsletter/index.htm
National Transit Institute (NTI), at Rutgers, The State University of New Jersey	Funded by FTA, NTI provides training, education, and clearinghouse services in support of U.S. public transportation and quality of life. NTI courses on transportation planning, environmental review, transit-oriented development, and transportation and land use are particularly relevant to climate change issues.	Professionals (transit agency staff, public transportation, transit industry, private companies)	Classroom and online courses	http://www.ntionline.com/
Outreach through conferences	FTA organizes, sponsors, and participates in numerous conferences as part of its outreach efforts, including conferences and sessions geared toward education on environmental and climate change issues. In the last year, FTA sponsored and participated in climate change panels at the annual Transportation Research Board conference, the Rail-volution conference, the American Public Transportation Association sustainability workshop, and the New Partners for Smart Growth Conference.	Professionals (transit agencies, state and local governments, academics)	Conferences	http://www.fta.dot.gov/news/news_events_415.html
Partnership for AiR Transportation Noise & Emissions Reduction (PARTNER)	A leading aviation cooperative research organization and an FAA/NASA/Transport Canada-sponsored Center of Excellence, PARTNER fosters breakthrough technological, operational, policy, and workforce advances for the betterment of mobility, economy, national security, and the environment. PARTNER comprises nine universities and 51 advisory board members. Many of its efforts have led to outreach and educational initiatives. PARTNER has funded the research of more than 200 master's and Ph.D. students, many in climate research. Partners: NASA, Transport Canada	Professionals (aviation stakeholders, including airlines, airports, manufacturers, the public, and government organizations)	Formal/informal	http://www.partner.aero

Table 9-1 (Continued) **Federal Climate Change Programs Grouped by Primary Audience**

Program Name	Description	Audiences	Learning Setting	Web Site
Peer Exchanges on Transportation and Climate Change	FHWA is hosting peer exchanges for information sharing among 19 climate resilience pilots at state departments of transportation (DOTs) and metropolitan planning organizations (MPOs).	Professionals (state DOTs, local transportation agencies, MPOs, public)	Formal/informal	http://www.fhwa.dot.gov/environment/climate_change/adaptation/ongoing_and_current_research/vulnerability_assessment_pilots/index.cfm; http://www.fhwa.dot.gov/environment/climate_change/adaptation/workshops_and_peer_exchanges/
Systematic Impacts of Climate Change Conference (October 11-12, 2012)	This two-day workshop examined the systematic effects of climate change on the national transportation systems, and identified what previous and current research has identified about climate change, what gaps exist in the research, and what researchers want to explore further.	Professionals from the transportation sector, and academics	Classroom/briefing style	N/A
Transit and Climate Change Adaptation (August 8, 2011)	Discussed how climate change has implications for the planning process and asset management programs, as well as project-level design considerations in the transit realm. Guest speakers included representatives from the New York and Los Angeles Metropolitan Transportation Authorities.	Professionals (transit agencies, state and local governments)	Webinar	http://www.fta.dot.gov/documents/FTA_Climate_Change_Adaptation_Webinar_Notes_AUgust_8.pdf; http://www.fta.dot.gov/sitemap_14078.html
Transit and Environmental Sustainability	Provides information on transit's role in environmental sustainability, FTA sustainability efforts, resources and tools, and a clearinghouse of transit agency practices.	Professionals (transit agencies, state and local governments)	Web site	http://www.fta.dot.gov/13835.html
Transportation and Climate Change	FHWA periodically hosts webinars on transportation and climate change adaptation and mitigation. FHWA is currently hosting a series of public webinars on adapting transportation systems to climate change impacts.	Professionals (state DOTs and MPOs)	Webinars	http://www.fhwa.dot.gov/environment/climate_change/mitigation/webinars/; http://www.fhwa.dot.gov/environment/climate_change/adaptation/webinars/
Using Asset Management to Adapt to Weather Extremes: Lessons Learned from Transport for London (TfL) (December 15, 2011)	Transportation systems "on both sides of the pond" face challenges with bringing assets up to a state of good repair while dealing with extreme weather and changing climates. Flooding and heat waves further stress aging assets. TfL manages London's buses, road network, underground rail, and above-ground rail. TfL engineers and specialists describe how their agency has integrated climate impacts into asset management systems to better adapt transportation infrastructure and operations to risks. Presenters explain TfL risk assessments, asset management processes, highways drainage hotspot identification, and adaptive design of future assets, such as floodproofing for a major new construction project.	Professionals (transit agencies, state and local governments)	Webinar	http://www.fta.dot.gov/sitemap_14127.html

Table 9-1 (Continued)   **Federal Climate Change Programs Grouped by Primary Audience**

Program Name	Description	Audiences	Learning Setting	Web Site
U.S. Environmental Protection Agency (EPA)				
Climate Change Indicators in the United States	In December 2012, EPA updated *Climate Change Indicators in the United States*. Available in print and online, this popular report presents 26 indicators that track observed signs of climate change. The indicators focus primarily on the U.S., but in some cases global trends are presented to provide context or a basis for comparison. The indicators are divided into five chapters: Greenhouse Gases, Weather and Climate, Oceans, Snow and Ice, and Human Society and Ecosystems. The Indicators are based on peer-reviewed data from various government agencies, academic institutions, and other organizations.	Public/ professionals	Formal/ informal	http://epa.gov/ climatechange/ science/indicators/
Climate Ready Estuaries (CRE) Program	CRE works with the National Estuary Program (NEP) and the coastal management community to assess climate change vulnerabilities, develop and implement adaptation strategies, and engage and educate stakeholders. CRE shares NEP examples to help other coastal managers, and provides technical guidance and assistance about climate change adaptation. The CRE Web site offers information on climate change impacts to different estuary regions, access to tools and resources to monitor changes, and information to help managers develop adaptation plans for estuaries and coastal communities.	Public/ professionals	Training	www.epa.gov/cre
ENERGY STAR for Existing Residential Homes	ENERGY STAR educates and empowers American homeowners with information about the actions they can take to reduce GHG emissions by improving the energy efficiency of their homes. Since 2009, EPA has offered two online tools for home energy savings: the Home Energy Yardstick, which allows homeowners to compare their homes' energy use with others across the country; and the interactive ENERGY STAR Home Advisor, which provides homeowners customized recommendations for improving the energy efficiency of their homes.	Public, home improvement contractors	Online tools, Web site, written collateral (factsheets, brochures, etc.)	http://www. energystar.gov/ homeimprovement
EPA/Institute for Tribal Environmental Professionals (ITEP)	Through a cooperative agreement with ITEP at Northern Arizona University, EPA has supported development of a national climate change adaptation planning training program and online resources for tribes. In the first two years of the agreement, 87 people from 62 tribes or tribal organizations have been trained in developing adaptation plans to prepare for the expected impacts of climate change.	Tribes and tribal organizations, public officials	Formal/ informal	http://www4.nau. edu/tribal climatechange/
Public				
National Oceanic and Atmospheric Administration (NOAA)				
NOAA@NSIDC (National Snow and Ice Data Center)	NSIDC manages about 60 NOAA data sets, and publishes several new data sets each year, with an emphasis on *in situ* data, digitizing old and sometimes forgotten but valuable analog data, and data sets from operational communities, such as the U.S. Navy. NSIDC also helps develop educational pages, created Google Earth™ files that enable the public to overlay data-based images on a virtual globe, and houses many photographic prints of glaciers, taken from the air and the ground. Partners: NSF, NASA	Public	Informal	http://nsidc.org/ data/virtual_ globes/

Table 9-1 (Continued)   **Federal Climate Change Programs Grouped by Primary Audience**

Program Name	Description	Audiences	Learning Setting	Web Site
Sea Grant Office	Administered through NOAA, the National Sea Grant Program is a nationwide network of 32 university-based programs that work with coastal communities. Sea Grant College engages this network of the nation's top universities in conducting scientific research, education, training, and extension projects designed to foster science-based decisions about the use and conservation of natural resources and to increase coastal resiliency. The Sea Grant network is engaged in a multifaceted and diverse series of programs to address climate change in coastal and Great Lakes regions.	Public	Formal/ informal	http://www. seagrant.noaa.gov/
National Science Foundation (NSF)				
Antarctic Artists and Writers	This program supports writing and artistic projects specifically designed to increase understanding and appreciation of the Antarctic and of human activities on the southernmost continent.	Public	Informal	http://www.nsf. gov/funding/pgm_ summ.jsp?pims_ id=503518
Smithsonian Institution				
GEO-Panamá	The Smithsonian Tropical Research Institute (STRI) contributed to GEO-Panamá, a series of publications that appeared in *La Prensa*, a major newspaper in Panama. This series touched on issues related to climate change, including ocean level rise in Panama City.	Public	Informal	http://www.stri. si.edu/english/ about_stri/ headline_news/ news/article. php?id=684
Green Revolution	The Smithsonian Institution Traveling Exhibition Service has partnered with Chicago's Museum of Science and Industry to present Green Revolution, a fully digital exhibition that gives host organizations the power to build (and control) their own "eco-zibit." Green Revolution is a multiplatform initiative that focuses on several major themes: waste, energy, green pioneers, gardening and composting, green construction, and our carbon footprint.	Public	Informal	http://www. sites.si.edu/ greenRevolution/ index.htm
Nuestra casa en el universo	STRI communication associate Jorge Ventocilla and Catherine Potvin, from McGill University, edited a 44-page book *Nuestra casa en el universo* (Our Home in the Universe), an educational tool on climate change, and the Reducing Emissions from Deforestation and Forest Degradation proposal for indigenous communities in the Latin American tropics.	Public	Informal	http://www.stri. si.edu/nuestracasa/ nuestra_casa.pdf
Ocean Month Annual Forum	STRI hosts an annual forum in celebration of Ocean Month. Climate change has been included in the program for this forum since 2005.	Public	Informal	https://www.stri. si.edu/english/ education_ fellowships/index. php
Public Outreach Program	STRI's public outreach program includes the "Smithsonian Talk of the Month" in Colón, Panamá. It provides STRI researchers and those from other academic institutions working at STRI an opportunity to share the results of their studies with the people from Colón. Prior to each presentation, STRI guides and volunteers go to four local radio stations to invite the community to attend the talk. In several of the monthly talks, researchers approach issues of climate change and its impact on countries such as Panamá, where the bulk of the population lives along the coast. These talks and other public outreach efforts stress that the Galeta Point Laboratory's instruments show rising sea levels, and it is essential to protect the local coastal and marine habitats, including coral reefs, seagrass beds, mangroves, wetlands, and lowland forests.	Public	Informal	https://www.stri. si.edu/english/ education_ fellowships/index. php

Table 9-1 (Continued)   **Federal Climate Change Programs Grouped by Primary Audience**

Program Name	Description	Audiences	Learning Setting	Web Site
Salamander Lab	Outreach for zoo visitors is conducted through the Salamander Lab, located at the Reptile Discovery Center, on research projects relating to climate change, specifically on the hellbender salamander and the Shenandoah salamander.	Public	Informal	http://nationalzoo.si.edu/ActivitiesAnd Events/Celebrations ambassadors.cfm
U.S. Department of Agriculture (USDA)				
Climate Change in the Southern Region	This Web site provides information on upcoming climate change seminars, climate-related reading materials, regional and agency climate initiatives, and tips for reducing one's carbon footprint. Leaders from various resource areas participate in regionwide climate change seminars, whose topics include region-specific information, adaptation, carbon, and planning.	Public	Formal/ informal	http://fsweb.r8.fs.fed.us/climate/index.php
Forest Service Research Web Site	This Web site provides online access to U.S. Forest Service (USFS) climate change research.	Public	Formal/ informal	http://www.fs.fed.us/research/climate/usfs-cc-research.shtml
Treesearch	This online search engine provides access to almost 30,000 USFS publications, including more than 4,500 climate change-related publications for the general public and land managers.	Public	Formal/ informal	http://www.treesearch.fs.fed.us
U.S. Department of Energy (DOE)				
Atmospheric Research Measurement (ARM) Climate Research Facility	Through DOE's Office of Science, ARM provides online materials to develop basic science awareness related to climate change and supports community outreach in ARM site regions.	Public	Formal/ informal	education.arm.gov
Energy Literacy: Essential Principles and Fundamental Concepts for Energy Education	Intended to bolster energy literacy for all citizens, this document serves as a framework to teach energy using science, technology, and social science principles. Led by DOE, it was agreed to by 13 federal agencies, with significant public input.	K to gray	Online pamphlet; also, an alignment tool for educators to ensure all principles used	http://www1.eere.energy.gov/education/energy_literacy.html
National Training & Education Resource (NTER)	This DOE-created online, open-source training platform allows anyone to upload course materials or create content using state-of-the art tools to create immersive content.	All ages, but expected to be used for adults primarily	Online, open-source training platform	http://nterlearning.org
Solar Career Map	Explores a range of solar energy occupations, describing diverse jobs across the industry, charting possible progression between them, and identifying the high-quality training necessary to do them well.	Public	Career visualization online tool	http://www1.eere.energy.gov/solar/careermap/
U.S. Department of Transportation (DOT)				
University Transportation Centers Program	This program awards grants to universities across the United States to advance the state of the art in transportation research and to develop the next generation of transportation professionals.	Public	University	http://utc.dot.gov/

Table 9-1 (Continued)   **Federal Climate Change Programs Grouped by Primary Audience**

Program Name	Description	Audiences	Learning Setting	Web Site
Transportation and Climate Change Clearinghouse	Designed as a one-stop source of information on transportation and climate change issues, this clearinghouse includes information on GHG inventories, analytic methods and tools, GHG reduction strategies, potential impacts of climate change on transportation infrastructure, and approaches for integrating climate change considerations into transportation decision making. The clearinghouse is funded jointly through the National Cooperative Highway Research Program and DOT's Center for Climate Change and Environmental Forecasting.	Public	Web site	http://www.climate.dot.gov

U.S. Environmental Protection Agency (EPA)

Program Name	Description	Audiences	Learning Setting	Web Site
Climate "Back to Basics" Informational Materials	Among the resources available on the Climate Change Web site and in print form is a series of "What You Can Do" fact sheets and Web pages that provide more than 25 easy steps to reduce GHG emissions and also increase energy efficiency and save resources. This information features actions that readers can take at home, at the office, on the road, and at school. A related science education resource for adults and students is the brochure "Frequently Asked Questions about Global Warming and Climate Change: Back to Basics," available in print and at http://epa.gov/climatechange/science/multimedia.html. The brochure addresses key questions asked by the public about this issue by restating in easy-to-understand language the most current climate science from widely accepted, peer-reviewed scientific literature.	Public	Formal/informal	www.epa.gov/climatechange/wycd
EPA Climate Change and Health Effects on Older Adults Web Site	EPA's Aging Initiative has created a Web page that contains a fact sheet entitled "It's Too Darn Hot: Planning for Excessive Heat Events." The fact sheet has been widely disseminated throughout aging and public health networks. In an effort to reach people for whom English is not their first language, this fact sheet was translated into 15 languages. "Beat the Heat" posters highlighting key messages about steps to take during extreme heat are available in English and Spanish and have been shared in senior centers around the country.	Public	Formal/informal	http://www.epa.gov/aging/resources/climatechange/index.htm
EPA Climate Change Web Site	Managed by EPA's Climate Change Division, the site is among the top results for "climate change" across search engines and averages more than 200,000 unique visitors a month. Updated in 2012, the site features information about climate change science, greenhouse gas (GHG) emissions and inventories, health and environmental impacts, adaptation activities and opportunities, EPA's varied activities on the issue, what individuals can do, frequent questions, and other educational resources.	Public	Formal/informal	www.epa.gov/climatechange
EPA's Online Tools for Accessing Facility-Level Greenhouse Gas Data	EPA's Greenhouse Gas Reporting Program collects GHG data from large sources of GHG emissions and suppliers of products that release GHGs when released or combusted. EPA has developed easy-to-use online tools to share publicly available information gathered annually since 2010. The Facility-Level Information on GreenHouse gases Tool (FLIGHT) allows users to filter and view emissions data by facility, industry, location, or gas, and to create pie charts and other graphics based on custom searches. In the spring of 2013, FLIGHT will also be available as an application for mobile devices. The full set of nonconfidential GHG data collected through the program is also available through Envirofacts, EPA's one-stop-shop for environmental information.	Public	Formal/informal	http://www.epa.gov/ghgreporting/

Table 9-1 (Continued)   **Federal Climate Change Programs Grouped by Primary Audience**

Program Name	Description	Audiences	Learning Setting	Web Site
General Audiences (K–12, Undergraduate, and Graduate Students; K–12 Teachers; Informal Educators; Professionals; Public				
National Aeronautics and Space Administration (NASA)				
Earth Climate Course: What Determines a Planet's Climate?	This set of student activities and teachers' guides connects NASA Earth science research with the teaching and learning of core science and mathematics concepts and skills, while addressing national education standards. The four modules cover (1) Temperature Variations and Habitability, (2) Modeling Hot and Cold Planets, (3) Using Mathematical Models to Investigate Planetary Habitability, and (4) How Atmospheres Affect Planetary Atmospheres. Scientific inquiry and research tools play a major role in the lessons. Presented with a science problem, students seek answers and consensus by experimenting with physical and computer models, collecting and analyzing their own measurements, and conducting comparisons with real-world data from satellites and ground-based observations.	K–12 students, K–12 teachers, undergraduate students	Formal	http://icp.giss.nasa. gov/education/ modules/eccm/
Earth Observatory	Earth Observatory shares the images, stories, and discoveries about climate and the environment that emerge from NASA research, including NASA's satellite missions, in-the-field research, and climate models.	Public, K–12 students, K–12 teachers, informal educators	Informal	http://earth observatory. nasa.gov/
Global Climate Change	This Web resource includes the planet's vital signs, feature stories, visualizations, and links to NASA missions involved in investigation of climate change. The interactive tool, Earth on the Earth 3D, provides near-real-time depiction of important climate variables from NASA Earth-observing satellites.	Public, K–12 students, K–12 teachers, informal educators	Informal	http://climate.nasa. gov/
MyNASAData	Working to make NASA Earth science data accessible to the K–12 and citizen scientist communities, the project's principal activity is to create "microsets" from large scientific data sets, and to wrap these with tools, lesson plans, and supporting documentation so that teachers can use the information in the classroom. Climate change-related lesson plans are available for middle and high schools.	K–12 teachers, K–12 students, citizen scientists	Formal/ informal	http://mynasadata. larc.nasa.gov/
NASA Minority University Research and Education Innovations in Climate Education (NICE)	NICE was created in fiscal year (FY) 2011 to extend the results of NASA's Earth Science Program to the education community by sponsoring unique and stimulating opportunities for global climate and Earth system science education at minority-serving institutions. NICE is designed to improve the quality of the nation's STEM education and enhance faculty, student, and teacher access to NASA-unique content related to global climate and Earth system change. In FY 2013, NICE is focusing on tribal colleges and universities. Partners: NSF, NOAA, ICE-t green team	K–12 teachers, undergraduate students	Formal/ informal	http://www. nasa.gov/offices/ education/ programs/ descriptions/ NASA_Innovations_ in_Climate_ Education.html
National Oceanic and Atmospheric Administration (NOAA)				
Climate.gov	This source of timely and authoritative scientific data and information about climate works to promote public understanding of climate science and climate-related events, to make its data products and services easy to access and use, to provide climate-related support to the private sector and the nation's economy, and to serve people making climate-related decisions with tools and resources that help them answer specific questions.	Public, professionals, K–12 teachers, informal educators, graduate students, undergraduate students, K–12 students	Informal	http://www.climate. gov/

Table 9-1 (Continued)   **Federal Climate Change Programs Grouped by Primary Audience**

Program Name	Description	Audiences	Learning Setting	Web Site
National Science Foundation (NSF)				
Antarctic Integrated System Science (AISS)	The discoveries of disciplinary science increasingly highlight the need for integrative approaches to forge new understanding of the complex interactions that govern Antarctica and its past, present, and future roles in the Earth system. AISS was established to respond to this need and foster progress on some of the most pressing issues on a planet subject to potentially accelerated change. AISS administers projects that transcend disciplinary boundaries, are highly integrated, and address questions broader in scope than those typically supported by the disciplinary programs described above.	Undergraduate students, graduate students, K–12 teachers, professionals, public	Formal, informal	http://www.nsf.gov/funding/pgm_summ.jsp?pims_id=503240
Arctic Research and Education Program	This program supports activities that bridge research and education in concert with funded research grants and agreements through supplement requests or as separate proposal requests to support new ventures. Arctic research spans the major STEM fields and is often multi- or interdisciplinary. Research in the Arctic has clear applications for education and outreach at many levels. The region itself is an interesting hook for teaching about life, physical, and social sciences, and such concepts as ocean and atmosphere circulation, climate, Earth system science, animal migrations, and life in extreme environments.	Undergraduate students, graduate students, K–12 teachers, professionals	Formal (K-12, undergrad, grad)	http://www.nsf.gov/funding/pgm_summ.jsp?pims_id=13448
Arctic Research Opportunities	The goal of NSF's Arctic Sciences Section is to gain a better understanding of the Arctic's physical, biological, geological, chemical, social, and cultural processes; the interactions of oceanic, terrestrial, atmospheric, biological, social, cultural, and economic systems; and the connections that define the Arctic. This umbrella solicitation provides detailed information on research opportunities to be supported by the Arctic Natural Sciences, Arctic System Science, Arctic Social Sciences, Arctic Observing Network, and Advanced Cyberinfrastructure programs.	Undergraduate students, graduate students, K–12 teachers, professionals, public	Formal, informal	http://www.nsf.gov/funding/pgm_summ.jsp?pims_id=5521
Climate Change Education Partnership (CCEP)	CCEP seeks to establish a coordinated national network of regionally or thematically based partnerships devoted to increasing the adoption of effective, high-quality educational programs and resources related to the science of climate change and its impacts. Each CCEP is required to be of a large enough scale that it will have catalytic or transformative impacts that cannot be achieved through other core NSF program awards.	Undergraduate students, graduate students, K–12 teachers, professionals	Formal (K-12, undergrad, grad)	http://www.nsf.gov/funding/pgm_summ.jsp?pims_id=503477
Coastal SEES	Coastal SEES (Science, Engineering and Education for Sustainability) is focused on the sustainability of coastal systems. For this solicitation, NSF defines coastal systems as the swath of land closely connected to the sea, including barrier islands, wetlands, mudflats, beaches, estuaries, cities, towns, recreational areas, and maritime facilities; the continental seas and shelves; and the overlying atmosphere. These systems are subject to complex and dynamic interactions among natural and human-driven processes.	Undergraduate students, graduate students, K–12 teachers, professionals	Formal (K-12, undergrad, grad)	http://www.nsf.gov/funding/pgm_summ.jsp?pims_id=504816
Cyber-Enabled Sustainability Science and Engineering (CyberSEES)	CyberSEES's goal is to advance interdisciplinary research in which the science and engineering of sustainability are enabled by advances in computing, and where computational innovation is grounded in the context of sustainability problems.	Undergraduate students, graduate students, K–12 teachers, professionals	Formal (K-12, undergrad, grad)	http://www.nsf.gov/funding/pgm_summ.jsp?pims_id=504829

Table 9-1 (Continued)   **Federal Climate Change Programs Grouped by Primary Audience**

Program Name	Description	Audiences	Learning Setting	Web Site
Dynamics of Coupled Natural and Human Systems (CNH)	The CNH competition promotes quantitative, interdisciplinary analyses of relevant human and natural system processes and complex interactions among human and natural systems at diverse scales.	Undergraduate students, graduate students, K-12 teachers	Formal (K-12, undergrad, grad)	http://www.nsf.gov/funding/pgm_summ.jsp?pims_id=13681
Long-Term Ecological Research (LTER)	Research at LTER sites provides experiments, databases, and research programs for use by other scientists. It must test important ecological or ecosystem theories, including ecosystem stability, biodiversity, community structure, and energy flow. LTER currently supports 26 active sites representing major biotic regions of the continental U.S. and Alaska, the marine environment, and the Antarctic continent. Its disciplinary scope includes population and community ecology, ecosystem science, evolutionary biology, phylogenetic systematics, social and economic sciences, urban ecology, oceanography, mathematics, computer science, and science education.	Undergraduate students, graduate students, K-12 teachers, professionals	Formal (K-12, undergrad, grad)	http://www.nsf.gov/funding/pgm_summ.jsp?pims_id=7671
Ocean Acidification	The need for understanding the potential adverse impacts of a slowly acidifying sea upon marine ecosystems is widely recognized and included as a priority objective in the new National Ocean Policy. The effects of ocean acidification could significantly affect strategies for developing practices enhancing the sustainability of ocean resources. This program supports basic research concerning the nature, extent, and impact of ocean acidification on oceanic environments in the past, present, and future.	Undergraduate students, graduate students, K-12 teachers, professionals	Formal (K-12, undergrad, grad)	
Sustainability Research Networks (SRNs)	SRNs will engage and explore fundamental theoretical issues and empirical questions in sustainability science, engineering, and education that will increase understanding of the ultimate sustainability challenge—maintaining and improving the quality of life for the nation within a healthy Earth system. SRNs will link scientists, engineers, and educators, at existing institutions, centers, and networks, and will also develop new research efforts and collaborations.	Undergraduate students, graduate students, K-12 teachers, professionals	Formal (K-12, undergrad, grad)	http://www.nsf.gov/funding/pgm_summ.jsp?pims_id=503645
U.S. Department of Energy (DOE)				
Collegiate Wind Competition	This competition is designed to strengthen the future wind workforce by connecting industry to young innovators and inspiring career choices in wind energy. The competition is a forum for college students from multiple disciplines to investigate innovative wind energy concepts; gain experience designing, building, and testing a wind turbine to perform according to a customized market data-derived business plan; and increase their knowledge of wind industry barriers.	Undergraduate students and faculty; professionals in the wind industry	Competition event	http://www.windpoweringamerica.gov/filter_detail.asp?itemid=3777
Student & Educator Resources Web Page	Provides age-appropriate educational resources for K-12 and higher-education students looking to learn more about the biomass field, a list of biomass-related academic institutions for students interested in pursuing higher education in the field, and resources for educators teaching bioenergy-related lessons at the K-12 level.	K-12 students, undergraduate and graduate students, educators	Web site	http://www1.eere.energy.gov/biomass/for_students.html
Wind for Schools	This program installs small wind turbines at rural elementary and secondary schools and develops Wind Application Centers at higher-education institutions. Wind Application Centers provide technical assistance in all aspects of wind energy to rural schools and communities.	K-12 students and teachers; undergraduate students and faculty; general public	Classroom; on-site training; Web site; software	http://www.windpoweringamerica.gov/schools/projects.asp

Appendices

Appendix A: Emission Trends

Appendix B: Bibliography

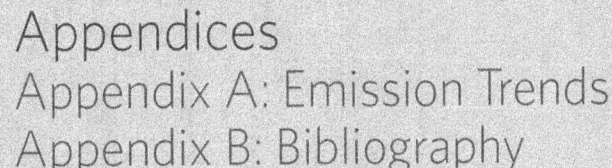

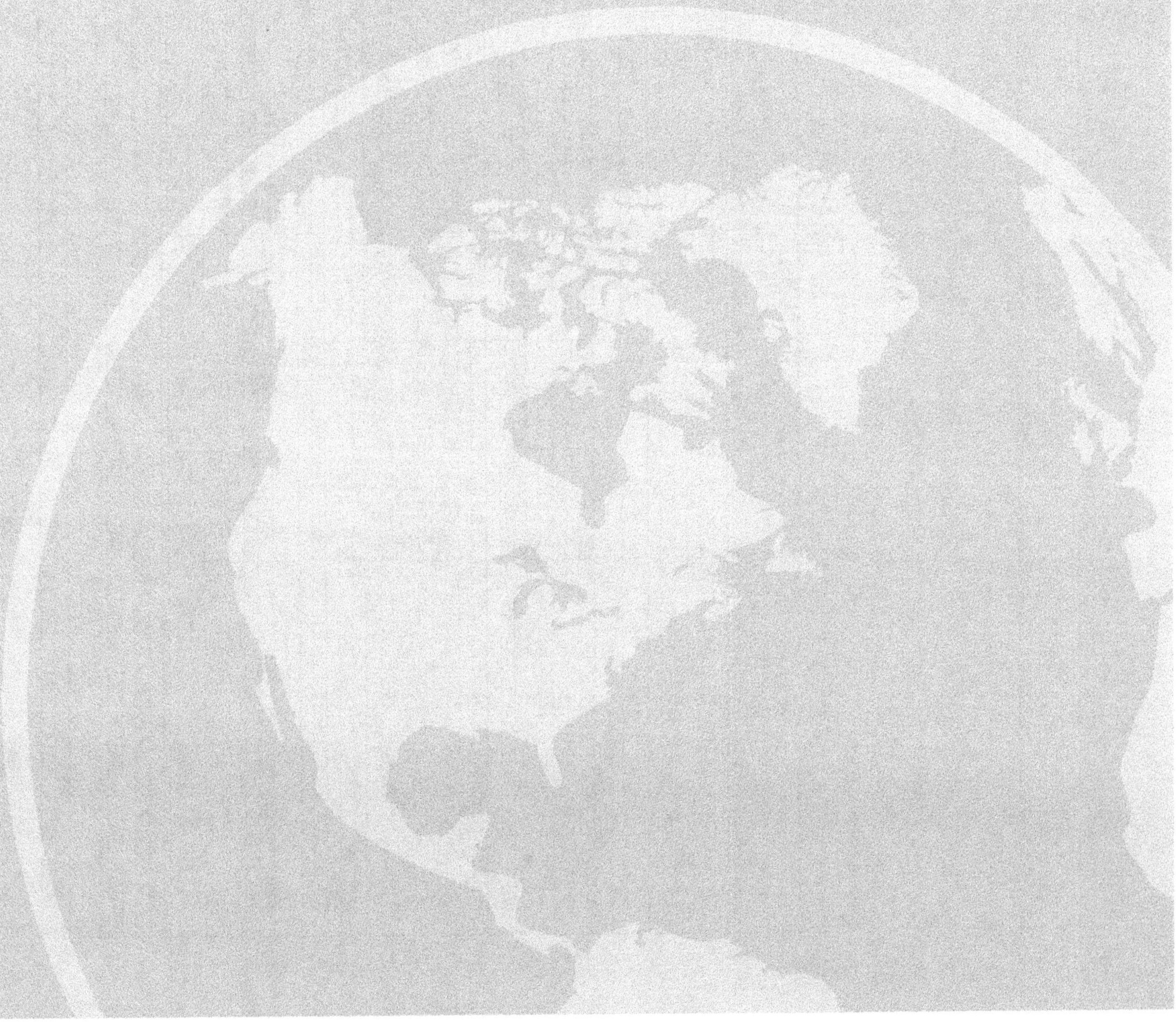

A

Appendix A: Emission Trends

Inventory 2011
Submission 2013 v1.1
UNITED STATES OF AMERICA

TABLE 10 EMISSION TRENDS
CO₂
(Part 1 of 3)

GREENHOUSE GAS SOURCE AND SINK CATEGORIES	Base year (1990)	1991	1992	1993	1994	1995	1996	1997	1998	1999
	(Gg)	(Gg)	(Gg)	(Gg)	(Gg)	(Gg)	(Gg)	(Gg)	(Gg)	(Gg)
1. Energy	4,911,976.68	4,872,772.30	4,976,519.13	5,089,990.69	5,170,741.85	5,226,111.72	5,412,334.61	5,484,245.98	5,525,875.69	5,601,482.77
A. Fuel Combustion (Sectoral Approach)	4,873,918.02	4,834,499.06	4,938,526.88	5,048,656.81	5,129,366.13	5,183,184.46	5,371,912.32	5,444,270.86	5,495,895.65	5,570,473.87
1. Energy Industries	1,820,817.12	1,818,191.70	1,831,538.76	1,906,903.88	1,931,238.84	1,947,924.74	2,020,993.05	2,083,398.69	2,177,387.99	2,190,523.00
2. Manufacturing Industries and Construction	848,555.99	828,273.81	859,446.33	858,303.65	866,789.18	870,390.21	906,648.20	907,158.15	868,801.78	844,615.05
3. Transport	1,445,418.12	1,401,573.35	1,463,597.87	1,499,200.33	1,543,917.64	1,577,849.60	1,621,928.25	1,638,364.16	1,674,047.95	1,730,141.07
4. Other Sectors	557,309.28	571,993.32	574,119.16	585,859.54	579,027.15	578,216.84	617,416.71	598,167.15	546,978.38	569,130.78
5. Other	201,817.51	214,466.87	209,824.76	198,389.41	208,393.33	208,803.06	204,926.12	212,182.71	228,679.54	236,063.98
B. Fugitive Emissions from Fuels	38,058.66	38,273.24	37,992.25	41,333.89	41,375.71	42,927.26	40,422.28	39,975.12	29,980.04	31,008.90
1. Solid Fuels	IE,NE,NO	IE,NE,NO	IE,NE,NO	IE,NE,NO	IE,NE,NO	IE,NE,NO	IE,NE,NO	IE,NE,NO	IE,NE,NO	IE,NE,NO
2. Oil and Natural Gas	38,058.66	38,273.24	37,992.25	41,333.89	41,375.71	42,927.26	40,422.28	39,975.12	29,980.04	31,008.90
2. Industrial Processes	188,717.28	177,814.47	180,398.17	177,914.37	183,846.08	190,043.98	190,110.57	193,360.11	189,884.98	197,404.23
A. Mineral Products	54,030.69	52,404.62	53,112.52	54,947.18	57,390.83	60,444.12	61,873.96	62,914.07	64,348.48	65,310.42
B. Chemical Industry	24,774.71	24,610.50	25,905.75	26,167.76	27,364.96	26,832.24	26,818.70	27,464.43	28,695.84	27,270.83
C. Metal Production	109,911.89	100,799.35	101,379.89	96,799.43	99,090.28	102,767.62	101,417.90	102,981.61	96,840.66	94,822.95
D. Other Production	NE	NE	NE	NE	NE	NE	NE	NE	NE	NE
E. Production of Halocarbons and SF₆										
F. Consumption of Halocarbons and SF₆										
G. Other	NA,NO	NA,NO	NA,NO	NA,NO	NA,NO	NA,NO	NA,NO	NA,NO	NA,NO	NA,NO
3. Solvent and Other Product Use	NA,NE	NA,NE	NA,NE	NA,NE	NA,NE	NA,NE	NA,NE	NA,NE	NA,NE	NA,NE
4. Agriculture										
A. Enteric Fermentation										
B. Manure Management										
C. Rice Cultivation										
D. Agricultural Soils										
E. Prescribed Burning of Savannas										
F. Field Burning of Agricultural Residues										
G. Other										
5. Land Use, Land-Use Change and Forestry(a)	-786,411.70	-791,029.77	-776,901.37	-776,683.37	-825,292.74	-789,863.62	-818,112.84	-781,566.48	-720,426.70	-651,652.02
A. Forest Land	-565,052.14	-564,307.51	-558,757.27	-564,024.58	-587,063.90	-597,140.53	-608,292.76	-570,477.04	-517,297.45	-448,056.65
B. Cropland	-5,985.06	-16,809.84	-7,639.18	-10,023.61	-20,435.67	2,768.90	-5,248.04	-7,746.33	-1,296.97	-2,703.32
C. Grassland	-12,948.83	-17,019.71	-16,557.72	-13,367.45	-27,733.65	-11,327.25	-27,705.00	-19,540.19	-20,244.47	-14,002.29
D. Wetlands	1,033.48	962.28	919.55	980.58	937.87	1,017.94	872.25	1,037.48	1,084.22	1,154.91
E. Settlements	-47,495.47	-48,589.18	-49,682.89	-50,776.60	-51,870.31	-52,964.02	-54,057.73	-55,151.44	-56,245.15	-57,338.86
F. Other Land	NE	NE	NE	NE	NE	NE	NE	NE	NE	NE
G. Other	-155,963.69	-145,265.81	-145,183.86	-139,471.70	-139,127.09	-132,218.67	-123,681.56	-129,688.95	-126,426.89	-130,675.83
6. Waste	IE,NA,NE	IE,NA,NE	IE,NA,NE	IE,NA,NE	IE,NA,NE	IE,NA,NE	IE,NA,NE	IE,NA,NE	IE,NA,NE	IE,NA,NE
A. Solid Waste Disposal on Land	NA,NE	NA,NE	NA,NE	NA,NE	NA,NE	NA,NE	NA,NE	NA,NE	NA,NE	NA,NE
B. Waste-water Handling										
C. Waste Incineration	IE	IE	IE	IE	IE	IE	IE	IE	IE	IE
D. Other	NA	NA	NA	NA	NA	NA	NA	NA	NA	NA
7. Other (as specified in Summary 1.4)	NA	NA	NA	NA	NA	NA	NA	NA	NA	NA
Total CO₂ emissions including net CO₂ from LULUCF	4,314,282.25	4,259,557.00	4,380,015.92	4,491,221.69	4,529,295.18	4,626,292.07	4,784,332.33	4,896,039.61	4,995,333.97	5,137,264.98
Total CO₂ emissions excluding net CO₂ from LULUCF	5,100,693.96	5,050,586.77	5,156,917.29	5,267,905.06	5,354,587.92	5,416,155.70	5,602,445.17	5,677,606.09	5,715,760.67	5,788,887.01
Memo Items:										
International Bunkers	103,462.57	117,569.49	107,862.97	97,829.15	96,689.41	98,491.64	99,749.73	106,960.91	110,490.71	102,733.04
Aviation	38,033.60	46,339.14	46,769.35	46,889.85	48,342.47	49,903.00	51,029.10	54,485.17	54,080.46	57,557.15
Marine	65,428.97	71,230.35	61,093.62	50,939.30	48,346.94	48,588.64	48,720.63	52,475.74	56,410.25	45,175.90
Multilateral Operations	IE	IE	IE	IE	IE	IE	IE	IE	IE	IE
CO₂ Emissions from Biomass	218,636.81	219,424.05	229,781.83	224,870.28	231,324.16	236,105.48	240,451.49	234,653.56	217,304.31	220,560.72

Note: All footnotes for this table are given at the end of the table on sheet 5.

TABLE 10 EMISSION TRENDS
CO$_2$
(Part 2 of 3)

Inventory 2011
Submission 2013 v1.1
UNITED STATES OF AMERICA

GREENHOUSE GAS SOURCE AND SINK CATEGORIES	2000	2001	2002	2003	2004	2005	2006	2007	2008	2009
	(Gg)	(Gg)	(Gg)	(Gg)	(Gg)	(Gg)	(Gg)	(Gg)	(Gg)	(Gg)
1. Energy	5,777,296.93	5,695,045.39	5,736,040.05	5,789,216.83	5,902,457.68	5,934,056.22	5,883,547.56	5,946,413.61	5,774,919.28	5,390,590.53
A. Fuel Combustion (Sectoral Approach)	5,747,185.88	5,665,482.42	5,705,694.84	5,760,046.39	5,873,577.83	5,903,827.51	5,823,193.81	5,915,251.43	5,741,997.16	5,358,083.23
1. Energy Industries	2,296,890.10	2,257,925.58	2,272,680.88	2,304,169.42	2,337,043.46	2,402,142.06	2,346,406.47	2,412,826.58	2,360,919.64	2,146,415.03
2. Manufacturing Industries and Construction	844,268.07	837,047.12	824,031.80	822,784.48	846,630.63	823,408.24	848,133.70	844,420.34	802,039.69	722,627.08
3. Transport	1,775,023.92	1,759,576.67	1,802,183.86	1,793,353.31	1,839,740.82	1,864,177.09	1,866,595.77	1,879,300.98	1,790,964.97	1,726,751.85
4. Other Sectors	601,487.60	586,888.75	584,713.51	613,792.81	602,363.03	581,411.81	530,091.37	560,523.05	570,720.39	560,392.34
5. Other	229,516.18	224,044.30	222,084.78	225,946.36	247,799.90	232,688.31	231,966.50	218,180.48	217,352.47	201,896.93
B. Fugitive Emissions from Fuels	30,111.05	29,562.96	30,345.21	29,170.44	28,879.84	30,228.71	30,353.76	31,162.18	32,922.12	32,507.31
1. Solid Fuels	IE,NE,NO	IE,NE,NO	IE,NE,NO	IE,NE,NO	IE,NE,NO	IE,NE,NO	IE,NE,NO	IE,NE,NO	IE,NE,NO	IE,NE,NO
2. Oil and Natural Gas	30,111.05	29,562.96	30,345.21	29,170.44	28,879.84	30,228.71	30,353.76	31,162.18	32,922.12	32,507.31
2. Industrial Processes	185,404.56	167,395.45	167,352.75	162,296.03	167,867.98	166,346.91	170,567.12	172,904.22	160,264.44	119,010.83
A. Mineral Products	63,673.15	63,022.08	64,894.72	64,256.63	69,396.86	70,746.96	73,069.12	70,954.35	65,245.87	51,378.21
B. Chemical Industry	25,844.70	21,597.44	22,945.80	21,397.67	22,591.98	21,816.66	21,185.12	23,283.67	20,415.81	18,657.03
C. Metal Production	95,886.71	83,175.93	79,512.23	76,641.72	75,879.15	73,783.29	76,312.88	78,666.19	74,602.77	48,975.59
D. Other Production	NE	NE	NE	NE	NE	NE	NE	NE	NE	NE
E. Production of Halocarbons and SF$_6$										
F. Consumption of Halocarbons and SF$_6$	NA,NO	NA,NO	NA,NO	NA,NO	NA,NO	NA,NO	NA,NO	NA,NO	NA,NO	NA,NO
G. Other	NA,NE	NA,NE	NA,NE	NA,NE	NA,NE	NA,NE	NA,NE	NA,NE	NA,NE	NA,NE
3. Solvent and Other Product Use										
4. Agriculture										
A. Enteric Fermentation										
B. Manure Management										
C. Rice Cultivation										
D. Agricultural Soils										
E. Prescribed Burning of Savannas										
F. Field Burning of Agricultural Residues										
G. Other										
5. Land Use, Land-Use Change and Forestry[a]	-673,063.43	-729,428.52	-850,533.24	-949,294.76	-991,490.63	-988,895.15	-953,053.88	-919,968.64	-892,975.15	-874,300.40
A. Forest Land	-431,111.76	-553,467.36	-679,349.21	-791,020.11	-817,448.71	-799,624.91	-764,068.13	-757,035.42	-757,052.84	-757,472.84
B. Cropland	-11,157.97	6,979.89	20,948.07	22,769.56	14,106.08	1,055.40	17,893.21	16,127.35	17,974.03	17,163.32
C. Grassland	-47,433.73	-18,554.93	-22,414.32	-15,212.38	-11,208.67	-11,248.52	-24,831.23	-1,884.33	-1,768.53	-1,651.68
D. Wetlands	1,227.28	1,140.27	1,000.95	983.07	1,077.08	1,078.91	878.94	1,011.63	992.14	1,088.63
E. Settlements	-58,432.57	-59,377.38	-60,322.19	-61,267.00	-62,211.82	-63,156.63	-64,101.44	-65,046.25	-65,991.06	-66,935.87
F. Other Land	NE	NE	NE	NE	NE	NE	NE	NE	NE	NE
G. Other	-126,094.68	-106,149.02	-110,396.53	-105,547.89	-115,804.60	-116,999.40	-118,825.23	-113,141.62	-87,128.89	-66,491.96
6. Waste	IE,NA,NE	IE,NA,NE	IE,NA,NE	IE,NA,NE	IE,NA,NE	IE,NA,NE	IE,NA,NE	IE,NA,NE	IE,NA,NE	IE,NA,NE2
A. Solid Waste Disposal on Land	NA,NE	NA,NE	NA,NE	NA,NE	NA,NE	NA,NE	NA,NE	NA,NE	NA,NE	NA,NE
B. Waste-water Handling										
C. Waste Incineration	IE	IE	IE	IE	IE	IE	IE	IE	IE	IE
D. Other	NA	NA	NA	NA	NA	NA	NA	NA	NA	NA
7. Other (as specified in Summary 1.A)	NA	NA	NA	NA	NA	NA	NA	NA	NA	NA
Total CO$_2$ emissions including net CO$_2$ from LULUCF	5,289,698.06	5,133,412.32	5,052,859.56	5,002,218.10	5,078,835.02	5,111,507.97	5,071,060.80	5,199,349.19	5,042,208.58	4,635,300.96
Total CO$_2$ emissions excluding net CO$_2$ from LULUCF	5,962,701.49	5,862,840.84	5,903,392.80	5,951,512.86	6,070,325.66	6,100,403.13	6,024,114.68	6,119,317.83	5,935,183.73	5,509,601.36
Memo Items:										
International Bunkers	101,726.18	93,731.32	94,442.98	98,309.91	108,391.00	113,139.25	114,115.98	115,345.34	114,341.85	106,410.32
Aviation	62,029.31	56,384.52	54,626.24	55,196.36	56,239.23	60,125.45	60,283.69	61,489.49	56,145.71	52,785.00
Marine	39,696.86	37,346.80	39,816.74	43,113.55	52,151.77	53,013.80	53,832.30	53,855.85	58,196.14	53,625.32
Multilateral Operations	IE	IE	IE	IE	IE	IE	IE	IE	IE	IE
CO$_2$ Emissions from Biomass	226,555.50	202,498.82	203,559.77	208,724.36	224,089.19	228,651.10	232,668.86	238,307.61	251,734.38	245,057.04

Note: All footnotes for this table are given at the end of the table on sheet 5.

TABLE 10 EMISSION TRENDS
CO$_2$
(Part 3 of 3)

Inventory 2011
Submission 2013 v1.1
UNITED STATES OF AMERICA

GREENHOUSE GAS SOURCE AND SINK CATEGORIES	2010 (Gg)	2011 (Gg)	Change from base to latest reported year %
1. Energy	5,585,641.74	5,452,528.41	11.00
A. Fuel Combustion (Sectoral Approach)	5,552,996.00	5,419,837.29	11.20
1. Energy Industries	2,259,189.96	2,158,510.32	18.55
2. Manufacturing Industries and Construction	780,239.67	773,192.26	-8.88
3. Transport	1,742,149.61	1,725,577.55	19.38
4. Other Sectors	555,204.37	550,857.14	-1.16
5. Other	216,212.39	211,700.02	4.90
B. Fugitive Emissions from Fuels	IE,NE,NO	IE,NE,NO	-14.10
1. Solid Fuels	IE,NE,NO	IE,NE,NO	0.00
2. Oil and Natural Gas	32,645.75	32,691.12	-14.10
2. Industrial Processes	141,396.86	151,292.18	-19.83
A. Mineral Products	57,806.43	58,590.21	8.44
B. Chemical Industry	21,736.70	21,664.69	-12.55
C. Metal Production	61,853.72	71,037.27	-35.37
D. Other Production	NE	NE	0.00
E. Production of Halocarbons and SF$_6$			
F. Consumption of Halocarbons and SF$_6$			
G. Other	NA,NO	NA,NO	0.00
3. Solvent and Other Product Use	NA,NE	NA,NE	0.00
4. Agriculture			
A. Enteric Fermentation			
B. Manure Management			
C. Rice Cultivation			
D. Agricultural Soils			
E. Prescribed Burning of Savannas			
F. Field Burning of Agricultural Residues			
G. Other			
5. Land Use, Land-Use Change and Forestry[a]	-879,410.48	-806,007.06	13.94
A. Forest Land	-758,184.94	-761,804.08	34.82
B. Cropland	19,884.34	19,765.20	-40.24
C. Grassland	-1,502.25	-1,354.10	-89.54
D. Wetlands	1,009.91	917.70	-11.20
E. Settlements	-67,880.69	-68,825.50	44.91

TABLE 10 EMISSION TRENDS

CH₄

(Part 1 of 3)

Inventory 2011
Submission 2013 v1.1
UNITED STATES OF AMERICA

GREENHOUSE GAS SOURCE AND SINK CATEGORIES	Base year (1990)	1991	1992	1993	1994	1995	1996	1997	1998	1999
	(Gg)	(Gg)	(Gg)	(Gg)	(Gg)	(Gg)	(Gg)	(Gg)	(Gg)	(Gg)
1. Energy	14,219.52	14,148.50	13,971.66	13,558.34	13,663.08	13,452.42	13,572.07	13,580.26	13,212.73	12,867.55
A. Fuel Combustion (Sectoral Approach)	573.69	575.41	588.06	562.63	549.98	540.73	545.98	504.48	468.42	463.49
1. Energy Industries	15.77	15.75	15.86	16.54	16.87	17.02	17.47	18.08	19.27	19.42
2. Manufacturing Industries and Construction	85.56	83.47	86.05	87.11	90.66	91.58	93.51	95.22	90.33	89.62
3. Transport	207.57	201.91	201.03	197.98	193.94	187.48	178.86	170.82	162.22	150.83
4. Other Sectors	262.02	271.36	282.15	258.04	245.41	241.83	253.44	217.50	193.73	200.75
5. Other	2.76	2.91	2.97	2.97	3.09	2.82	2.70	2.85	2.87	2.89
B. Fugitive Emissions from Fuels	13,645.82	13,573.09	13,383.59	12,995.70	13,113.10	12,911.69	13,026.09	13,075.77	12,744.31	12,404.06
1. Solid Fuels	4,290.85	4,155.70	4,077.32	3,551.44	3,631.23	3,585.27	3,583.39	3,521.73	3,508.75	3,328.70
2. Oil and Natural Gas	9,354.98	9,417.39	9,306.27	9,444.27	9,481.87	9,326.41	9,442.70	9,554.04	9,235.56	9,075.36
2. Industrial Processes	155.63	160.01	165.08	169.19	179.65	187.71	192.89	199.45	201.13	208.64
A. Mineral Products	NA	NA	NA	NA	NA	NA	NA	NA	NA	NA
B. Chemical Industry	109.40	118.96	121.03	124.40	133.76	140.43	146.96	153.36	156.34	165.60
C. Metal Production	46.24	41.04	44.05	44.79	45.90	47.28	45.93	46.09	44.80	43.04
D. Other Production										
E. Production of Halocarbons and SF₆										
F. Consumption of Halocarbons and SF₆										
G. Other	NA,NO	NA,NO	NA,NO	NA,NO	NA,NO	NA,NO	NA,NO	NA,NO	NA,NO	NA,NO
3. Solvent and Other Product Use										
4. Agriculture	8,168.68	8,242.82	8,436.47	8,548.35	8,832.44	9,012.37	8,909.17	8,887.63	8,998.06	9,044.59
A. Enteric Fermentation	6,320.86	6,333.24	6,540.08	6,621.59	6,741.35	6,806.84	6,852.70	6,721.20	6,649.14	6,652.64
B. Manure Management	1,498.82	1,567.79	1,511.61	1,583.25	1,689.78	1,743.42	1,715.17	1,799.34	1,962.08	1,986.67
C. Rice Cultivation	339.21	333.19	374.79	334.24	391.13	362.90	331.75	356.24	376.26	394.87
D. Agricultural Soils	NA	NA	NA	NA	NA	NA	NA	NA	NA	NA
E. Prescribed Burning of Savannas	NA	NA	NA	NA	NA	NA	NA	NA	NA	NA
F. Field Burning of Agricultural Residues	9.78	8.61	9.99	9.17	10.18	9.22	9.56	10.25	10.58	10.40
G. Other	NA	NA	NA	NA	NA	NA	NA	NA	NA	NA
5. Land Use, Land-Use Change and Forestry	118.37	103.50	154.56	96.28	285.72	160.34	455.04	98.14	120.87	428.88
A. Forest Land	118.37	103.50	154.56	96.28	285.72	160.34	455.04	90.14	120.87	428.88
B. Cropland	NA	NA	NA	NA	NA	NA	NA	NA	NA	NA
C. Grassland	NA	NA	NA	NA	NA	NA	NA	NA	NA	NA
D. Wetlands	NE	NE	NE	NE	NE	NE	NE	NE	NE	NE
E. Settlements	NE	NE	NE	NE	NE	NE	NE	NE	NE	NE
F. Other Land	NE	NE	NE	NE	NE	NE	NE	NE	NE	NE
G. Other	NE	NE	NE	NE	NE	NE	NE	NE	NE	NE
6. Waste	7,810.46	7,870.33	7,918.24	7,878.41	7,820.60	7,484.08	7,321.62	6,915.28	6,569.63	6,355.75
A. Solid Waste Disposal on Land	7,037.07	7,083.45	7,110.88	7,065.09	6,988.68	6,643.42	6,476.41	6,054.10	5,703.75	5,478.19
B. Waste-water Handling	758.15	769.46	787.76	788.28	801.15	805.82	805.63	817.38	818.20	824.15
C. Waste Incineration	IE,NE	IE,NE	IE,NE	IE,NE	IE,NE	IE,NE	IE,NE	IE,NE	IE,NE	IE,NE
D. Other	15.24	17.42	19.60	25.04	30.77	34.84	39.59	43.80	47.68	53.42
7. Other (as specified in Summary 1.A)	NA	NA	NA	NA	NA	NA	NA	NA	NA	NA
Total CH₄ emissions including CH₄ from LULUCF	30,472.66	30,525.16	30,646.02	30,250.46	30,781.48	30,296.92	30,450.80	29,672.15	29,102.43	28,905.42
Total CH₄ emissions excluding CH₄ from LULUCF	30,354.28	30,421.66	30,491.46	30,154.19	30,495.77	30,136.58	29,995.76	29,582.01	28,981.56	28,476.54
Memo Items:										
International Bunkers	6.53	7.11	6.10	5.08	4.82	4.85	4.86	5.24	5.63	4.51
Aviation	NA	NA	NA	NA	NA	NA	NA	NA	NA	NA
Marine	6.53	7.11	6.10	5.08	4.82	4.85	4.86	5.24	5.63	4.51
Multilateral Operations	IE	IE	IE	IE	IE	IE	IE	IE	IE	IE
CO₂ Emissions from Biomass										

Note: All footnotes for this table are given at the end of the table on sheet 5.

TABLE 10 EMISSION TRENDS
CH₄
(Part 2 of 3)

Inventory 2011
Submission 2013 v1.1
UNITED STATES OF AMERICA

GREENHOUSE GAS SOURCE AND SINK CATEGORIES	2000	2001	2002	2003	2004	2005	2006	2007	2008	2009
	(Gg)	(Gg)	(Gg)	(Gg)	(Gg)	(Gg)	(Gg)	(Gg)	(Gg)	(Gg)
1. Energy	13,068.78	13,153.53	12,735.36	12,801.27	12,639.94	12,364.71	12,852.31	12,852.66	13,066.72	12,610.43
A. Fuel Combustion (Sectoral Approach)	468.60	443.90	429.71	432.24	431.63	428.47	399.91	404.35	405.00	385.39
1. Energy Industries	20.46	20.34	20.55	20.66	21.14	22.02	21.72	22.66	22.13	20.67
2. Manufacturing Industries and Construction	89.64	85.33	83.34	83.20	87.79	87.03	88.69	85.76	81.26	73.81
3. Transport	142.26	138.17	121.38	111.82	105.41	98.17	91.74	84.42	77.13	72.15
4. Other Sectors	213.41	196.53	201.23	212.89	213.40	217.58	194.17	208.22	221.50	215.61
5. Other	2.83	3.53	3.21	3.68	3.89	3.67	3.59	3.29	2.99	3.15
B. Fugitive Emissions from Fuels	12,600.18	12,709.62	12,305.65	12,369.03	12,208.32	11,936.25	12,452.40	12,448.31	12,661.72	12,225.04
1. Solid Fuels	3,227.05	3,195.35	2,998.34	2,993.23	3,041.32	2,973.49	3,034.31	3,009.45	3,448.93	3,592.13
2. Oil and Natural Gas	9,373.13	9,514.28	9,307.31	9,375.80	9,166.99	8,962.75	9,418.09	9,438.86	9,212.79	8,632.91
2. Industrial Processes	207.94	186.43	191.48	187.02	205.14	184.45	188.45	189.29	169.15	155.06
A. Mineral Products	NA	NA	NA	NA	NA	NA	NA	NA	NA	NA
B. Chemical Industry	163.55	147.46	154.26	149.46	166.13	149.98	153.42	155.66	137.88	138.18
C. Metal Production	44.39	38.98	37.22	37.56	39.01	34.46	35.03	33.63	31.27	17.78
D. Other Production										
E. Production of Halocarbons and SF₆										
F. Consumption of Halocarbons and SF₆										
G. Other	NA,NO	NA,NO	NA,NO	NA,NO	NA,NO	NA,NO	NA,NO	NA,NO	NA,NO	NA,NO
3. Solvent and Other Product Use	NA,NO	NA,NO	NA,NO	NA,NO	NA,NO	NA,NO	NA,NO	NA,NO	NA,NO	NA,NO
4. Agriculture	8,958.33	9,013.04	9,042.54	9,092.63	8,944.38	9,120.61	9,211.39	9,549.52	9,536.99	9,455.63
A. Enteric Fermentation	6,578.49	6,540.42	6,551.79	6,563.92	6,440.03	6,521.73	6,631.12	6,751.21	6,731.45	6,693.01
B. Manure Management	2,012.78	2,098.87	2,156.31	2,187.96	2,134.59	2,264.89	2,287.72	2,493.05	2,452.15	2,402.82
C. Rice Cultivation	356.84	363.78	325.20	328.37	360.22	326.10	281.97	294.56	342.73	349.06
D. Agricultural Soils	NA	NA	NA	NA	NA	NA	NA	NA	NA	NA
E. Prescribed Burning of Savannas	NA	NA	NA	NA	NA	NA	NA	NA	NA	NA
F. Field Burning of Agricultural Residues	10.22	9.97	9.24	11.79	9.54	7.89	10.58	10.70	10.67	10.74
G. Other	NA	NA	NA	NA	NA	NA	NA	NA	NA	NA
5. Land Use, Land-Use Change and Forestry	544.45	320.46	484.43	312.20	177.96	382.55	843.33	683.82	412.64	271.48
A. Forest Land	544.45	320.46	484.43	312.20	177.96	382.55	843.33	683.82	412.64	271.48
B. Cropland	NA	NA	NA	NA	NA	NA	NA	NA	NA	NA
C. Grassland	NA	NA	NA	NA	NA	NA	NA	NA	NA	NA
D. Wetlands	NE	NE	NE	NE	NE	NE	NE	NE	NE	NE
E. Settlements	NE	NE	NE	NE	NE	NE	NE	NE	NE	NE
F. Other Land	NE	NE	NE	NE	NE	NE	NE	NE	NE	NE
G. Other	NE	NE	NE	NE	NE	NE	NE	NE	NE	NE
6. Waste	6,217.01	5,994.52	6,037.95	6,292.84	6,108.99	6,216.65	6,179.92	6,183.23	6,280.18	6,257.81
A. Solid Waste Disposal on Land	5,336.76	5,132.72	5,176.99	5,430.09	5,239.73	5,357.07	5,310.99	5,313.86	5,408.68	5,396.60
B. Waste-water Handling	820.56	801.74	800.21	793.51	794.98	785.01	793.52	790.60	791.31	785.91
C. Waste Incineration	IE,NE	IE,NE	IE,NE	IE,NE	IE,NE	IE,NE	IE,NE	IE,NE	IE,NE	IE,NE
D. Other	59.69	60.06	60.75	69.24	74.28	74.57	75.41	78.78	80.20	75.30
7. Other (as specified in Summary 1.A)	NA	NA	NA	NA	NA	NA	NA	NA	NA	NA
Total CH₄ emissions including CH₄ from LULUCF	28,996.52	28,667.99	28,491.75	28,685.36	28,076.42	28,268.97	29,275.40	29,458.53	29,465.69	28,751.31
Total CH₄ emissions excluding CH₄ from LULUCF	28,452.08	28,347.53	28,007.32	28,373.16	27,898.45	27,886.41	28,432.08	28,774.71	29,053.05	28,479.83
Memo Items:										
International Bunkers	3.96	3.73	3.98	4.31	5.21	5.29	5.38	5.38	5.81	5.30
Aviation	NA	NA	NA	NA	NA	NA	NA	NA	NA	NA
Marine	3.96	3.73	3.98	4.31	5.21	5.29	5.38	5.38	5.81	5.30
Multilateral Operations	IE	IE	IE	IE	IE	IE	IE	IE	IE	IE
CO₂ Emissions from Biomass										

Note: All footnotes for this table are given at the end of the table on sheet 5.

TABLE 10 EMISSION TRENDS
CH_4
(Part 3 of 3)

Inventory 2011
Submission 2013 v1.1
UNITED STATES OF AMERICA

GREENHOUSE GAS SOURCE AND SINK CATEGORIES	2010	2011	Change from base to latest reported year
	(Gg)	(Gg)	%
1. Energy	12,375.39	12,015.78	-15.50
A. Fuel Combustion (Sectoral Approach)	386.00	382.05	-33.41
1. Energy Industries	21.85	21.24	34.65
2. Manufacturing Industries and Construction	79.52	78.83	-7.87
3. Transport	69.10	66.26	-68.08
4. Other Sectors	211.91	212.19	-19.02
5. Other	3.62	3.53	27.62
B. Fugitive Emissions from Fuels	11,989.39	11,633.73	-14.75
1. Solid Fuels	3,684.25	3,242.22	-24.44
2. Oil and Natural Gas	8,305.14	8,391.51	-10.30
2. Industrial Processes	171.59	176.96	13.70
A. Mineral Products	NA	NA	0.00
B. Chemical Industry	146.61	148.89	36.10
C. Metal Production	24.98	28.08	-39.28
D. Other Production			
E. Production of Halocarbons and SF_6			
F. Consumption of Halocarbons and SF_6			
G. Other	NA,NO	NA,NO	0.00
3. Solvent and Other Product Use			
4. Agriculture	9,519.01	9,345.30	14.40
A. Enteric Fermentation	6,632.37	6,541.59	3.49
B. Manure Management	2,466.09	2,478.01	65.33
C. Rice Cultivation	409.72	315.96	-6.86
D. Agricultural Soils	NA	NA	0.00
E. Prescribed Burning of Savannas	NA	NA	0.00
F. Field Burning of Agricultural Residues	10.82	9.74	-0.46
G. Other	NA	NA	0.00
5. Land Use, Land-Use Change and Forestry	222.33	674.85	470.12
A. Forest Land	222.33	674.85	470.12
B. Cropland	NA	NA	0.00
C. Grassland	NA	NA	0.00
D. Wetlands	NE	NE	0.00
E. Settlements	NE	NE	0.00
F. Other Land	NE	NE	0.00

TABLE 10 EMISSION TRENDS
N₂O
(Part 1 of 3)

Inventory 2011
Submission 2013 v1.1
UNITED STATES OF AMERICA

GREENHOUSE GAS SOURCE AND SINK CATEGORIES	Base year (1990)	1991	1992	1993	1994	1995	1996	1997	1998	1999
	(Gg)	(Gg)	(Gg)	(Gg)	(Gg)	(Gg)	(Gg)	(Gg)	(Gg)	(Gg)
1. Energy	183.10	188.97	199.76	207.13	213.56	217.34	220.93	222.84	222.53	219.34
A. Fuel Combustion (Sectoral Approach)	183.10	188.97	199.76	207.13	213.56	217.34	220.93	222.84	222.53	219.34
1. Energy Industries	23.76	23.72	23.90	24.91	25.41	25.64	26.32	26.82	28.01	28.60
2. Manufacturing Industries and Construction	13.89	13.58	14.01	14.15	14.70	14.81	15.16	15.42	14.72	14.64
3. Transport	137.37	143.72	153.84	160.50	166.08	169.66	172.01	173.80	173.56	169.77
4. Other Sectors	4.87	5.02	5.11	4.77	4.59	4.51	4.77	4.23	3.78	3.95
5. Other	3.20	2.93	2.91	2.80	2.78	2.72	2.68	2.57	2.47	2.38
B. Fugitive Emissions from Fuels	IE,NA,NE	IE,NA,NE	IE,NA,NE	IE,NA,NE	IE,NA,NE	IE,NA,NE	IE,NA,NE	IE,NA,NE	IE,NA,NE	IE,NA,NE
1. Solid Fuels	IE,NE	IE,NE	IE,NE	IE,NE	IE,NE	IE,NE	IE,NE	IE,NE	IE,NE	IE,NE
2. Oil and Natural Gas	IE,NA,NE	IE,NA,NE	IE,NA,NE	IE,NA,NE	IE,NA,NE	IE,NA,NE	IE,NA,NE	IE,NA,NE	IE,NA,NE	IE,NA,NE
2. Industrial Processes	109.54	108.49	103.63	107.60	109.80	121.90	124.04	101.34	85.81	81.87
A. Mineral Products	NA	NA	NA	NA	NA	NA	NA	NA	NA	NA
B. Chemical Industry	109.54	108.49	103.63	107.60	109.80	121.90	124.04	101.34	85.81	81.87
C. Metal Production	NA	NA	NA	NA	NA	NA	NA	NA	NA	NA
D. Other Production										
E. Production of Halocarbons and SF₆										
F. Consumption of Halocarbons and SF₆										
G. Other	NA,NO	NA,NO	NA,NO	NA,NO	NA,NO	NA,NO	NA,NO	NA,NO	NA,NO	NA,NO
3. Solvent and Other Product Use	14.21	13.81	13.02	14.80	14.80	14.80	14.80	15.74	15.74	15.74
4. Agriculture	781.67	805.82	816.64	937.09	844.85	871.75	906.60	891.28	830.77	832.29
A. Enteric Fermentation										
B. Manure Management	46.32	47.07	47.21	46.53	48.96	50.24	50.05	50.68	51.44	53.44
C. Rice Cultivation										
D. Agricultural Soils	735.09	758.52	769.16	890.31	795.61	821.25	856.28	840.31	779.03	778.56
E. Prescribed Burning of Savannas	NA	NA	NA	NA	NA	NA	NA	NA	NA	NA
F. Field Burning of Agricultural Residues	0.26	0.24	0.27	0.25	0.29	0.26	0.27	0.29	0.29	0.29
G. Other	NA	NA	NA	NA	NA	NA	NA	NA	NA	NA
5. Land Use, Land-Use Change and Forestry	9.93	9.19	12.32	9.78	20.53	13.29	29.59	9.46	10.58	27.84
A. Forest Land	6.76	5.93	8.84	5.75	16.23	9.51	26.09	5.99	7.80	25.25
B. Cropland	IE,NE	IE,NE	IE,NE	IE,NE	IE,NE	IE,NE	IE,NE	IE,NE	IE,NE	IE,NE
C. Grassland	IE,NE	IE,NE	IE,NE	IE,NE	IE,NE	IE,NE	IE,NE	IE,NE	IE,NE	IE,NE
D. Wetlands	0.02	0.02	0.02	0.01	0.01	0.02	0.01	0.02	0.01	0.02
E. Settlements	3.16	3.25	3.46	4.01	4.28	3.76	3.48	3.46	2.77	2.58
F. Other Land	NE	NE	NE	NE	NE	NE	NE	NE	NE	NE
G. Other	IE,NA,NO	IE,NA,NO	IE,NA,NO	IE,NA,NO	IE,NA,NO	IE,NA,NO	IE,NA,NO	IE,NA,NO	IE,NA,NO	IE,NA,NO
6. Waste	12.30	12.71	13.29	13.87	14.59	14.93	15.55	15.90	16.56	17.50
A. Solid Waste Disposal on Land										
B. Waste-water Handling	11.16	11.41	11.82	12.00	12.28	12.31	12.58	12.62	12.98	13.49
C. Waste Incineration	IE	IE	IE	IE	IE	IE	IE	IE	IE	IE
D. Other	1.14	1.31	1.47	1.88	2.31	2.61	2.97	3.28	3.58	4.01
7. Other (as specified in Summary 1.A)	NA	NA	NA	NA	NA	NA	NA	NA	NA	NA
Total N₂O emissions including N₂O from LULUCF	1,110.75	1,139.00	1,158.67	1,290.26	1,218.13	1,254.00	1,311.51	1,256.56	1,181.99	1,194.58
Total N₂O emissions excluding N₂O from LULUCF	1,100.82	1,129.81	1,146.35	1,280.48	1,197.60	1,240.71	1,281.92	1,247.09	1,171.41	1,166.74
Memo Items:										
International Bunkers	4.06	4.26	4.06	3.79	3.90	3.75	3.97	4.04	4.29	3.87
Aviation	2.41	2.46	2.51	2.50	2.67	2.52	2.74	2.71	2.86	2.72
Marine	1.66	1.81	1.55	1.29	1.22	1.23	1.23	1.33	1.43	1.14
Multilateral Operations	IE	IE	IE	IE	IE	IE	IE	IE	IE	IE
CO₂ Emissions from Biomass										

Note: All footnotes for this table are given at the end of the table on sheet 5.

TABLE 10 EMISSION TRENDS

N₂O

(Part 2 of 3)

Inventory 2011
Submission 2013 v1.1
UNITED STATES OF AMERICA

GREENHOUSE GAS SOURCE AND SINK CATEGORIES	2000	2001	2002	2003	2004	2005	2006	2007	2008	2009
	(Gg)	(Gg)	(Gg)	(Gg)	(Gg)	(Gg)	(Gg)	(Gg)	(Gg)	(Gg)
1. Energy	**218.94**	**211.01**	**203.69**	**196.45**	**193.47**	**186.78**	**177.11**	**163.17**	**151.36**	**141.18**
A. Fuel Combustion (Sectoral Approach)	218.94	211.01	203.69	196.45	193.47	186.78	177.11	163.17	151.36	141.18
1. Energy Industries	31.04	32.79	38.70	42.61	47.55	51.60	52.30	53.90	54.27	54.21
2. Manufacturing Industries and Construction	14.66	14.39	14.13	14.22	14.95	14.91	15.26	14.81	14.12	13.05
3. Transport	166.60	157.30	144.42	132.79	124.06	113.52	103.37	88.14	76.58	67.69
4. Other Sectors	4.20	4.00	3.97	4.23	4.26	4.24	3.78	3.99	4.15	4.05
5. Other	2.44	2.54	2.47	2.59	2.64	2.51	2.40	2.33	2.24	2.18
B. Fugitive Emissions from Fuels	IE,NA,NE	IE,NA,NE	IE,NA,NE	IE,NA,NE	IE,NA,NE	IE,NA,NE	IE,NA,NE	IE,NA,NE	IE,NA,NE	IE,NA,NE
1. Solid Fuels	IE,NE	IE,NE	IE,NE	IE,NE	IE,NE	IE,NE	IE,NE	IE,NE	IE,NE	IE,NE
2. Oil and Natural Gas	IE,NA,NE	IE,NA,NE	IE,NA,NE	IE,NA,NE	IE,NA,NE	IE,NA,NE	IE,NA,NE	IE,NA,NE	IE,NA,NE	IE,NA,NE
2. Industrial Processes	**81.92**	**67.17**	**73.04**	**71.37**	**65.01**	**78.58**	**82.21**	**82.21**	**62.72**	**54.23**
A. Mineral Products	NA	NA	NA	NA	NA	NA	NA	NA	NA	NA
B. Chemical Industry	81.92	67.17	73.04	71.37	65.01	78.58	82.21	82.21	62.72	54.23
C. Metal Production	NA	NA	NA	NA	NA	NA	NA	NA	NA	NA
D. Other Production										
E. Production of Halocarbons and SF₆										
F. Consumption of Halocarbons and SF₆										
G. Other	NA,NO	NA,NO	NA,NO	NA,NO	NA,NO	NA,NO	NA,NO	NA,NO	NA,NO	NA,NO
3. Solvent and Other Product Use	**15.74**	**15.74**	**14.15**	**14.15**	**14.15**	**14.15**	**14.15**	**14.15**	**14.15**	**14.15**
4. Agriculture	**787.26**	**820.57**	**832.21**	**810.24**	**856.38**	**821.47**	**842.52**	**872.13**	**849.38**	**840.72**
A. Enteric Fermentation										
B. Manure Management	54.85	54.50	55.71	56.16	54.10	55.08	57.60	57.96	57.31	57.13
C. Rice Cultivation										
D. Agricultural Soils	732.13	765.78	776.24	753.78	802.01	766.14	784.62	813.84	791.76	783.27
E. Prescribed Burning of Savannas	NA	NA	NA	NA	NA	NA	NA	NA	NA	NA
F. Field Burning of Agricultural Residues	0.29	0.29	0.27	0.31	0.27	0.25	0.30	0.33	0.30	0.31
G. Other	NA	NA	NA	NA	NA	NA	NA	NA	NA	NA
5. Land Use, Land-Use Change and Forestry	**35.12**	**23.46**	**32.66**	**23.28**	**16.00**	**27.08**	**52.65**	**44.06**	**28.71**	**20.70**
A. Forest Land	31.43	19.04	28.21	18.60	11.00	22.32	47.81	38.98	23.98	16.17
B. Cropland	IE,NE	IE,NE	IE,NE	IE,NE	IE,NE	IE,NE	IE,NE	IE,NE	IE,NE	IE,NE
C. Grassland	IE,NE	IE,NE	IE,NE	IE,NE	IE,NE	IE,NE	IE,NE	IE,NE	IE,NE	IE,NE
D. Wetlands	0.02	0.02	0.02	0.02	0.02	0.02	0.01	0.02	0.02	0.02
E. Settlements	3.66	4.40	4.43	4.66	4.98	4.74	4.83	5.06	4.71	4.51
F. Other Land	NE	NE	NE	NE	NE	NE	NE	NE	NE	NE
G. Other	IE,NA,NO	IE,NA,NO	IE,NA,NO	IE,NA,NO	IE,NA,NO	IE,NA,NO	IE,NA,NO	IE,NA,NO	IE,NA,NO	IE,NA,NO
6. Waste	**18.35**	**18.85**	**18.71**	**19.53**	**20.25**	**20.62**	**20.99**	**21.49**	**21.81**	**21.64**
A. Solid Waste Disposal on Land										
B. Waste-water Handling	13.87	14.34	14.16	14.34	14.68	15.03	15.33	15.59	15.80	16.00
C. Waste Incineration	IE	IE	IE	IE	IE	IE	IE	IE	IE	IE
D. Other	4.48	4.50	4.56	5.19	5.57	5.59	5.66	5.91	6.01	5.65
7. Other (as specified in Summary 1.A)	**NA**	**NA**	**NA**	**NA**	**NA**	**NA**	**NA**	**NA**	**NA**	**NA**
Total N₂O emissions including N₂O from LULUCF	**1,157.33**	**1,156.81**	**1,174.46**	**1,135.03**	**1,165.27**	**1,148.68**	**1,189.62**	**1,213.14**	**1,128.13**	**1,092.63**
Total N₂O emissions excluding N₂O from LULUCF	**1,122.21**	**1,133.35**	**1,141.80**	**1,111.74**	**1,149.27**	**1,121.61**	**1,136.97**	**1,169.08**	**1,099.42**	**1,071.92**
Memo Items:										
International Bunkers	2.23	2.43	2.50	2.59	2.87	2.94	3.00	3.08	3.18	3.17
Aviation	1.22	1.48	1.49	1.50	1.55	1.60	1.63	1.71	1.70	1.81
Marine	1.01	0.95	1.01	1.09	1.32	1.34	1.37	1.37	1.48	1.36
Multilateral Operations	IE	IE	IE	IE	IE	IE	IE	IE	IE	IE
CO₂ Emissions from Biomass										

Note: All footnotes for this table are given at the end of the table on sheet 5.

TABLE 10 EMISSION TRENDS
N₂O
(Part 3 of 3)

Inventory 2011
Submission 2013 v1.1
UNITED STATES OF AMERICA

GREENHOUSE GAS SOURCE AND SINK CATEGORIES	2010 (Gg)	2011 (Gg)	Change from base to latest reported year (%)
1. Energy	140.62	131.74	-28.05
A. Fuel Combustion (Sectoral Approach)	140.62	131.74	-28.05
1. Energy Industries	59.54	57.88	143.59
2. Manufacturing Industries and Construction	13.94	13.84	-0.41
3. Transport	60.96	53.92	-60.75
4. Other Sectors	3.97	3.96	-18.78
5. Other	2.21	2.15	-32.96
B. Fugitive Emissions from Fuels	IE,NA,NE	IE,NA,NE	0.00
1. Solid Fuels	IE,NE	IE,NE	0.00
2. Oil and Natural Gas	IE,NA,NE	IE,NA,NE	0.00
2. Industrial Processes	68.22	84.21	-23.13
A. Mineral Products	NA	NA	0.00
B. Chemical Industry	68.22	84.21	-23.13
C. Metal Production	NA	NA	0.00
D. Other Production			
E. Production of Halocarbons and SF₆			
F. Consumption of Halocarbons and SF₆			
G. Other	NA,NO	NA,NO	0.00
3. Solvent and Other Product Use	14.15	14.15	-0.38
4. Agriculture	846.36	855.63	9.46
A. Enteric Fermentation			
B. Manure Management	57.29	58.01	25.25
C. Rice Cultivation			
D. Agricultural Soils	788.75	797.34	8.47
E. Prescribed Burning of Savannas	NA	NA	0.00
F. Field Burning of Agricultural Residues	0.32	0.28	7.31
G. Other	NA	NA	0.00
5. Land Use, Land-Use Change and Forestry	18.22	43.29	335.76
A. Forest Land	13.45	38.49	469.65
B. Cropland	IE,NE	IE,NE	0.00
C. Grassland	IE,NE	IE,NE	0.00
D. Wetlands	0.02	0.01	-14.10

TABLE 10 EMISSION TRENDS
HFCs, PFCs and SF₆
(Part 1 of 3)

Inventory 2011
Submission 2013 v1.1
UNITED STATES OF AMERICA

GREENHOUSE GAS SOURCE AND SINK CATEGORIES	Base year (1990) (Gg)	1991 (Gg)	1992 (Gg)	1993 (Gg)	1994 (Gg)	1995 (Gg)	1996 (Gg)	1997 (Gg)	1998 (Gg)	1999 (Gg)
Emissions of HFCs[a] (Gg CO₂ equivalent)	36,924.10	33,540.69	38,282.65	39,503.73	45,592.64	64,035.14	73,986.13	84,503.54	101,185.43	99,920.63
HFC-23	3.13	2.81	3.12	2.85	2.72	2.84	2.69	2.60	3.41	2.64
HFC-32	CJE,NA,NE,NO	CJE,NA,NE,NO	CJE,NA,NE,NO	CJE,NA,NE,NO	CJE,NA,NE,NO	IE,NA,NO	IE,NA,NO	IE,NA,NO	IE,NA,NO	0.00
HFC-41	CJE,NA,NO	IE,NA,NO	IE,NA,NO	IE,NA,NO	CJE,NA,NO	CJE,NA,NO	IE,NA,NO	IE,NA,NO	IE,NA,NO	IE,NA,NO
HFC-43-10mee	CJE,NA,NO	CJE,NA,NO	IE,NA,NO	IE,NA,NO	CJE,NA,NO	CJE,NA,NO	CJE,NA,NO	IE,NA,NO	IE,NA,NO	IE,NA,NO
HFC-125	CJE,NA,NO	CJE,NA,NO	IE,NA,NO	0.17	0.35	0.72	1.11	1.54	1.81	2.10
HFC-134	CJE,NA,NO	CJE,NA,NO	CJE,NA,NO	CJE,NA,NO	CJE,NA,NO	CJE,NA,NO	CJE,NA,NO	CJE,NA,NO	CJE,NA,NO	CJE,NA,NO
HFC-134a	CJE,NA,NO	IE,NA,NO	0.83	3.63	8.78	19.86	26.63	33.51	37.43	42.00
HFC-152a	CJE,NA,NO	CJE,NA,NO	CJE,NA,NO	CJE,NA,NO	CJE,NA,NO	CJE,NA,NO	CJE,NA,NO	CJE,NA,NO	CJE,NA,NO	CJE,NA,NO
HFC-143	IE,NA,NO	IE,NA,NO	IE,NA,NO	CJE,NA,NO	CJE,NA,NO	CJE,NA,NO	CJE,NA,NO	CJE,NA,NO	CJE,NA,NO	CJE,NA,NO
HFC-143a	CJE,NA,NO	CJE,NA,NO	CJE,NA,NO	0.07	0.16	0.29	0.44	0.63	0.81	1.04
HFC-227ea	CJE,NA,NO	IE,NA,NO	IE,NA,NO	CJE,NA,NO	0.01	0.02	0.04	0.05	0.06	0.08
HFC-236fa	CJE,NA,NO	CJE,NA,NO	IE,NA,NO	CJE,NA,NO	0.01	0.02	0.04	0.04	0.05	0.06
HFC-245ca	CJE,NA,NO	CJE,NA,NO	CJE,NA,NO	CJE,NA,NO	CJE,NA,NO	CJE,NA,NO	CJE,NA,NO	CJE,NA,NO	CJE,NA,NO	CJE,NA,NO
Unspecified mix of listed HFCs[a] (Gg CO₂ equivalent)	331.04	640.05	648.53	658.43	661.37	1,594.51	2,785.64	3,449.74	4,066.32	4,184.66
Emissions of PFCs[a] (Gg CO₂ equivalent)	20,645.87	17,774.74	16,539.87	16,507.74	15,167.42	15,587.02	16,606.19	15,222.69	14,029.04	13,961.47
CF₄	2.54	2.16	1.99	1.96	1.96	1.75	1.86	1.68	1.47	1.48
C₂F₆	0.45	0.40	0.39	0.41	0.41	0.45	0.49	0.47	0.49	0.49
C₃F₈	0.00	0.00	0.00	0.00	0.00	0.00	0.00	0.00	0.00	0.00
C₄F₁₀	CJE,NA,NO	CJE,NA,NO	CJE,NA,NO	CJE,NA,NO	CJE,NA,NO	CJE,NA,NO	CJE,NA,NO	CJE,NA,NO	CJE,NA,NO	CJE,NA,NO
c-C₄F₈	CJE,NA,NO	CJE,NA,NO	CJE,NA,NO	CJE,NA,NO	CJE,NA,NO	CJE,NA,NO	CJE,NA,NO	CJE,NA,NO	CJE,NA,NO	CJE,NA,NO
C₅F₁₂	CJE,NA,NO	CJE,NA,NO	CJE,NA,NO	CJE,NA,NO	CJE,NA,NO	CJE,NA,NO	CJE,NA,NO	CJE,NA,NO	CJE,NA,NO	CJE,NA,NO
C₆F₁₄	CJE,NA,NO	CJE,NA,NO	CJE,NA,NO	CJE,NA,NO	CJE,NA,NO	CJE,NA,NO	CJE,NA,NO	CJE,NA,NO	CJE,NA,NO	CJE,NA,NO
Unspecified mix of listed PFCs[a] (Gg CO₂ equivalent)	NA,NE,NO	NA,NE,NO	NA,NE,NO	NA,NE,NO	NA,NE,NO	NA,NE,NO	NA,NE,NO	NA,NE,NO	NA,NE,NO	NA,NE,NO
Emissions of SF₆[b] (Gg CO₂ equivalent)	32,634.53	31,252.92	31,446.62	30,902.91	29,402.59	27,959.51	27,202.99	25,449.29	22,849.19	22,904.73
SF₆	1.37	1.31	1.32	1.29	1.23	1.17	1.14	1.06	0.94	0.95

Note: All footnotes for this table are given at the end of the table on sheet 5.

TABLE 10 EMISSION TRENDS
HFCs, PFCs and SF₆
(Part 2 of 3)

Inventory 2011
Submission 2013 v1.1
UNITED STATES OF AMERICA

GREENHOUSE GAS SOURCE AND SINK CATEGORIES	2000 (Gg)	2001 (Gg)	2002 (Gg)	2003 (Gg)	2004 (Gg)	2005 (Gg)	2006 (Gg)	2007 (Gg)	2008 (Gg)	2009 (Gg)
Emissions of HFCs[a][b] - (Gg CO₂ equivalent)	**104,964.81**	**101,417.37**	**108,117.90**	**103,719.25**	**113,176.87**	**115,002.68**	**115,974.25**	**119,973.45**	**117,451.89**	**111,949.05**
HFC-23	2.47	1.70	1.82	1.07	1.49	1.37	1.21	1.48	1.19	0.48
HFC-32	0.03	0.07	0.13	0.22	0.34	0.50	0.97	1.49	2.02	2.61
HFC-41	IE,NA,NO	IE,NA,NO	IE,NA,NO	IE,NA,NO	IE,NA,NO	IE,NA,NO	IE,NA,NO	IE,NA,NO	IE,NA,NO	IE,NA,NO
HFC-43-10mee	C,IE,NA,NO	C,IE,NA,NO	C,IE,NA,NO	C,IE,NA,NO	C,IE,NA,NO	C,IE,NA,NO	C,IE,NA,NO	C,IE,NA,NO	C,IE,NA,NO	C,IE,NA,NO
HFC-125	2.32	2.44	2.56	2.69	2.86	3.05	3.58	4.30	5.12	6.18
HFC-134	C,IE,NA,NO	C,IE,NA,NO	C,IE,NA,NO	C,IE,NA,NO	C,IE,NA,NO	C,IE,NA,NO	C,IE,NA,NO	C,IE,NA,NO	C,IE,NA,NO	C,IE,NA,NO
HFC-134a	46.41	49.46	52.54	54.53	56.62	57.64	57.57	55.52	53.27	51.33
HFC-152a	C,IE,NA,NO	C,IE,NA,NO	C,IE,NA,NO	C,IE,NA,NO	C,IE,NA,NO	C,IE,NA,NO	C,IE,NA,NO	C,IE,NA,NO	C,IE,NA,NO	C,IE,NA,NO
HFC-143	C,IE,NA,NO	C,IE,NA,NO	C,IE,NA,NO	C,IE,NA,NO	C,IE,NA,NO	C,IE,NA,NO	C,IE,NA,NO	C,IE,NA,NO	C,IE,NA,NO	C,IE,NA,NO
HFC-143a	1.23	1.42	1.63	1.84	2.06	2.29	2.51	2.72	2.91	3.32
HFC-227ea	C,IE,NA,NO	C,IE,NA,NO	C,IE,NA,NO	C,IE,NA,NO	C,IE,NA,NO	C,IE,NA,NO	C,IE,NA,NO	C,IE,NA,NO	C,IE,NA,NO	C,IE,NA,NO
HFC-236fa	0.09	0.09	0.10	0.11	0.12	0.12	0.13	0.14	0.14	0.14
HFC-245ca	C,IE,NA,NO	C,IE,NA,NO	C,IE,NA,NO	C,IE,NA,NO	C,IE,NA,NO	C,IE,NA,NO	C,IE,NA,NO	C,IE,NA,NO	C,IE,NA,NO	C,IE,NA,NO
Unspecified mix of listed HFCs[a][b] - (Gg CO₂ equivalent)	4,017.97	4,005.65	4,436.48	4,956.68	5,324.50	5,649.54	5,986.58	6,321.77	6,665.90	7,045.25
Emissions of PFCs[a][b] - (Gg CO₂ equivalent)	**13,473.80**	**6,979.60**	**8,711.06**	**7,988.60**	**6,125.08**	**6,394.63**	**6,030.44**	**7,670.73**	**6,607.08**	**4,458.52**
CF₄	1.48	0.67	0.88	0.67	0.55	0.56	0.51	0.70	0.55	0.37
C₂F₆	0.41	0.27	0.31	0.28	0.27	0.26	0.28	0.33	0.31	0.22
C₃F₈	0.02	0.01	0.01	0.01	0.01	0.00	0.01	0.01	0.01	0.00
C₄F₁₀	C,NA,NE,NO	C,NA,NE,NO	C,NA,NE,NO	C,NA,NE,NO	C,NA,NE,NO	C,NA,NE,NO	C,NA,NE,NO	C,NA,NE,NO	C,NA,NE,NO	C,NA,NE,NO
c-C₄F₈	C,NA,NE,NO	C,NA,NE,NO	C,NA,NE,NO	C,NA,NE,NO	C,NA,NE,NO	C,NA,NE,NO	C,NA,NE,NO	C,NA,NE,NO	C,NA,NE,NO	C,NA,NE,NO
C₅F₁₂	C,IE,NA,NO	C,IE,NA,NO	C,IE,NA,NO	C,IE,NA,NO	C,IE,NA,NO	C,NA,NE,NO	C,NA,NE,NO	C,NA,NE,NO	C,NA,NE,NO	C,NA,NE,NO
C₆F₁₄	NA,NE,NO	NA,NE,NO	NA,NE,NO	NA,NE,NO	NA,NE,NO	NA,NE,NO	NA,NE,NO	NA,NE,NO	NA,NE,NO	NA,NE,NO
Unspecified mix of listed PFCs[a][b] - (Gg CO₂ equivalent)	NA,NE,NO	NA,NE,NO	NA,NE,NO	NA,NE,NO	NA,NE,NO	NA,NE,NO	NA,NE,NO	NA,NE,NO	NA,NE,NO	NA,NE,NO
Emissions of SF₆[a][b] - (Gg CO₂ equivalent)	**18,827.49**	**18,009.80**	**17,006.25**	**16,681.60**	**15,498.42**	**14,986.64**	**13,684.57**	**12,287.30**	**11,391.23**	**9,815.90**
SF₆	0.79	0.75	0.71	0.70	0.65	0.63	0.57	0.51	0.48	0.41

Note: All footnotes for this table are given at the end of the table on sheet 5.

TABLE 10 EMISSION TRENDS
HFCs, PFCs and SF₆
(Part 3 of 3)

Inventory 2011
Submission 2013 v1.1
UNITED STATES OF AMERICA

GREENHOUSE GAS SOURCE AND SINK CATEGORIES	2010 (Gg)	2011 (Gg)	Change from base to latest reported year (%)
Emissions of HFCs[a][b] - (Gg CO₂ equivalent)	**121,275.07**	**128,951.68**	**2,69.23**
HFC-23	0.58	0.62	-80.05
HFC-32	3.86	4.94	100.00
HFC-41	IE,NA,NO	IE,NA,NO	0.00
HFC-43-10mee	C,IE,NA,NO	C,IE,NA,NO	0.00
HFC-125	7.93	9.51	100.00
HFC-134	C,IE,NA,NO	C,IE,NA,NO	100.00
HFC-134a	51.40	51.01	0.00
HFC-152a	C,IE,NA,NO	C,IE,NA,NO	0.00
HFC-143	C,IE,NA,NO	C,IE,NA,NO	0.00
HFC-143a	3.86	4.41	100.00
HFC-227ea	C,IE,NA,NO	C,IE,NA,NO	100.00
HFC-236fa	0.15	0.15	1,295.15
HFC-245ca	C,NA,NE,NO	C,NA,NE,NO	0.00
Unspecified mix of listed HFCs[a][b] - (Gg CO₂ equivalent)	7,419.32	7,807.86	2,258.60
Emissions of PFCs[a][b] - (Gg CO₂ equivalent)	**5,946.51**	**7,017.60**	**-66.01**
CF₄	0.45	0.61	-75.84
C₂F₆	0.32	0.32	-28.64
C₃F₈	0.00	0.01	0.01
C₄F₁₀	C,NA,NE,NO	C,NA,NE,NO	100.00
c-C₄F₈	0.00	0.00	100.00
C₅F₁₂	C,NA,NE,NO	C,NA,NE,NO	0.00
C₆F₁₄	NA,NE,NO	NA,NE,NO	100.00
Unspecified mix of listed PFCs[a][b] - (Gg CO₂ equivalent)	NA,NE,NO	NA,NE,NO	0.00
Emissions of SF₆[a][b] - (Gg CO₂ equivalent)	**10,078.11**	**9,379.53**	**-71.26**
SF₆	0.42	0.39	-71.26

Note: All footnotes for this table are given at the end of the table on sheet 5.

TABLE 10 EMISSION TRENDS
SUMMARY
(Part 1 of 3)

Inventory 2011
Submission 2013 v.1.1
UNITED STATES OF AMERICA

GREENHOUSE GAS EMISSIONS	Base year (1990) CO₂ equivalent (Gg)	1991 CO₂ equivalent (Gg)	1992 CO₂ equivalent (Gg)	1993 CO₂ equivalent (Gg)	1994 CO₂ equivalent (Gg)	1995 CO₂ equivalent (Gg)	1996 CO₂ equivalent (Gg)	1997 CO₂ equivalent (Gg)	1998 CO₂ equivalent (Gg)	1999 CO₂ equivalent (Gg)
CO₂ emissions including net CO₂ from LULUCF	4,314,282.25	4,259,557.00	4,380,015.92	4,491,221.69	4,529,395.18	4,626,292.07	4,784,332.33	4,896,039.61	4,995,333.97	5,137,264.98
CO₂ emissions excluding net CO₂ from LULUCF	5,100,693.96	5,050,586.77	5,156,917.29	5,267,905.06	5,354,387.92	5,416,155.70	5,602,445.17	5,677,606.09	5,715,760.67	5,788,887.01
CH₄ emissions including CH₄ from LULUCF	639,925.78	641,028.36	643,566.42	635,259.72	646,441.15	636,235.27	639,466.70	623,115.10	611,150.90	607,013.82
CH₄ emissions excluding CH₄ from LULUCF	637,429.98	638,854.76	640,320.57	633,237.91	640,441.10	632,868.13	629,910.00	621,222.24	608,612.76	598,007.25
N₂O emissions including N₂O from LULUCF	344,333.11	353,089.78	359,188.93	399,980.72	377,621.32	388,740.62	406,568.71	389,532.64	366,415.37	370,320.15
N₂O emissions excluding N₂O from LULUCF	341,253.70	350,239.87	355,368.75	396,950.15	371,255.88	384,621.23	397,395.46	386,598.96	363,135.88	361,689.53
HFCs	36,924.10	33,540.69	38,282.65	39,503.73	45,592.64	64,035.14	73,986.13	84,503.54	101,185.43	99,929.63
PFCs	20,645.87	17,774.74	16,339.87	16,507.74	15,167.42	15,587.02	16,600.19	15,222.69	14,029.04	13,961.47
SF₆	32,634.53	31,252.92	31,446.62	30,902.91	29,402.59	27,959.51	27,202.99	25,449.29	22,449.19	22,804.73
Total (including LULUCF)	5,388,745.64	5,336,243.49	5,469,040.41	5,613,376.51	5,643,490.31	5,758,849.63	5,948,157.05	6,033,862.86	6,110,563.96	6,251,294.78
Total (excluding LULUCF)	6,169,592.14	6,122,249.75	6,238,875.76	6,385,907.49	6,456,417.56	6,541,226.73	6,747,540.84	6,810,602.80	6,825,172.96	6,885,279.62

GREENHOUSE GAS SOURCE AND SINK CATEGORIES	Base year (1990) CO₂ equivalent (Gg)	1991 CO₂ equivalent (Gg)	1992 CO₂ equivalent (Gg)	1993 CO₂ equivalent (Gg)	1994 CO₂ equivalent (Gg)	1995 CO₂ equivalent (Gg)	1996 CO₂ equivalent (Gg)	1997 CO₂ equivalent (Gg)	1998 CO₂ equivalent (Gg)	1999 CO₂ equivalent (Gg)
1. Energy	5,267,347.08	5,228,471.08	5,331,849.06	5,438,924.70	5,523,870.28	5,575,988.21	5,765,836.29	5,838,510.93	5,872,327.30	5,939,695.89
2. Industrial Processes	316,147.45	297,374.07	302,260.37	301,737.95	311,819.03	339,357.00	350,404.17	354,139.01	358,373.87	353,861.72
3. Solvent and Other Product Use	4,404.02	4,281.69	4,037.02	4,587.52	4,587.52	4,587.52	4,587.52	4,879.50	4,879.50	4,879.50
4. Agriculture	413,861.23	422,904.55	430,324.88	470,009.72	447,386.14	459,501.19	468,138.22	462,923.43	446,497.03	447,946.62
5. Land Use, Land-Use Change and Forestry[2]	-780,846.50	-786,006.26	-769,835.35	-771,630.98	-812,927.25	-782,377.09	-799,383.79	-776,739.94	-714,609.00	-633,984.84
6. Waste	167,832.35	169,218.36	170,404.44	169,747.60	168,754.59	161,792.82	158,574.64	150,149.94	143,095.27	138,895.89
7. Other	NA	NA	NA	NA	NA	NA	NA	NA	NA	NA
Total (including LULUCF)[5]	5,388,745.64	5,336,243.49	5,469,040.41	5,613,376.51	5,643,490.31	5,758,849.63	5,948,157.05	6,033,862.86	6,110,563.96	6,251,294.78

TABLE 10 EMISSION TRENDS
SUMMARY
(Part 2 of 3)

Inventory 2011
Submission 2013 v.1.1
UNITED STATES OF AMERICA

GREENHOUSE GAS EMISSIONS	2000 CO₂ equivalent (Gg)	2001 CO₂ equivalent (Gg)	2002 CO₂ equivalent (Gg)	2003 CO₂ equivalent (Gg)	2004 CO₂ equivalent (Gg)	2005 CO₂ equivalent (Gg)	2006 CO₂ equivalent (Gg)	2007 CO₂ equivalent (Gg)	2008 CO₂ equivalent (Gg)	2009 CO₂ equivalent (Gg)
CO₂ emissions including net CO₂ from LULUCF	5,289,698.06	5,133,412.32	5,052,859.56	5,002,218.10	5,078,835.02	5,111,507.97	5,071,060.80	5,199,349.19	5,042,208.58	4,635,300.96
CO₂ emissions excluding net CO₂ from LULUCF	5,962,701.49	5,862,840.84	5,903,392.80	5,951,512.86	6,070,325.66	6,100,403.13	6,024,114.68	6,119,317.83	5,935,183.73	5,509,601.36
CH₄ emissions including CH₄ from LULUCF	608,926.99	602,027.78	598,326.85	602,392.54	589,604.73	593,648.28	614,783.46	618,629.08	618,779.46	603,777.56
CH₄ emissions excluding CH₄ from LULUCF	597,493.59	595,298.05	588,153.81	595,836.33	585,867.49	585,614.69	597,073.58	604,268.86	610,114.02	598,076.49
N₂O emissions including N₂O from LULUCF	358,771.16	358,610.42	364,082.85	351,857.91	361,232.39	356,091.39	368,782.10	376,073.19	349,720.57	338,714.51
N₂O emissions excluding N₂O from LULUCF	347,885.07	351,337.85	353,958.68	344,640.75	356,272.68	347,697.61	352,460.59	362,415.56	340,820.96	332,296.71
HFCs	104,964.81	101,117.37	108,117.90	103,719.25	113,176.87	115,002.68	115,974.25	119,973.45	117,451.89	111,949.05
PFCs	13,473.80	6,979.60	8,711.06	7,080.60	6,125.08	6,194.63	6,030.44	7,670.73	6,607.08	4,458.52
SF₆	18,827.49	18,009.80	17,006.25	16,681.60	15,498.42	14,986.61	13,684.57	12,287.30	11,391.23	9,815.90
Total (including LULUCF)	6,394,662.32	6,220,157.29	6,149,104.47	6,083,950.01	6,164,472.52	6,197,431.56	6,190,315.62	6,333,082.94	6,146,158.81	5,704,016.51
Total (excluding LULUCF)	7,045,346.25	6,935,583.51	6,979,340.50	7,019,471.39	7,147,266.20	7,169,899.34	7,489,338.11	7,225,933.72	7,021,568.90	6,566,198.03

GREENHOUSE GAS SOURCE AND SINK CATEGORIES	2000 CO₂ equivalent (Gg)	2001 CO₂ equivalent (Gg)	2002 CO₂ equivalent (Gg)	2003 CO₂ equivalent (Gg)	2004 CO₂ equivalent (Gg)	2005 CO₂ equivalent (Gg)	2006 CO₂ equivalent (Gg)	2007 CO₂ equivalent (Gg)	2008 CO₂ equivalent (Gg)	2009 CO₂ equivalent (Gg)
1. Energy	6,119,611.96	6,036,684.00	6,066,626.69	6,118,042.18	6,227,873.36	6,251,617.44	6,178,349.41	6,266,903.09	6,096,242.58	5,699,176.63
2. Industrial Processes	352,433.19	318,640.90	327,850.26	315,830.55	327,129.19	330,765.41	335,697.57	347,231.31	318,710.08	265,319.84
3. Solvent and Other Product Use	4,879.50	4,879.50	4,387.15	4,387.15	4,387.15	4,387.15	4,387.15	4,387.15	4,387.15	4,387.15
4. Agriculture	432,176.83	443,651.11	447,878.18	442,108.05	453,308.70	446,188.00	454,620.11	470,900.79	463,583.76	459,190.37
5. Land Use, Land-Use Change and Forestry[2]	-650,683.94	-715,426.23	-830,236.03	-935,521.38	-982,793.68	-972,467.78	-919,022.50	-891,950.78	-875,410.10	-862,181.53
6. Waste	136,244.78	131,728.00	132,598.22	138,203.47	134,941.34	136,941.34	136,283.88	136,531.38	138,645.34	138,123.85
7. Other	NA	NA	NA	NA	NA	NA	NA	NA	NA	NA
Total (including LULUCF)[5]	6,394,662.32	6,220,157.29	6,149,104.47	6,083,950.01	6,164,472.52	6,197,431.56	6,190,315.62	6,333,082.94	6,146,158.81	5,704,016.51

[1] The column "Base year" should be filled in only by those Parties with economies in transition that use a base year different from 1990 in accordance with the relevant decisions of the COP. For these Parties, this different base year is used to calculate the percentage change in the final column of this table.

[2] Fill in net emissions/removals as reported in table Summary 1.A. For the purposes of reporting, the signs for removals are always negative (-) and for emissions positive (+).

[3] Enter actual emissions estimates. If only potential emissions estimates are available, these should be reported in this table and an indication for this be provided in the documentation box. Only in these rows are the emissions expressed as CO₂ equivalent emissions.

[4] In accordance with the UNFCCC reporting guidelines, HFC and PFC emissions should be reported for each relevant chemical. However, if it is not possible to report values for each chemical (i.e. mixtures, confidential data, lack of disaggregation), this row could be used for reporting aggregate figures for HFCs and PFCs, respectively. Note that the unit used for this row is Gg of CO₂ equivalent and that appropriate notation keys should be entered in the cells for the individual chemicals.

[5] Includes net CO₂, CH₄, and N₂O from LULUCF.

TABLE 10 EMISSION TRENDS
SUMMARY
(Part 3 of 3)

Inventory 2011
Submission 2013 v1.1
UNITED STATES OF AMERICA

GREENHOUSE GAS EMISSIONS	2010	2011	Change from base to latest reported year
	CO$_2$ equivalent (Gg)	CO$_2$ equivalent (Gg)	(%)
CO$_2$ emissions including net CO$_2$ from LULUCF	4,847,628.12	4,707,813.53	9.12
CO$_2$ emissions excluding net CO$_2$ from LULUCF	5,727,038.60	5,663,820.59	9.86
CH$_4$ emissions including CH$_4$ from LULUCF	592,710.43	587,235.17	-8.23
CH$_4$ emissions excluding CH$_4$ from LULUCF	588,041.55	573,063.23	-10.10
N$_2$O emissions including N$_2$O from LULUCF	343,917.52	356,886.99	3.65
N$_2$O emissions excluding N$_2$O from LULUCF	338,270.27	343,468.24	0.65
HFCs	121,275.07	128,951.68	249.23
PFCs	5,946.51	7,017.60	-66.01
SF$_6$	10,070.11	9,379.53	-71.28
Total (including LULUCF)	**5,921,547.77**	**5,797,284.50**	**7.58**
Total (excluding LULUCF)	**6,790,642.12**	**6,665,700.87**	**8.04**

GREENHOUSE GAS SOURCE AND SINK CATEGORIES	2010	2011	Change from base to latest reported year
	CO$_2$ equivalent (Gg)	CO$_2$ equivalent (Gg)	(%)
1. Energy	5,889,117.78	5,745,698.03	9.08
2. Industrial Processes	303,439.65	326,461.30	3.26
3. Solvent and Other Product Use	4,387.15	4,387.15	-0.38
4. Agriculture	462,269.97	461,496.95	11.51
5. Land Use, Land-Use Change and Forestry$^{(2)}$	-869,094.35	-868,416.37	11.21
6. Waste	131,427.57	127,657.44	-23.94
7. Other	NA	NA	0.00
Total (including LULUCF)$^{(3)}$	**5,921,547.77**	**5,797,284.50**	**7.58**

(1) The column "Base year" should be filled in only by those Parties with economies in transition that use a base year different from 1990 in accordance with the relevant decisions of the COP. For these Parties, this different base year is used to calculate the percentage change in the final column of this table.

(2) Fill in net emissions/removals as reported in table Summary 1.A. For the purposes of reporting, the signs for removals are always negative (-) and for emissions positive (+).

(3) Enter actual emissions estimates. If only potential emissions estimates are available, these should be reported in this table and an indication for this be provided in the documentation box. Only in these rows are the emissions expressed as CO$_2$ equivalent emissions.

**SUMMARY 2   SUMMARY REPORT FOR CO_2 EQUIVALENT EMISSIONS
(Sheet 1 of 1)**

Inventory 2011
Submission 2013 v1.1
UNITED STATES OF AMERICA

GREENHOUSE GAS SOURCE AND SINK CATEGORIES	CO_2 [1]	CH_4	N_2O	HFCs [2]	PFCs [2]	SF_6 [2]	Total
				CO_2 equivalent (Gg)			
Total (Net Emissions) [1]	4,707,813.53	587,235.17	356,886.99	128,951.68	7,017.60	9,379.53	5,797,284.50
1. Energy	5,452,528.41	252,331.28	40,838.34				5,745,698.03
A. Fuel Combustion (Sectoral Approach)	5,419,837.29	8,023.01	40,838.34				5,468,698.64
1. Energy Industries	2,158,510.32	446.03	17,941.39				2,176,897.73
2. Manufacturing Industries and Construction	773,192.26	1,655.48	4,289.45				779,137.19
3. Transport	1,725,577.55	1,391.54	16,714.47				1,743,683.56
4. Other Sectors	550,857.14	4,455.93	1,227.07				556,540.15
5. Other	211,700.02	74.03	665.96				212,440.01
B. Fugitive Emissions from Fuels	32,691.12	244,308.27	IE,NA,NE				276,999.40
1. Solid Fuels	IE,NE,NO	68,086.52	IE,NE				68,086.52
2. Oil and Natural Gas	32,691.12	176,221.76	IE,NA,NE				208,912.88
2. Industrial Processes	151,292.18	3,716.20	26,104.11	128,951.68	7,017.60	9,379.53	326,461.30
A. Mineral Products	58,590.21	NA	NA				58,590.21
B. Chemical Industry	21,664.69	3,126.59	26,104.11	NA	NA	NA	50,895.39
C. Metal Production	71,037.27	589.61	NA	NA	2,942.43	1,407.30	75,976.62
D. Other Production	NE						NE
E. Production of Halocarbons and SF_6				6,934.00	NA,NE	NA,NE	6,934.00
F. Consumption of Halocarbons and SF_6 [2]				122,017.68	4,075.17	7,972.23	134,065.08
G. Other	NA,NO	NA,NO	NA,NO	NA	NA	NA	NA,NO
3. Solvent and Other Product Use	NA,NE		4,387.15				4,387.15
4. Agriculture		196,251.27	265,245.68				461,496.95
A. Enteric Fermentation		137,373.39					137,373.39
B. Manure Management		52,038.31	17,984.12				70,022.43
C. Rice Cultivation		6,635.07					6,635.07
D. Agricultural Soils [3]		NA	247,173.95				247,173.95
E. Prescribed Burning of Savannas		NA	NA				NA
F. Field Burning of Agricultural Residues		204.50	87.61				292.11
G. Other		NA	NA				NA
5. Land Use, Land-Use Change and Forestry [1]	-896,007.06	14,171.94	13,418.75				-868,416.37
A. Forest Land	-761,804.08	14,171.94	11,931.26				-735,700.88
B. Cropland	19,765.20	NA	IE,NE				19,765.20
C. Grassland	-1,354.10	NA	IE,NE				-1,354.10
D. Wetlands	917.70	NE	4.48				922.18
E. Settlements	-68,825.50	NE	1,483.01				-67,342.49
F. Other Land	NE	NE	NE				NE
G. Other	-84,706.28	NA,NO	IE,NA,NO				-84,706.28
6. Waste	IE,NA,NE	120,764.47	6,892.97				127,657.44
A. Solid Waste Disposal on Land	NA,NE	103,046.71					103,046.71
B. Waste-water Handling		16,168.08	5,177.24				21,345.32
C. Waste Incineration	IE	IE,NE	IE				IE,NE
D. Other	NA	1,549.69	1,715.73				3,265.42
7. Other (as specified in Summary 1.A)	NA	NA	NA	NA	NA	NA	NA

Memo Items: [4]							
International Bunkers	111,315.70	97.41	928.16				112,341.26
Aviation	64,856.50	NA	562.98				65,419.48
Marine	46,459.20	97.41	365.18				46,921.78
Multilateral Operations	IE	IE	IE				IE
CO_2 Emissions from Biomass	264,527.22						264,527.22

Total CO_2 Equivalent Emissions without Land Use, Land-Use Change and Forestry	6,665,700.87
Total CO_2 Equivalent Emissions with Land Use, Land-Use Change and Forestry	5,797,284.50

[1] For CO_2 from Land Use, Land-use Change and Forestry the net emissions/removals are to be reported. For the purposes of reporting, the signs for removals are always negative (-) and for emissions positive (+).

[2] Actual emissions should be included in the national totals. If no actual emissions were reported, potential emissions should be included.

[3] Parties which previously reported CO_2 from soils in the Agriculture sector should note this in the NIR.

[4] See footnote 8 to table Summary 1.A.

AAAS and NSTA (American Association for the Advancement of Science Project 2061 and National Science Teachers Association). 2007. *Atlas of Science Literacy.* Vols. 1 and 2. Washington, DC. <http://www.project2061.org/publications/atlas/default.htm>

AASHE (Association for the Advancement of Sustainability in Higher Education). 2010. *Sustainability Curriculum in Higher Education: A Call to Action.* Denver, CO. <http://www.aashe.org/files/A_Call_to_Action_final%282%29.pdf>

Alig, R.J., A.J. Plantinga, D. Haim, and M. Todd. 2010. *Area Changes in U.S. Forests and Other Major Land Uses, 1982 to 2002, with Projections to 2062.* Gen. Tech. Rep. PNW-GTR-815. Portland, OR: U.S. Department of Agriculture, Forest Service, Pacific Northwest Research Station. 98 p. <http://www.fs.fed.us/pnw/pubs/pnw_gtr815.pdf>

Angel, J.R., and K.E. Kunkel. 2010. The response of Great Lakes water levels to future climate scenarios with an emphasis on Lake Michigan-Huron. *Journal of Great Lakes Research* 36 (supplement 2): 51–58.

Beatty, A.; Steering Committee on Climate Change Education in Formal Settings, K-14; Board on Science Education; and Division of Behavior and Social Sciences and Education. 2013. *National Research Council Climate Change Education: Formal Settings, K-14: A Workshop Summary.* National Research Council of The National Academies. Washington, DC: National Academies Press. <http://www.nap.edu/catalog.php?record_id=13435>

Bierbaum, R., J.B. Smith, A. Lee, M. Blair, L. Carter, F.S. Chapin III, P. Fleming, S. Ruffo, M. Stults, and S. McNeeley. 2013. A comprehensive review of climate adaptation in the United States: more than before, but less than needed. *Mitigation and Adaptation Strategies for Global Change* 18(3): 361–406. <http://link.springer.com/article/10.1007%2Fs11027-012-9423-1>

Boyle, P., and W. Mott. 2009. *America, the Ocean, and Climate Change: New Research Insights for Conservation, Awareness, and Action.* Providence, RI: The Ocean Project. June. <theoceanproject.org/download-reports/>

Burkett, V., and M. Davidson. 2013. *Coastal Impacts, Adaptation, and Vulnerabilities: A Technical Input to the 2013 National Climate Assessment. Cooperative Report to the 2013 National Climate Assessment.* Washington, DC: Island Press. <http://www.coastalstates.org/wp-content/uploads/2011/03/Coastal-Impacts-Adaptation-Vulnerabilities-Oct-2012.pdf>

Congressional Budget Office. 2013. *The Budget and Economic Outlook: Fiscal Years 2013 to 2023.* February 5. <http://www.cbo.gov/publication/43907>

Cochran, P., O.H. Huntington, C. Pungowiyi, S. Tom, F.S. Chapin III, H.P. Huntington, N.G. Maynard, and S.F. Trainor. 2013. Indigenous frameworks for observing and responding to climate change in Alaska. *Climatic Change* 120(March): 557–567. <http://dx.doi.org/10.1007/s10584-013-0735-2>

COEA/DOE/OST (President's Council of Economic Advisors, U.S. Department of Energy, White House Office of Science and Technology). 2013. *Economic Benefits of Increasing Electric Grid Resilience to Weather Outages.* August. <http://energy.gov/sites/prod/files/2013/08/f2/Grid%20Resiliency%20Report_FINAL.pdf>

Coyle, K.J. 2010. *Back to School: Back Outside! How Outdoor Education and Outdoor School Time Create High Performance Students.* Reston, VA: National Wildlife Federation. September. <http://www.peecworks.org/PEEC/PEEC_Research/01798BF4-001D0211.1/Coyle%202010%20Back%20to%20School%20web.pdf>

CPRAL (State of Louisiana, Coastal Protection and Restoration Authority of Louisiana). 2012. *Louisiana's Comprehensive Master Plan for a Sustainable Coast: Committed to Our Coast.* Baton Rouge, LA. <http://issuu.com/coastalmasterplan/docs/coastal_master_plan-v2?e=3722998/2447530>

Dittmer, Kyle. 2013. Changing streamflow on Columbia basin tribal lands—climate change and salmon. *Climatic Change* 120(April): 627–641. <http://dx.doi.org/10.1007/s10584-013-0745-0>

Appendix B:
Bibliography

Doyle, J.T., M.H. Redsteer, and M.J. Eggers. 2013. Exploring effects of climate change on Northern Plains American Indian health. *Climatic Change* 120(June): 643–645. <http://dx.doi.org/10.1007/s10584-013-0799-z>

Eddy, W.F., and K. Marton. 2012. *Effective Tracking of Building Energy Use: Improving the Commercial Buildings and Residential Energy Consumption Surveys (2012)*. Washington, DC: The National Academies Press. <http://www.nap.edu/catalog.php?record_id=13360>

EESI (Environmental and Energy Study Institute). 2013. "Polling the American Public on Climate Change." April. <http://www.usclimatenetwork.org/resource-database/polling-the-american-public-on-climate-change>

EOP (Executive Office of the President). 2009. *Federal Leadership in Environmental, Energy, and Economic Performance.* Executive Order 13514. *Federal Register* 74(194, Oct. 8). <http://www.gpo.gov/fdsys/pkg/FR-2009-10-08/pdf/E9-24518.pdf>

———. 2010. *Stewardship of the Ocean, Our Coasts, and the Great Lakes.* Executive Order 13547. *Federal Register* 75(140, July 22). <http://www.gpo.gov/fdsys/pkg/FR-2010-07-22/pdf/2010-18169.pdf>

———. 2012. *Establishing the Hurricane Sandy Rebuilding Task Force.* Executive Order 13632. *Federal Register* 77(241, Dec. 7). <http://www.gpo.gov/fdsys/pkg/FR-2012-12-14/pdf/2012-30310.pdf>

———. 2013a. *The President's Climate Action Plan.* Washington, DC. June. <http://www.whitehouse.gov/sites/default/files/image/president27sclimateactionplan.pdf>

———. 2013b. *Preparing the United States for the Impacts of Climate Change.* Executive Order 13653. *Federal Register* 78(215, Nov. 6). <http://www.gpo.gov/fdsys/pkg/FR-2013-11-06/pdf/2013-26785.pdf>

EOP/CEQ (Executive Office of the President, Council on Environmental Quality). 2011. *Instructions for Implementing Climate Change Adaptation Planning in Accordance with Executive Order 13514.* March 4. <http://www.whitehouse.gov/sites/default/files/microsites/ceq/adaptation_final_implementing_instructions_3_3.pdf>

Feely, R.A., S.C. Doney, and S.R. Cooley. 2009. Ocean acidification: present conditions and future changes in a high-CO_2 world. *Oceanography* 22(4): 36–47.

Fisher, D. 2013. "Climate Coverage Falls Further in 2012." *The Daily Climate.* Climate Central. January 2. <http://www.climatecentral.org/news/climate-coverage-falls-further-in-2012-15424>

Forest, S., and M.A. Feder. 2013. *Climate Change Education: Goals, Audiences, and Strategies: A Workshop Summary.* National Research Council of The National Academies. Washington, DC: National Academies Press. <http://www.nap.edu/catalog.php?record_id=13224>

Garfin, G., A. Jardine, R. Merideth, M. Black, and S. LeRoy, eds. 2013. *Assessment of Climate Change in the Southwest United States: A Report Prepared for the National Climate Assessment.* A report by the Southwest Climate Alliance. Washington, DC: Island Press. <http://islandpress.org/ip/books/book/distributed/A/bo9199001.html>

Gautam, M.R., K. Chief, and W.J. Smith, Jr. 2013. Climate change in arid lands and Native American socioeconomic vulnerability: the case of the Pyramid Lake Paiute Tribe. *Climatic Change* 120(April): 585–599. <http://dx.doi.org/10.1007/s10584-013-0737-0>

Grajal, A., and S.R. Goldman, eds. 2012. *Climate Change Education: A Primer for Zoos and Aquariums.* Brookfield, IL: Chicago Zoological Society. September. <http://www.clizen.org/files/ClimateChangeEducationEbookFirstEditionRevised.pdf>

Gregg R.M., L.J. Hansen, K.M. Feifel, J.L. Hitt, J.M. Kershner, A. Score, and J.R. Hoffman. 2011. *The State of Marine and Coastal Adaptation in North America: A Synthesis of Emerging Ideas.* A report for the Gordon and Betty Moore Foundation. Bainbridge Island, WA: EcoAdapt. <http://ecoadapt.org/documents/marine-adaptation-report.pdf>

Griffis, R., and J. Howard, eds. 2012. *Oceans and Marine Resources in a Changing Climate: Technical Input to the 2013 National Climate Assessment.* Washington, DC: United States Global Change Research Program. <http://downloads.usgcrp.gov/NCA/technicalinputreports/Griffis_Howard_Ocean_Marine_Resources.pdf>

Haim, D., R.J. Alig, A.J. Plantinga, and B. Sohngen. 2011. Climate change and future land use in the United States: an economic approach. *Climatic Change Economics* 2(1): 27-51. <http://treesearch.fs.fed.us/pubs/40838>

Hampton Roads Planning District Commission. 2013. *Coastal Resiliency: Adapting to Climate Change in Hampton Roads.* July. <http://www.hrpdc.org/uploads/docs/07182013-PDC-E9I.pdf>

Hayhoe, K., D. Wuebbles, and C.S. Team. 2008. *Climate Change and Chicago: Projections and Potential Impacts.* <http://www.chicagoclimateaction.org/>

Haynes, R.W., D.M. Adams, R.J. Alig, P.J. Ince, R. John, and X. Zhou. 2007. *The 2005 RPA Timber Assessment Update*. Gen. Tech. Rep. PNW-GTR-699. Portland, OR: U.S. Department of Agriculture, Forest Service, Pacific Northwest Research Station. 212 pp. <http://www.fs.fed.us/pnw/publications/gtr699/>

Hinzman, L.D., N.D. Bettez, W.R. Bolton, F.S. Chapin, M.B. Dyurgerov, C.L. Fastie, B. Griffith, R.D. Hollister, A. Hope, H.P. Huntington, A.M. Jensen, G.J. Jia, T. Jorgenson, D.L. Kane, D.R. Klein, G. Kofinas, A.H. Lynch, A.H. Lloyd, A.D. McGuire, F.E. Nelson, W.C. Oechel, T.E. Osterkamp, C.H. Racine, V.E. Romanovsky, R.S. Stone, D.A. Stow, M. Sturm, C.E. Tweedie, G.L. Vourlitis, M.D. Walker, D.A. Walker, P.J. Webber, J.M. Welker, K.S. Winker, and K. Yoshikawa. 2005. Evidence and implications of recent climate change in northern Alaska and other Arctic regions. *Climatic Change* 72: 251–298. <http://www.treesearch.fs.fed.us/pubs/25527>

Horton, R., W. Solecki, and C. Rosenzweig. 2011. *Climate Change in the Northeast—A Sourcebook*. Draft Technical Input Report Prepared for the U.S. National Climate Assessment. November 18. <http://downloads.usgcrp.gov/NCA/Activities/nca_ne_full_report_v2.pdf>

HSRTF (Hurricane Sandy Rebuilding Task Force). 2013. *Hurricane Sandy Rebuilding Strategy: Stronger Communities, A Resilient Region*. Presented to the President of the United States. <http://portal.hud.gov/hudportal/HUD?src=/press/press_releases_media_advisories/2013/HUDNo.13-125>

ICCATF (Interagency Climate Change Adaptation Task Force). 2010. *Progress Report of the Interagency Climate Change Adaptation Task Force: Recommended Actions in Support of a National Climate Change Adaptation Strategy*. Washington, DC: Executive Office of the President, Council on Environmental Policy. October 5. <http://www.whitehouse.gov/sites/default/files/microsites/ceq/Interagency-Climate-Change-Adaptation-Progress-Report.pdf>

———. 2011a. *Federal Actions for a Climate Resilient Nation: Progress Report*. Washington, DC: Executive Office of the President, Council on Environmental Policy. October 28. <http://www.whitehouse.gov/sites/default/files/microsites/ceq/2011_adaptation_progress_report.pdf>

———. 2011b. *National Action Plan: Priorities for Managing Freshwater Resources in a Changing Climate*. October. <http://www.whitehouse.gov/sites/default/files/microsites/ceq/2011_national_action_plan.pdf>

IPCC (Intergovernmental Panel on Climate Change). 1996. *IPCC Second Assessment—Climate Change 1995. A Report of the Intergovernmental Panel on Climate Change*. Geneva, Switzerland: United Nations Environment Programme, World Meteorological Organization, IPCC. <http://www.ipcc.ch/pdf/climate-changes-1995/ipcc-2nd-assessment/2nd-assessment-en.pdf>

———. 2000. *Good Practice Guidance and Uncertainty Management in National Greenhouse Gas Inventories*. Ed. J. Penman et al. <http://www.ipcc-nggip.iges.or.jp/public/gp/english/>

———. 2001. Climate Change 2001: *The Scientific Basis. Contribution of Working Group I to the Third Assessment Report of the Intergovernmental Panel on Climate Change*. Ed. J.T. Houghton, Y. Ding, D.J. Griggs, M. Noguer, P.J. van der Linden, X. Dai, K. Maskell, and C.A. Johnson. Cambridge, United Kingdom, and New York, NY: Cambridge University Press. <http://www.grida.no/publications/other/ipcc_tar/>

———. 2003. *Good Practice Guidance for Land Use, Land-Use Change and Forestry*. Ed. J. Penman, M. Gytarsky, T. Hiraishi, T. Krug, D. Kruger, R. Pipatti, L. Buendia, K. Miwa, T. Ngara, K. Tanabe, and F. Wagner. IPCC National Greenhouse Gas Inventories Programme. Hayama, Japan: Institute for Global Environmental Strategies. <http://www.ipcc-nggip.iges.or.jp/public/gpglulucf/ gpglulucf.html>

———. 2006. *2006 IPCC Guidelines for National Greenhouse Gas Inventories*. Ed. S. Eggleston, L. Buendia, K. Miwa, T. Ngara, and K. Tanabe. Prepared by the IPCC Task Force on National Greenhouse Gas Inventories. Hayama, Japan: Institute for Global Environmental Strategies. <http://www.ipcc-nggip.iges.or.jp/public/2006gl/index.html>

———. 2007a. *Climate Change 2007: The Physical Science Basis. Contribution of Working Group I to the Fourth Assessment Report of the Intergovernmental Panel on Climate Change*. Ed. S. Solomon, D. Qin, M. Manning, Z. Chen, M. Marquis, K.B. Averyt, M. Tignor, and H.L. Miller. Cambridge, UK: Cambridge University Press. <http://www.ipcc.ch/publications_and_data/publications_ipcc_ fourth_assessment_report_synthesis_report.htm>

———. 2007b. *Climate Change 2007: Synthesis Report. Contribution of Working Groups I, II, and III to the Fourth Assessment Report of the Intergovernmental Panel on Climate Change*. Ed. R.K. Pachauri and A. Reisinger. Geneva, Switzerland: IPCC. <http://www.ipcc.ch/publications_and_data/ar4/syr/en/contents.html>

————. 2011. *Renewable Energy Sources and Climate Change Mitigation.* Special Report of the Intergovernmental Panel on Climate. Prepared by Working Group III of the Intergovernmental Panel on Climate Change. Ed. O. Edenhofer, R. Pichs-Madruga, Y. Sokona, K. Seyboth, P. Matschoss, S. Kadner, T. Zwickel, P. Eickemeier, G. Hansen, S. Schlömer, C. von Stechow. Cambridge, United Kingdom, and New York, NY: Cambridge University Press. <http://srren.ipcc-wg3.de/>

————. 2012. *Renewable 2012. Managing the Risks of Extreme Events and Disasters to Advance Climate Change Adaptation.* Special Report of the Intergovernmental Panel on Climate. Prepared by Working Group III of the Intergovernmental Panel on Climate Change. Ed. O.C.B. Field, V. Barros, T.F. Stocker, Q.D. Qin, D.J. Dokken, K.L. Ebi, M.D. Mastrandrea, K.J. Mach, G.-K. Plattner, S.K. Allen, M. Tignor, and P.M. Midgley. Cambridge, United Kingdom, and New York, NY: Cambridge University Press. <http://ipcc-wg2.gov/SREX/report/>

————. 2013. *Climate Change 2013: The Physical Science Basis.* Contribution of Working Group I to the Fifth Assessment Report of the Intergovernmental Panel on Climate Change. Ed. T.F. Stocker, D. Qin, G.-K. Plattner, M. Tignor, S.K. Allen, J. Boschung, A. Nauels, Y. Xia, V. Bex, and P.M. Midgley. Cambridge, United Kingdom, and New York, NY: Cambridge University Press.

IPCC/UNEP/OECD/IEA (Intergovernmental Panel on Climate Change, United Nations Environment Programme, Organisation for Economic Co-operation and Development, International Energy Agency). 1997. *Revised 1996 IPCC Guidelines for National Greenhouse Gas Inventories.* Ed. J.T. Houghton, L.G. Meira Filho, B. Lim, K. Treanton, I. Mamaty, Y. Bonduki, D.J. Griggs, and B.A. Callender. Paris: OECD. <http://www.ipcc-nggip.iges. or.jp/public/gl/invs1.html>

ISSE (Integrative Science for Society and Environment). 2010. *Integrative Science for Society and Environment: A Strategic Research Initiative.* Research Initiatives Subcommittee of the LTER Planning Process Conference Committee and the Cyberinfrastructure Core Team. April. <http://www.csrc.sr.unh.edu/~lammers/MacroscaleHydrology/Papers/ISSE_complete_10April.pdf>

IWGCCH (Interagency Working Group on Climate Change and Health). 2010. *A Human Health Perspective on Climate Change. A Report Outlining the Research Needs on the Human Health Effects of Climate Change.* Environmental Health Perspectives and National Institute of Environmental Health Sciences. April. <http://downloads.globalchange.gov/cchhg/climatereport2010.pdf>

Izaurralde, R.C., R. Sahajpal, X. Zhang, and D.H. Manowitz. 2012. *National Geo-Database for Biofuel Simulations and Regional Analysis of Biorefinery Siting Based on Cellulosic Feedstock Grown on Marginal Lands.* PNNL-21283. Richland, WA: Pacific Northwest National Laboratory. April. <http://www.pnnl.gov/main/publications/external/technical_reports/PNNL-21283.pdf>

Janetos, A., L. Hansen, D. Inouye, B. P. Kelly, L. Meyerson, B. Peterson, and R. Shaw. 2008. "Biodiversity." In *The Effects of Climate Change on Agriculture, Land Resources, Water Resources, and Biodiversity in the United States.* Vol. Synthesis and Assessment Product 4.3. Ed. P. Backlund, A. Janetos, D. Schimel, J. Hatfield, K. Boote, P. Fay, L. Hahn, C. Izaurralde, B.A. Kimball, T. Mader, et al. Washington, D.C.: U.S. Department of Agriculture, pp. 151–181.

Janetos, A.C., R.S. Chen, D. Arndt, and M.A. Kenney. 2012. *National Climate Assessment Indicators: Background, Development, & Examples.* Washington, DC: National Research Council. February. <http://downloads.usgcrp.gov/NCA/Activities/NCA-Indicators-Technical-Input-Report-FINAL--3-1-12.pdf>

Kalra, N., L. Ecola, R. Keefe, B. Weatherford, M. Wachs, P. Plumeau, S. Lawe, and C. Smith. 2012. *Reference Sourcebook for Reducing Greenhouse Gas Emissions from Transportation Sources.* FHWA Project DTHF61-09-F-00117. Washington, DC: U.S. Department of Transportation, Federal Highway Administration. February. <http://www.fhwa.dot.gov/environment/climate_change/mitigation/resources_and_publications/reference_sourcebook/>

Karl, T.R., S.J. Hassol, C.D. Miller, and W.L. Murray, eds. 2009. *Global Climate Change Impacts in the United States: A State of Knowledge Report from the U.S. Global Change Research Program.* Cambridge, UK, and New York, NY: Cambridge University Press. <http://downloads.globalchange.gov/usimpacts/pdfs/climate-impacts-report.pdf>

KDFWR (State of Kentucky, Department of Fish and Wildlife Resources). 2010. *Action Plan to Respond to Climate Change in Kentucky: A Strategy of Resilience.* Frankfort, KY. <http://fw.ky.gov/kfwis/stwg/2010Update/Climate_Change_Chapter.pdf>

Keener, V.W., J.J. Marra, M.L. Finucane, D. Spooner, and M.H. Smith, eds. 2012. *Climate Change and Pacific Islands: Indicators and Impacts. Report for the 2012 Pacific Islands Regional Climate Assessment.* Washington, DC: Island Press. <http://www.cakex.org/virtual-library/climate-change-and-pacific-islands-indicators-and-impacts>

Kunkel, K.E, L.E. Stevens, S.E. Stevens, L. Sun, E. Janssen, D. Wuebbles, C.E. Konrad II, C.M. Fuhrman, B.D. Keim, M.C. Kruk, A. Billet, H. Needham, M. Schafer, and J.G. Dobson. 2013a. *Regional Climate Trends and Scenarios for the U.S. National Climate Assessment. Part 2. Climate of the Southeast U.S.* NOAA Technical Report NESDIS 142-2. Washington, DC: National Oceanic and Atmospheric Administration, National Environmental Satellite, Data, and Information Service. January. <http://www.sercc.com/NOAA_NESDIS_Tech_Report_Climate_of_the_Southeast_U.S.pdf>

Kunkel, K.E, L.E. Stevens, S.E. Stevens, L. Sun, E. Janssen, D. Wuebbles, S.D. Hilberg, M.S. Timlin, L. Stoecker, N.E. Westcott, and J.G. Dobson. 2013b. *Regional Climate Trends and Scenarios for the U.S. National Climate Assessment. Part 3. Climate of the Midwest U.S.* NOAA Technical Report NESDIS 142-3. Washington, DC: National Oceanic and Atmospheric Administration, National Environmental Satellite, Data, and Information Service. January. <http://www.nesdis.noaa.gov/technical_reports/ NOAA_NESDIS_Tech_Report_142-3-Climate_of_the_Midwest_U.S.pdf>

Kunkel, K.E, L.E. Stevens, S.E. Stevens, L. Sun, E. Janssen, D. Wuebbles, M.C. Kruk, D.P. Thomas, M. Shulski, N. Umphlett, K. Hubbard, K. Robbins, L. Romolo, A. Akyuz, T. Pathak, T. Bergantino, and J.G. Dobson. 2013c. *Regional Climate Trends and Scenarios for the U.S. National Climate Assessment. Part 4. Climate of the U.S. Great Plains.* NOAA Technical Report NESDIS 142-4. Washington, DC: National Oceanic and Atmospheric Administration, National Environmental Satellite, Data, and Information Service. January. <http://www.nesdis.noaa.gov/technical_reports/NOAA_NESDIS_Tech_Report_142-4-Climate_of_the_U.S.%20Great_Plains.pdf>

Kunkel, K.E, L.E. Stevens, S.E. Stevens, L. Sun, E. Janssen, D. Wuebbles, K.T. Redmond, and J.G. Dobson. 2013d. *Regional Climate Trends and Scenarios for the U.S. National Climate Assessment. Part 6. Climate of the Northwest U.S.* NOAA Technical Report NESDIS 142-6. Washington, DC: National Oceanic and Atmospheric Administration, National Environmental Satellite, Data, and Information Service. January. <http://www.nesdis.noaa.gov/technical_reports/ NOAA_NESDIS_Tech_Report_142-6-Climate_of_the_Northwest_U.S.pdf>

Leiserowitz, A., E. Maibach, C. Roser-Renouf, G. Feinberg, and P. Howe. 2013. *Extreme Weather and Climate Change in the American Mind: April 2013.* Yale University and George Mason University. New Haven, CT: Yale Project on Climate Change Communication. <http://www.climatechangecommunication.org/sites/default/files/reports/ Extreme-Weather-Public-Opinion-April-2013_0.pdf>

Lewes Mitigation Planning Team and Lewes Pilot Project Subcommittee. 2011. *The City of Lewes Hazard Mitigation and Climate Adaptation Action Plan.* Delaware Sea Grant College Program, ICLEI-Local Governments for Sustainability, University of Delaware Sustainable Coastal Communities Program. June. <http://www.ci.lewes.de.us/pdfs/Lewes_Hazard_Mitigation_and_CLimate_Adaptation_Action_Plan_FinalDraft_8-2011.pdf>

Loveland, T., R. Mahmood, T. Patel-Weynand, K. Karstensen, K. Beckendorf, N. Bliss, and A. Carleton. 2012. *National Climate Assessment Technical Report on the Impacts of Climate and Land Use and Land Cover Change.* U.S. Geological Survey Open-File Report 2012–1155. Reston, VA: U.S. Department of the Interior, U.S. Geological Survey. <http://pubs.usgs.gov/of/2012/1155/of2012-1155.pdf>

Lynn, K., J. Daigle, J. Hoffman, F. Lake, N. Michelle, D. Ranco, C. Viles, G. Voggesser, and P. Williams. 2013. The impacts of climate change on tribal traditional foods. *Climatic Change* 120(March): 545–556. <http://dx.doi.org/10.1007/s10584-013-0736-1>

Maldonado, J.K., C. Shearer, R. Bronen, K. Peterson, and H. Lazrus. 2013. The impact of climate change on tribal communities in the US: displacement, relocation, and human rights. *Climatic Change* 120(April): 601–614. <http://dx.doi.org/10.1007/s10584-013-0746-z>

Markon, C.J., S.F. Trainor, and F.S. Chapin, III, eds. 2012. *The United States National Climate Assessment—Alaska Technical Regional Report.* USGS Circular 1379. Reston, VA: U.S. Department of the Interior, U.S. Geological Survey. <http://pubs.usgs.gov/circ/1379/pdf/circ1379.pdf>

McKenzie, B.S. 2010. *Public Transportation Usage Among U.S. Workers: 2008 and 2009.* American Community Survey Reports. ACSBR/090-5. Washington, DC: U.S. Department of Commerce, Census Bureau. October. <http://www.census.gov/prod/2010pubs/acsbr09-5.pdf>

McKenzie, B.S., and M. Rapino. 2011. *Commuting in the United States: 2009.* American Community Survey Reports. ACS-15. Washington, DC: U.S. Department of Commerce, Census Bureau. September. <http://www.census.gov/prod/2011pubs/acs-15.pdf>

MDEP (State of Maine, Department of Environmental Protection). 2012. *Chapter 355: Coastal Sand Dune Rules.* 06-096. <http://www.maine.gov/sos/cec/rules/06/096/096c355.doc>

Michalak, A.M., R.B. Jackson, G. Marland, C.L. Sabine, and the Carbon Cycle Science Working Group. 2011. *A U.S. Carbon Cycle Science Plan.* Boulder, CO: University Corporation for Atmospheric Research. August. <http://carboncycle.joss.ucar.edu/sites/default/files/documents/USCarbonCycleSciencePlan-2011.pdf>

Milly, P.C.D., and K.A.Dunne, 2011. On the hydrologic adjustment of climate-model projections: the potential pitfall of potential evapotranspiration. *Earth Interactions* 15: 1–14. doi: 10.1175/2010ei363.1.

MMC (Multihazard Mitigation Council). 2005. *Natural Hazard Mitigation Saves: An Independent Study to Assess the Future Savings from Mitigation Activities.* Washington, DC: U.S. Department of Homeland Security, Federal Emergency Management Agency, and National Institute of Building Sciences.
<http://www.floods.org/PDF/MMC_Volume1_FindingsConclusionsRecommendations.pdf>

Nakićenović, N., O. Davidson, G. Davis, A. Grübler, T. Kram, E. Lebre La Rovere, B. Metz, T. Morita, W. Pepper, H. Pitcher, A. Sankovski, P. Shukla, R. Swart, R. Watson, and Z. Dad. 2000. *Summary for Policymakers: Emission Scenarios. A Special Report of IPCC Working Group III of the Intergovernmental Panel on Climate Change.* World Meteorological Organization and United Nations Environment Programme. <https://www.ipcc.ch/pdf/special-reports/spm/sres-en.pdf>

NAS (National Academy of Sciences), Achieve, the American Association for the Advancement of Science, and the National Science Teachers Association. 2013. *Next Generation Science Standards for Today's Students and Tomorrow's Workforce.* Washington, DC: Achieve, Inc. June. <http://www.nextgenscience.org/next-generation-science-standards>

NCADAC (National Climate Assessment Development Advisory Committee). 2013. *Third National Climate Assessment: Draft Report.* Washington, DC: U.S. Global Change Research Program. May. <http://www.globalchange.gov/what-we-do/assessment>

NFWPCAP (National Fish, Wildlife, and Plants Climate Adaptation Partnership). 2012. *National Fish, Wildlife, and Plants Climate Adaptation Strategy.* Washington, DC: Association of Fish and Wildlife Agencies, Council of Environmental Quality, Great Lakes Indian Fish and Wildlife Commission, National Oceanic and Atmospheric Administration, and U.S. Fish and Wildlife Service. <http://www.wildlifeadaptationstrategy.gov/pdf/NFWPCAS-Final.pdf>

NOAA (National Oceanic and Atmospheric Administration). 2011a. The U.S. Population Living at the Coast, 1970–2030. <http://stateofthecoast.noaa.gov/population/welcome.html>

———. 2011b. Spatial Trends in Coastal Socioeconomics. Demographic Trends (1970–2011) from the Census Bureau.<http://coastalsocioeconomics.noaa.gov/>

———. 2012. *NOAA's List of Coastal Counties for the Bureau of the Census.* Statistical Abstract Series. <http://www.census.gov/geo/landview/lv6help/coastal_cty.pdf>

NOAA/ESRL (National Oceanic and Atmospheric Administration, Earth System Research Laboratory). 2009. "Trends in Atmospheric Carbon Dioxide." <http://www.esrl.noaa.gov/gmd/ccgg/trends>

NOAA/NCDC (National Oceanic and Atmospheric Administration, National Climatic Data Center). 2012a. *Billion-Dollar U.S. Weather/Climate Disasters: 1980–2012.* Asheville, NC. December. <http://www.ncdc.noaa.gov/billions/events>

———. 2012b. *State of the Climate: National Overview—Annual 2012.* Asheville, NC. December. <http://www.ncdc.noaa.gov/sotc/national/2012/13>

NOC (National Ocean Council). 2013. *National Ocean Policy Implementation Plan.* Washington, DC. April. <http://www.whitehouse.gov//sites/default/files/national_ocean_policy_implementation_plan.pdf>

NPCC2 (New York City Panel on Climate Change). 2010. *Climate Change Adaptation in New York City: Building a Risk Management Response: New York City Panel on Climate Change 2010 Report.* Vol. 1196. New York, NY. May. <http://onlinelibrary.wiley.com/doi/10.1111/nyas.2010.1196.issue-1/issuetoc>

———. 2013. *Climate Risk Information 2013: Observations, Climate Change Projections, and Maps.* Ed. C. Rosenzweig and W. Solecki. Prepared for use by the City of New York Special Initiative on Rebuilding and Resiliency. New York, NY. June. <http://www.nyc.gov/html/planyc2030/downloads/pdf/npcc_climate_risk_information_2013_report.pdf>

NRC (National Research Council) Committee on America's Climate Choices. 2011. *America's Climate Choices.* Washington, DC: The National Academies. <http://nas-sites.org/americasclimatechoices/sample-page/panel-reports/americas-climate-choices-final-report/>

Orr, J. C., S. Pantoja, and H. O. Pörtner. 2005. Introduction to special section: The ocean in a high-CO_2 world. *Journal of Geophysical Research-Oceans* 110, C09S01.

Propst, S.C. 2012. *Innovative Approaches for Adapting to Water Variability in the West.* Washington, DC: Georgetown Climate Center. March. <http://www.georgetownclimate.org/sites/default/files/Water%20Variability%20in%20the%20West.pdf>

Quinn, H., H. Schweingruber, and T. Keller, eds. 2013. Committee on Conceptual Framework for the New K-12 Science Education Standards; Board on Science Education; Division of Behavioral and Social Sciences and Education; and National Research Council. *A Framework for New K-12 Science Education: Practices, Crosscutting Concepts, and Core Ideas.* National

Research Council of The National Academies. Washington, DC: National Academies Press. <http://www.nap.edu/catalog.php?record_id=13165#>

Skaggs, R., K. Hibbard, T.C. Janetos, and J.S. Rice. 2012. *Climate and Energy-Water-Land System Interactions. Technical Report to the U.S. Department of Energy in Support of the National Climate Assessment.* PNNL-21185. Richland, WA: U.S. Department of Energy, Pacific Northwest National Laboratory. March. <http://www.cakex.org/sites/default/files/documents/PNNL-21185.pdf>

Solecki, W., C. Rosenzweig, S. Hammer, and S. Mehrotra. 2012. Urbanization of climate change: responding to a new global challenge. In *The Urban Transformation: Health, Shelter and Climate Change.* Ed. E.D. Sclar, N. Volavka-Close, and P. Brown. London, United Kingdom: Routledge.

Southern Climate Impacts Planning Program. 2010. *SCIPP Drought Information Document.* University of Oklahoma and Louisiana State University. June 15. <http://www.southernclimate.org/publications/SCIPP_Drought_Final.pdf>

Staudinger, M.D., N.B. Grimm, A. Staudt, S.L. Carter, F.S. Chapin III, P. Kareiva, M. Ruckelshaus, and B.A. Stein. 2012. *Impacts of Climate Change on Biodiversity, Ecosystems, and Ecosystem Services: Technical Input to the 2013 National Climate Assessment. Cooperative Report to the 2013 National Climate Assessment.* Reston, VA: U.S. Department of the Interior, U.S. Geological Survey. July. <http://downloads.usgcrp.gov/NCA/Activities/Biodiversity-Ecosystems-and-Ecosystem-Services-Technical-Input.pdf>

Stults, M., and J. Pagach. 2011. *Preparing for Climate Change in Groton, Connecticut: A Model Process for Communities in the Northeast.* ICELI-Local Governments for Sustainability and Connecticut Department of Environmental Protection. April. <http://www.groton-ct.gov/depts/plandev/docs/Final%20Report_Groton%20Coastal%20Climate%20Change%20ProjectJP.pdf>

UGLSB (Upper Great Lakes Study Board). 2012. *Lake Superior Regulation: Addressing Uncertainty in Upper Great Lakes Water Levels. Summary of Findings and Recommendations.* Final Report to the International Joint Commission. <http://www.ijc.org/files/tinymce/uploaded/IUGLS_Summary_Report.pdf>

UNEP and WMO (United Nations Environment Programme and World Meteorological Organization). 2011. *Integrated Assessment of Black Carbon and Tropospheric Ozone: Summary for Decision Makers.* UNEP/GC/26/INF/20. Nairobi, Kenya: UNEP. <http://www.unep.org/dewa/Portals/67/pdf/Black_Carbon.pdf>

UNFCCC (United Nations Framework Convention on Climate Change). 2003. *Review of the Implementation of Commitments and of Other Provisions of the Convention. National Communications: Greenhouse Gas Inventories From Parties Included in Annex I to the Convention. UNFCCC Guidelines on Reporting and Review.* FCCC/CP/2002/8. March 28. <http://unfccc.int/resource/docs/cop8/08.pdf>

———. 2006. *Updated UNFCCC Reporting Guidelines on Annual Inventories Following Incorporation of the Provisions of Decision 14/CP.11.* FCCC/ SBSTA/2006/9. August 18. <http://unfccc.int/resource/docs/2006/sbsta/eng/09.pdf>

———. 2012. *Common Tabular Format for UUNFXXX Biennial Reporting Guidelines for Developed Country Parties.* FCCC/CP/2012/8/L.12. December 7. <http://unfccc.int/resource/docs/2012/cop18/eng/l12.pdf>

———. 2013. *Report of the Conference of the Parties on Its Eighteenth Session, Held in Doha from 26 November to 8 December 2012.* FCCC/ CP/2012/8/Add.1. February 28. <http://unfccc.int/resource/docs/2012/cop18/eng/08a01.pdf>

USACE (United States Army Corps of Engineers). 2011. *Sea-Level Change Considerations for Civil Works Programs.* Circular No. 1165-2-212. October 1. <http://planning.usace.army.mil/toolbox/library/ECs/EC11652212Nov2011.pdf>

USAID (United States Agency for International Development). 2012. *USAID Climate Change and Development Strategy: 2012–2016.* Washington, DC. January. <http://www.cgdev.org/doc/Rethinking%20Aid/Climate_Change_&_Dev_Strategy.pdf>

U.S. Congress (United States Congress). 2013. "American Taxpayer Relief Act of 2012." P.L. 112-240, H.R. 8, 126 Stat. 2313. January 2. <http://transportation.house.gov/hearing/overview-united-states%E2%80%99-freight-transportation-system>

USDA (United States Department of Agriculture). 2010. *Strategic Plan: FY 2010–2015.* Washington, DC. <http://www.ocfo.usda.gov/sp2010/sp2010.pdf>

USDA/FS (United States Department of Agriculture, Forest Service). 2012. *Future of America's Forest and Rangelands: Forest Service 2010 Resources Planning Act Assessment.* Gen. Tech. Rep. WO-87. Washington, DC. 198 p. <http://www.fs.fed.us/research/rpa/assessment/>

USDA/FSA (United States Department of Agriculture, Farm Service Agency). 2013. Conservation Reserve Program Reports and Statistics. <http://www.fsa.usda.gov/FSA/webapp?area=home&subject=copr&topic=rns-css>

USDA/GCTF (United States Department of Agriculture, Global Change Task Force). 2010. *USDA Climate Change Science Plan.* Washington, DC. December 8.
<http://www.usda.gov/oce/climate_change/science_plan2010/USDA_CCSPlan_120810.pdf>

USDA/NASS (United States Department of Agriculture, Natural Agricultural Statistics Service). 2009. *2007 Census of Agriculture. United States Summary and State Data.* Vol. 1. Geographic Area Series, Part 51. AC-07-A-51. Washington, DC. December. <http://www.agcensus.usda.gov/Publications/2007/Full_Report/usv1.pdf>

———. 2013. *Crop Production Historical Track Records.* ISSN: 2157-8990.
<http://usda01.library.cornell.edu/usda/current/htrcp/htrcp-04-12-2013.txt>

U.S. DOC/BEA ((United States Department of Commerce, Bureau of Economic Analysis). 2012. 2012 Comprehensive Revision of the National Income and Product Accounts: Current-dollar and "real" GDP, 1929–2012. Washington, DC. Last Modified January 2013. <http://www.bea.gov/national/index.htm#gdp>

U.S. DOC/Census (United States Department of Commerce, Census Bureau). 2010. Population of U.S. Cities.
<http://www.census.gov>

———. 2012. International Database. June. <http://www.census.gov/ipc/www/idbnew.html>

U.S. DHS (United States Department of Homeland Security). 2012. *Climate Change Adaptation Roadmap.* Washington, DC. June. <http://www.dhs.gov/sites/default/files/publications/Appendix%20A%20DHS%20FY2012%20Climate%20 Change%20Adaptation%20Plan_0.pdf>

U.S. DHS/FEMA (United States Department of Homeland Security, Federal Emergency Management Agency). 2011. *FEMA Mitigation and Insurance Strategic Plan: 2012–2014.* FEMA P-857. Washington, DC. September.
<http://www.fema.gov/media-library-data/20130726-1811-25045-8194/fema_mitigation_strategic_plan_508.pdf>

U.S. DOE (United States Department of Energy). 2011. *Report on the First Quadrennial Technology Review.* DOE/S-0001. Washington, DC. September. <http://energy.gov/sites/prod/files/QTR_report.pdf>

———. 2013. "Energy Dept. Reports: U.S. Wind Energy Production and Manufacturing Reaches Record High." August 6. <http://energy.gov/articles/energy-dept-reports-us-wind-energy-production-and-manufacturing-reaches-record-highs>

U.S. DOE/EIA (U.S. Department of Energy, Energy Information Administration). 2009. *Annual Energy Outlook 2009.* Washington, DC. <http://www.eia.doe.gov/oiaf/aeo/leg_reg.html>

———. 2011. *Direct Federal Financial Interventions and Subsidies in Energy in Fiscal Year 2010.* Washington, DC. July. <http://www.eia.gov/analysis/requests/subsidy/pdf/subsidy.pdf>

———. 2012. *Annual Energy Review 2011.* DOE/EIA-0384(2011). Washington, DC. September. <www.eia.gov/aer>

———. 2013a. *2012 Domestic Uranium Production Report.* Washington, DC. June.
<http://www.eia.gov/uranium/production/annual/pdf/dupr.pdf>

———. 2013b. *Annual Energy Outlook 2013 with Projections to 2040.* DOE/EIA-0383(2013). Washington, DC. April. <http://www.eia.gov/forecasts/aeo/>

———. 2013c. *Power Monthly with Data for December 2012.* Washington, DC. February.
<http://www.eia.gov/electricity/monthly/current_year/february2013.pdf>

———. 2013d. "Energy Sources Have Changed throughout the History of the United States." DOE/EIA-0035(2013/07). Washington, DC. July. <http://www.eia.gov/mer>

———. 2013e. "How Dependent Are We on Foreign Oil?" *Energy In Brief.* May 10.
<http://www.eia.gov/energy_in_brief/article/foreign_oil_dependence.cfm>

———. 2013f. "How Dependent Is the United States on Foreign Oil?" June 3.
<http://www.eia.gov/tools/faqs/faq.cfm?id=32&t=6>

———. 2013g. *Monthly Energy Review.* Supplemental Tables on Petroleum Product detail. DOE/EIA-0035(2013/02). Washington, DC. February. <http://www.eia.gov/mer>

———. 2013h. *Monthly Energy Review.* DOE/EIA-0035(2013/05). Washington, DC. May. <http://www.eia.gov/mer>

———. 2013i. *Monthly Energy Review.* DOE/EIA-0035(2013/06). Washington, DC. June. <http://www.eia.gov/mer>

———. 2013j. *Monthly Energy Review*. DOE/EIA-0035(2013/08). Washington, DC. August. <http://www.eia.gov/mer>

———. 2013k. *Monthly Energy Review*. DOE/EIA-0035(2013/09). Washington, DC. September. <http://www.eia.gov/mer>

———. 2013l. "U.S. Natural Gas Imports & Exports 2012." May 23. <http://www.eia.gov/naturalgas/importsexports/annual/>

———. 2013m. *State Energy Efficiency Program Evaluation Inventory*. July.
<http://www.eia.gov/efficiency/programs/inventory/pdf/inventory.pdf>

U.S. DOE and NREL (United States Department of Energy and National Renewable Energy Laboratory). 2013. *U.S. Energy Sector Vulnerabilities to Climate Change and Extreme Weather*. DOE/PI-0013. Washington, DC. July.
<http://energy.gov/sites/prod/files/2013/07/f2/20130716-Energy%20Sector%20Vulnerabilities%20Report.pdf>

U.S. DOI/BR (United States Department of the Interior, Bureau of Reclamation). 2012. *Reclamation: Managing Water in the West. Colorado River Basin Water Supply and Demand Study: Study Report*. December. <http://www.usbr.gov/lc/region/programs/crbstudy/finalreport/Study%20Report/CRBS_Study_Report_FINAL.pdf>

U.S.DOI/FWS (United States Department of the Interior, National Fish, Wildlife, and Plants Climate Adaptation Partnership). 2012. *National Fish, Wildlife, and Plants Climate Adaptation Strategy*. Washington, DC: Association of Fish and Wildlife Agencies, Council on environmental Quality, Great Lakes Indian Fish and Wildlife Commission, National Oceanic and Atmospheric Administration, and U.S. Fish and Wildlife Service. <http://www.nps.gov/orgs/ccrp/upload/NPS_CCActionPlan.pdf>

U.S.DOI/NPS (United States Department of Agriculture, National Park Service). 2010. *Climate Change Action Plan: 2012–2014*. Climate Change Response Program. Washington, DC. <http://www.nps.gov/orgs/ccrp/upload/NPS_CCActionPlan.pdf>

———. 2011. *A Call to Action: Preparing for a Second Century of Stewardship and Engagement*. Washington, DC. August.
<http://www.nps.gov/calltoaction/PDF/Directors_Call_to_Action_Report.pdf>

———. 2012. *Green Parks Plan: Advancing Our Mission through Sustainable Operations*. Washington, DC. April.
<http://www.nps.gov/greenparksplan/downloads/NPS_2012_Green_Parks_Plan.pdf>

U.S. DOS (United States Department of State). 2007. *U.S. Climate Action Report 2006: Fourth National Communication of the United States of America Under the United Nations Framework Convention on Climate Change*. Washington, DC. July.
<http://www.state.gov/g/oes/rls/rpts/car/>

———. 2010. *U.S. Climate Action Report 2010: Fifth National Communication of the United States of America Under the United Nations Framework Convention on Climate Change*. Washington, DC. June. <http://www.state.gov/e/oes/rls/rpts/car5/>

———. 2012. *Meeting the Fast Start Commitment: U.S. Climate Finance in Fiscal Year 2012*. Washington, DC.
<http://www.state.gov/documents/organization/201130.pdf>

U.S. DOS and USAID (United States Department of State and United States Agency for International Development). 2010. *Leading Through Civilian Power: The First Quadrennial Diplomacy and Development Review*. Washington, DC.
<http://www.state.gov/documents/organization/153108.pdf>

U.S. DOT (United States Department of Transportation). 2013. *U.S. Department of Transportation Climate Adaptation Plan: Ensuring Transportation Infrastructure and System Resilience*. Washington, DC. February.
<http://www.dot.gov/sites/dot.dev/files/docs/DOT%20Adaptation%20Plan.pdf>

U.S. DOT/OIG (United States Department of Transportation, Office of Inspector General). 2012. *Aviation Industry Performance: A Review of the Aviation Industry, 2008–2011*. CC-2012-029. Washington, DC. September 24.
<http://www.oig.dot.gov/sites/dot/files/Aviation%20Industry%20Performance%5E9-24-12.pdf>

U.S. DOT/RITA/BTS (United States Department of Transportation, Research and Innovative Technology Administration, Bureau of Transportation Statistics). 2013a. *National Transportation Statistics 2013*. Washington, DC.
<http://www.bts.gov/publications/national_transportation_statistics/>

———. 2013b. "Total Passengers on U.S Airlines and Foreign Airlines U.S. Flights Increased 1.3% in 2012 from 2011." BTS-16-13. Washington, DC. April 4. <http://www.rita.dot.gov/bts/press_releases/bts016_13>

U.S. EPA (United States Environmental Protection Agency). 2009. *Endangerment and Cause or Contribute Findings for Greenhouse Gases Under Section 202(a) o the Clean Air Act*. Final Rule. *Federal Register* 74(239, Dec. 15).
<http://www.epa.gov/climatechange/Downloads/endangerment/Federal_Register-EPA-HQ-OAR-2009-0171-Dec.15-09.pdf>

———. 2012. Cross-EPA Work Group on Climate Change Adaptation Planning. *U.S. Environmental Protection Agency Climate Change Adaptation Plan*. June.
<http://www.epa.gov/climatechange/pdfs/EPA-climate-change-adaptation-plan-final-for-public-comment-2-7-13.pdf>

———. 2013a. *Energy Star® Overview of 2012 Achievements.* March. <http://www.energystar.gov/ia/partners/publications/pubdocs/ES%20bi-fold%20031313%20FINAL%20for%20print%20rev.pdf?8254-b82b>

———. 2013b. *Methodologies for U.S. Greenhouse Gas Emissions Projections: Non-CO$_2$ and Non-Energy CO$_2$ Sources.* Washington, DC. August.

U.S. EPA/OAP (United States Environmental Protection Agency, Office of Atmospheric Programs). 2013. *Inventory of U.S. Greenhouse Gas Emissions and Sinks: 1990–2011.* EPA 430-R-13-001. Washington, DC. April. <http://www.epa.gov/climatechange/emissions/usinventoryreport.html>

U.S. EPA/OAQPS (United States Environmental Protection Agency, Office of Air Quality Planning and Standards). 2009. "1970–2008 Average annual emissions, all criteria pollutants in MS Excel." <http://www.epa.gov/ttn/chief/trends/index.html>.

———. 2010. National Emissions Inventory (NEI) Air Pollutant Emissions Trends Data. "2009 Average annual emissions, all criteria pollutants in MS Excel." <http://www.epa.gov/ttn/chief/trends/index.html>

U.S. EPA/OSW (United States Environmental Protection Agency, Office of Solid Waste). 2013a. *Municipal Solid Waste Generation, Recycling, and Disposal in the United States: Facts and Figures for 2011.* EPA-530-13-001. Washington, DC. May. <http://www.epa.gov/epawaste/nonhaz/municipal/pubs/MSWcharacterization_508_053113_fs.pdf>

———. 2013b. *Municipal Solid Waste in the United States: 2011 Facts and Figures.* EPA530-R-13-001. Washington, DC. May. <http://www.epa.gov/epawaste/nonhaz/municipal/pubs/MSWcharacterization_fnl_060713_2_rpt.pdf>

USGCRP (United States Global Change Research Program). 2009. *Climate Literacy: The Essential Principles of Climate Science. A Guide for Individuals and Communities.* Washington, DC. <http://downloads.globalchange.gov/Literacy/climate_literacy_lowres_english.pdf>

———. 2011. *Our Changing Planet: The U.S. Global Change Research Program for Fiscal Year 2011.* A Supplement to the President's Budget for Fiscal Year 2011. Washington, DC. <http://downloads.globalchange.gov/ocp/ocp2011/ocp2011.pdf>

———. 2012a. *Our Changing Planet: The U.S. Global Change Research Program for Fiscal Year 2012.* A Supplement to the President's Budget for Fiscal Year 2012. Washington, DC. <http://downloads.globalchange.gov/ocp/ocp2012/ocp2012.pdf>

———. 2012b. *The National Global Change Research Plan 2012–2021: A Strategic Plan for the U.S. Global Research Program.* USGCRP, National Coordination Office. <http://downloads.globalchange.gov/strategic-plan/2012/usgcrp-strategic-plan-2012.pdf>

———. 2013. *Our Changing Planet: The U.S. Global Change Research Program for Fiscal Year 2013.* A Supplement to the President's Budget for Fiscal Year 2013. Washington, DC. <http://downloads.globalchange.gov/ocp/ocp2013/ocp2013.pdf>

U.S. House (United States House of Representatives, Committee on Transportation and Infrastructure). 2013. "Overview of the United States' Freight Transportation System." April 2. <http://transportation.house.gov/hearing/overview-united-states%E2%80%99-freight-transportation-system>

Uzarski, D.G., T.M. Burton, R.E. Kolar, and M.J. Cooper. 2009. The ecological impacts of fragmentation and vegetation removal in Lake Huron's coastal wetlands. *Aquatic Ecosystem Health & Management* 12(1): 45–62.

Vogel, J.M., and J.B. Smith. 2010. *Climate Change Vulnerability Assessments: A Review of Water Utility Practices.* EPA 800-R-10-001. Washington, DC: U.S. Environmental Protection Agency, Office of Water. August. <http://water.epa.gov/scitech/climatechange/upload/Climate-Change-Vulnerability-Assessments-Sept-2010.pdf>

Vogel, J., V. Zoltay, J. Smith, D. Kemp, and T. Johnson. 2011. *Climate Change Vulnerability Assessments: Four Case Studies of Water Utility Practices.* EPA/600/R-10/077F: Washington, DC: U.S. Environmental Protection Agency, Global Change Research Program, National Center for Environmental Assessment. <http://cfpub.epa.gov/ncea/global/recordisplay.cfm?deid=233808>

Vose, J.M., D.L. Peterson, and T. Patel-Weynand, eds. 2012. *Effects of Climate Variability and Change on Forest Ecosystems: A Comprehensive Science Synthesis for the U.S. Forest Sector.* Gen. Tech. Rep. PNW-GTR-870. Portland, OR: U.S. Department of Agriculture, Forest Service, Pacific Northwest Research Station. 265 p. <http://www.treesearch.fs.fed.us/pubs/42610>

Walthall, C.L., J. Hatfield, P. Backlund, L. Lengnick, E. Marshall, M. Walsh, S. Adkins, M. Aillery, E.A. Ainsworth, C. Ammann, C.J. Anderson, I. Bartomeus, L.H. Baumgard, F. Booker, B. Bradley, D.M. Blumenthal, J. Bunce, K. Burkey, S.M. Dabney, J.A. Delgado, J. Dukes, A. Funk, K. Garrett, M. Glenn, D.A. Grantz, D. Goodrich, S. Hu, R.C. Izaurralde, R.A.C. Jones, S-H. Kim, A.D.B. Leaky, K. Lewers, T.L. Mader, A. McClung, J. Morgan, D.J. Muth, M. Nearing, D.M. Oosterhuis, D. Ort, C. Parmesan, W.T. Pettigrew, W. Polley, R. Rader, C. Rice, M. Rivington, E. Rosskopf, W.A. Salas, L.E. Sollenberger, R. Srygley, C. Stöckle, E.S. Takle, D. Timlin, J.W. White, R. Winfree, L. Wright-Morton, and L.H. Ziska. 2012. *Climate Change and Agriculture in the*

United States: Effects and Adaptation. USDA Technical Bulletin 1935. Washington, DC: U.S. Department of Agriculture, Agricultural Research Service. <http://www.usda.gov/oce/climate_change/effects_2012/CC%20and%20Agriculture%20 Report%20%2802-04-2013%29b.pdf>

Wear, D.N. 2011. *Forecasts of County-Level Land Uses under Three Future Scenarios: A Technical Document Supporting the Forest Service 2010 RPA Assessment.* Gen. Tech. Rep. SRS-141. Asheville, NC: U.S. Department of Agriculture Forest Service, Southern Research Station. 41 p. <http://www.fs.fed.us/research/rpa/assessment/>

Wear, D.N., R. Huggett, R. Li, B. Perryman, and S. Liu. 2013. *Forecasts of Forest Conditions in Regions of the United States under Future Scenarios: A Technical Document Supporting the Forest Service 2010 RPA Assessment.* Gen. Tech. Rep. SRS-170. Asheville, NC: U.S. Department of Agriculture Forest Service, Southern Research Station. 101 p. <http://www.srs.fs.usda.gov/pubs/43055>

Westcott, P., and R. Trostle. 2013. *USDA Agricultural Projections to 2022.* Long-Term Projections Report OCE-2013-1. Washington, DC: U.S. Department of Agriculture, Office of the Chief Economist. February. <http://www.ers.usda.gov/publications/oce-usda-agricultural-projections/oce131.aspx#.UXAmW7WG18E>.

White-Newsome, J.L., B.N. Sánchez, E.A. Parker, J.T. Dvonch, Z. Zhang, and M.S. O'Neill. 2011. Assessing heat-adaptive behaviors among older, urban-dwelling adults. *Maturitas* 70(1): 85–91. <http://www.ncbi.nlm.nih.gov/pubmed/21782363>

Wilbanks, T., D. Bilello, D. Schmalzer, M. Scott, D. Arent, J. Buizer, H. Chum, J. Dell, J. Edmonds, G. Franco, R. Jones, S. Rose, N. Roy, A. Sanstad, S. Seidel, J. Weyant, and D. Wuebbles. 2012a. *Climate Change and Energy Supply and Use. Technical Report for the U.S. Department of Energy in Support of the National Climate Assessment.* Oak Ridge, TN: U.S. Department of Energy, Oak Ridge National Laboratory. February 29. <http://www.esd.ornl.gov/eess/EnergySupplyUse.pdf>

Wilbanks, T., S. Fernandez. G. Backus, P. Garcia, K. Jonietz, P. Kirshen, M. Savonis, B. Solecki, L. Toole, M. Allen, R. Bierbaum, T. Brown, N. Brune, J. Buizer, J. Fu, O. Omitaomu, L. Scarlett, M. Susman, E. Vugrin, and R. Zimmerman. 2012b. *Climate Change and Infrastructure, Urban Systems, and Vulnerabilities: Technical Report for the U.S. Department of Energy in Support of the National Climate Assessment. Department of Energy.* Oak Ridge, TN: U.S. Department of Energy, Oak Ridge National Laboratory. February 29. <http://www.esd.ornl.gov/eess/Infrastructure.pdf>

WMO (World Meteorological Organization Secretariat). 2003. *The Second Report on the Adequacy of the Global Observing Systems for Climate in Support of the UNFCCC.* GCOS-82 (WMO/TD No. 1143). April. <https://www.wmo.int/pages/prog/gcos/Publications/gcos-82_2AR.pdf>

Woodall, C.W., K. Skog, J.E. Smith, and C.H. Perry. 2011. "Criterion 5: Climate Change and Global Carbon Cycles." In *National Report on Sustainable Forests—2010.* FS-979. Ed. G. Robertson, P. Gaulke, R. McWilliams, S. LaPlante, and R. Guldin. Washington DC: U.S. Department of Agriculture, Forest Service. 214 p. <http://www.fs.fed.us/research/sustain/docs/national-reports/2010/2010-sustainability-report.pdf>